What AI Can Do

The philosopher Spinoza once asserted that no one knows what a body can do, conceiving an intrinsic bodily power with unknown limits. Similarly, we can ask ourselves about Artificial Intelligence (AI): To what extent is the development of intelligence limited by its technical and material substrate? In other words, what can AI do? The answer is analogous to Spinoza's: Nobody knows the limit of AI.

Critically considering this issue from philosophical, interdisciplinary, and engineering perspectives, respectively, this book assesses the scope and pertinence of AI technology and explores how it could bring about both a better and more unpredictable future.

What AI Can Do highlights, at both the theoretical and practical levels, the cross-cutting relevance that AI is having on society, appealing to students of engineering, computer science, and philosophy, as well as all who hold a practical interest in the technology.

What AI Can Do:
Strengths and Limitations
of Artificial Intelligence

Edited by
Manuel Cebral-Loureda,
Elvira G. Rincón-Flores, and
Gildardo Sanchez-Ante

CRC Press
Taylor & Francis Group
Boca Raton London New York

CRC Press is an imprint of the
Taylor & Francis Group, an **informa** business
A CHAPMAN & HALL BOOK

First edition published 2024
by CRC Press
6000 Broken Sound Parkway NW, Suite 300, Boca Raton, FL 33487-2742

and by CRC Press
4 Park Square, Milton Park, Abingdon, Oxon, OX14 4RN

CRC Press is an imprint of Taylor & Francis Group, LLC

ISBN: 9781032396002 (hbk)
ISBN: 9781032395999 (pbk)
ISBN: 9781003350491 (ebk)

DOI: 10.1201/b23345

Typeset in Palatino
by codeMantra

Contents

Section I   Nature and Culture of the Algorithm

Section II   Knowledge Areas Facing AI

Section III Future Scenarios and Implications for the Application of AI

Editors

Manuel Cebral-Loureda holds a Ph.D. in Philosophy (Santiago de Compostela University) and a master's degree in Data Mining and Statistical Learning (UNED), and he is currently a professor and researcher at the Tecnológico de Monterrey, School of Humanities and Education. His interests focus on Digital Humanities, the Critical Reflection on Technology, AI, and Posthumanism. He is the Principal Investigator of the project "Neurohumanities Lab: Engaging Experiences for Human Flourishing." Some of his most recent articles include "Neural Deep Learning Models for Learning Analytics in a Digital Humanities Laboratory" (2021) or "Will and Desire in Modern Philosophy: A Computational Approach" (2020). Since 2021, he has been a member of the Mexican National System of Researchers (SNI).

Elvira G. Rincón-Flores holds a Ph.D. in Education Sciences from the University of Salamanca, Cum Laude thesis. Actually, she is an Impact Measurement Research Scientist at the Institute for the Future of Education of the Tecnológico de Monterrey, and she is also a professor at the same institution. She belongs to the National System of Researchers of Mexico as a Level 2, and she is the leader of the following research projects: Educational Spaces of the Tec 21 Model, Gamification in Higher Education, Student Mentoring, Well-being for Students, and Adaptive Learning. She also collaborates with the University of Lima in the development of a dynamic platform for gamification called Gamit! Her lines of research are the evaluation of educational innovation and educational gamification.

Gildardo Sanchez-Ante received the Ph.D. degree in Computer Science from Tecnológico de Monterrey in 2002. From 1999 to 2001, he was a Visiting Researcher at the Robotics Laboratory of Stanford University, and from 2004 to 2005, he was a Research Fellow at the National University of Singapore. He was the former Founding Rector of the Universidad Politécnica de Yucatan. Currently, he is a tenured research professor at Tecnológico de Monterrey. He is a senior member of the IEEE and the ACM. He is a regular member of the Academia Mexicana de la Computación and the Academia Jalisciense de Ciencias. He is a member of the National System of Researchers, Level 1. He develops research in automatic learning and pattern recognition, as well as its application in robotics. He has recently ventured into computational modeling of nanomaterial properties to optimize their performance.

Contributors

Mariángela Abbruzzese
Humanistic Studies Program
Tecnológico de Monterrey
CDMX, Mexico

Nora Aguirre-Celis
Department of Computer Science
University of Texas
Austin, Texas

Joanna Alvarado-Uribe
Institute for the Future of Education
Tecnológico de Monterrey
Monterrey, NL, Mexico

Vania Baldi
Digital Media and Interaction
  Centre of Universidade de Aveiro
Aveiro, Portugal
and
Departamento de Sociologia en
  ISCTE
CIES-ISCTE - Centre for Research
  and Studies in Sociology
Instituto Universitário de Lisboa
Lisboa, Portugal

Jacob Bañuelos
School of Humanities and Education
Tecnológico de Monterrey
CDMX, Mexico

Yamil Burguete
Estudios Humanísticos
Tecnológico de Monterrey
Puebla, Pue, Mexico

Eduardo Cascallar
Centre for Occupational &
  Organizational Psychology and
  Professional Learning
KU Leuven
Leuven, Belgium

Gerardo Castañeda-Garza
Institute for the Future of Education
Tecnológico de Monterrey
Monterrey, NL, Mexico

Héctor Gibrán Ceballos
Institute for the Future of Education
Tecnológico de Monterrey
Monterrey, NL, Mexico

Manuel Cebral-Loureda
Department of Humanistic Studies
Tecnológico de Monterrey
Monterrey, NL, Mexico

Juan Guadalupe Cristerna Villegas
Civil Engineering Department
Tecnológico de Monterrey
Monterrey, NL, Mexico

Francisco Díaz-Estrada
School of Humanities and Education
Tecnológico de Monterrey
Monterrey, NL, Mexico

Flavio Everardo
Medios y Cultura Digital
Tecnológico de Monterrey
Puebla, Pue, Mexico
and
University of Potsdam
Potsdam, Germany

Luis E. Falcon-Morales
Computer Science Department,
   School of Engineering and
   Sciences
Tecnológico de Monterrey
Guadalajara, Mexico

Alejandro García-González
National School of Medicine and
   Health Sciences
Tecnológico de Monterrey
Monterrey, NL, Mexico

Talía González-Cacho
School of Architecture, Art and
   Design
Tecnológico de Monterrey
Monterrey, NL, Mexico

Juan R. Lopez
Institute of Advanced Materials for
   Sustainable Manufacturing
Tecnológico de Monterrey
Monterrey, NL, Mexico

Karen Hinojosa-Hinojosa
School of Architecture, Art and
   Design
Tecnológico de Monterrey
Monterrey, NL, Mexico

Eva Kyndt
Centre for the New Workforce
Swinburne University of
   Technology
Melbourne, Australia

Diego López
Institute of Advanced Materials for
   Sustainable Manufacturing
Tecnológico de Monterrey
Monterrey, NL, Mexico

Eunice López-Camacho
Independent Researcher
Fort Payne, Alabama

Edgar R. López-Mena
Sciences Department, School of
   Engineering and Sciences
Tecnológico de Monterrey
Guadalajara, Mexico

Brecht De Man
PXL-Music
PXL University of Applied Sciences
   and Arts
Belgium

Omar Mata
Institute of Advanced Materials for
   Sustainable Manufacturing
Tecnológico de Monterrey
Monterrey, NL, Mexico

Adan Medina
Institute of Advanced Materials for
   Sustainable Manufacturing
Tecnológico de Monterrey
Monterrey, NL, Mexico

Paola Gabriela Mejía-Almada
Institute for the Future of Education
Tecnológico de Monterrey
Monterrey, NL, Mexico

Juanjo Mena
Department of Education
University of Salamanca
Salamanca, Spain

Andres Mendez-Vazquez
Department of Computer Science
CINVESTAV Unidad Guadalajara
Guadalajara, Mexico

Risto Miikkulainen
Department of Computer Science
University of Texas in Austin
Austin, Texas

Arturo Molina
Institute of Advanced Materials for
    Sustainable Manufacturing
Tecnológico de Monterrey
Monterrey, NL, Mexico

Luis Fernando Morán-Mirabal
Institute for the Future of Education
Tecnológico de Monterrey
Monterrey, NL, Mexico

Diego E. Navarro-López
Biotechnology Department, School
    of Engineering and Sciences
Tecnológico de Monterrey
Guadalajara, Mexico

Gilberto Ochoa-Ruiz
School of Engineering and Sciences
Tecnológico de Monterrey
Guadalajara, Mexico

Omar Olmos-López
Engineering and Science School
Tecnológico de Monterrey
Monterrey, NL, Mexico

Pedro Fonseca
Computer Science Department
Tecnológico de Monterrey
Monterrey, NL, Mexico

Pedro Ponce
Institute of Advanced Materials for
    Sustainable Manufacturing
Tecnológico de Monterrey
Monterrey, NL, Mexico

Carolina Sacristán-Ramírez
Estudios Humanísticos
Tecnológico de Monterrey
Monterrey, NL, Mexico

Ricardo Ramirez
School of Engineering and Sciences
Tecnológico de Monterrey
Monterrey, NL, Mexico

Ivan Reyes-Amezcua
Department of Computer Science
CINVESTAV Unidad Guadalajara
Guadalajara, Mexico

Paola Ricaurte
Department of Media and Digital
    Culture
Tecnológico de Monterrey
CDMX, Mexico

Florian Richter
AImotion Department
Technical University Ingolstadt of
    Applied Sciences
Ingolstadt, Germany

Elvira G. Rincón-Flores
Institute for the Future of Education
Tecnológico de Monterrey
Monterrey, NL, Mexico

**Carlos Felipe
Rodríguez-Hernández**
Institute for the Future of Education
Tecnológico de Monterrey
Monterrey, NL, Mexico

Olaf Román
Department of Educational
    Innovation and Digital Learning
Tecnológico de Monterrey
Monterrey, NL, Mexico

Gildardo Sanchez-Ante
Computer Science Department,
    School of Engineering and
    Sciences
Tecnológico de Monterrey
Guadalajara, Mexico

Juan Pablo Solís
Civil Engineering Department
Tecnológico de Monterrey
Monterrey, NL, Mexico

Juan Humberto Sossa-Azuela
Centro de Investigacion en
    Computacion
Instituto Politecnico Nacional
CDMX, Mexico

Gabriela Elisa Sued
Institute of Social Research
UNAM
CDMX, Mexico

Juliana Vivar-Vera
School of Social Science and
    Government
Tecnológico de Monterrey
Monterrey, NL, Mexico

Mariel Zasso
Department of Media and Digital
    Culture
Tecnológico de Monterrey
CDMX, Mexico

Section I

Nature and Culture of the Algorithm

The philosopher Spinoza stated, some centuries ago, that no one knows what a body can do, rejecting the primacy of the soul over it and conceiving an intrinsic bodily power with unknown limits. Similarly, we can ask ourselves about Artificial Intelligence (AI): to what extent is the development of intelligence limited by its technical and material substrate? Should this material substrate of intelligence be organic, or can it be inorganic and artificial? If so, what is the limit of this development, i.e., what can AI do? The answer is intended to be analogous to Spinoza's: nobody knows what AI can do…

In order to address this question, this first section develops a philosophical and more theoretical framework of the application of AI, understanding that any algorithm has its nature, i.e., a set of material conditionings and mathematical operations that define it; but it also has its culture, which has to do with the way it is integrated and assimilated by people and society. As for the relationship between body and mind in Spinoza, these two features of algorithms have no hierarchy: we have to understand how an algorithm works, how it has been constituted, etc… but we also have to see how it is being used, and how it is being assimilated and socialized. The power of an emancipated technological culture around AI will depend on the appropriateness of both approaches: the more engineering one together with the more humanistic.

More particularly, this section tries to show that AI is a major challenge for ethics as a whole, and not so much a specialized branch of it (Ritcher); the socio-technical construction of a gray eminence that induces us not to delve into the underlying issues of AI (Baldi); the critical aspects regarding the necessity of more decolonial and feminist approaches within AI

DOI: 10.1201/b23345-1

technologies (Ricaurte & Rosauro-Zaso); the cultural side that complements socio-algorithmic development (Sued); the socio-cultural reappropriation of deep learning technologies, first for *deepfake* practices, but also for *deeptruth* (Bañuelos & Abbruzzese); and how AI is challenging our own philosophical conceptions, even in a sense of transcendental philosophy (Cebral-Loureda).

Effectively, we found Ritcher's proposal extremely appropriate for the beginning of the section. It has all that a first or introductory section needs: it gives a strong conceptual framework, which, at the same time, opens many possibilities of development. Technology is not an artifact, because it is not something given, but something created as a means to achieve another thing. Under this point of view, even though we don't provide an end-to-end technology, when considering it a means for another thing, implies its ethical reflection, how it is being implemented and how reasonable is the end we pursue. Technology, and specially AI, becomes an essential field of research nowadays, constituting not an appendix or complement of what ethics is, but a core of fundamental interest that allows ethics itself to advance.

Following with the idea of conceiving technology as a means that needs ethical reflection and research, Baldi draws well the new scenario of social and political interactions, and how the technological structures are forcing us to accept a series of assumptions where the individual is exposed to questions without answers, an overwhelming network of quantifications where big platforms tend to win. In such a scenario, technology and AI are often used as a smokescreen—or gray eminence—appearing to have ends in itself, and allowing everything to change without anything changing.

Ricaurte and Rosauro-Zaso go deeper into the need of a cultural comprehension of AI, in order to promote and, even, guarantee a state of techno-diversity that is not feasible today. Intelligent systems tend to reproduce historical biases, demanding active answers and critical understandings that do not assume technology in an aseptic way. Because of the powerful emergence of AI technologies, it is easier to let go, not to ask why and how they are effectively operating, to bypass the matrix web of power that tends to prolong the oppressive structures of yesteryear. It is a very tough text, one that goes to the core of our ontological conception of reality, and how it is influenced by pre-understandings and patriarchal heritage.

Sued investigates algorithms as a social construct, something that goes beyond the technical, becoming a social entity to be considered under more complex criteria, such as the effects they have on our cultural and political ways of life. It is therefore necessary to compensate for the clear risks involved in the *algocratic* society with serious and constant socio-technical research. It is required to connect the cryptic processes involved within algorithmic culture—algorithmic black boxes—to a more accessible language that can be understood by the rest of the citizens, endowing them with greater technological knowledge along with a greater capacity for agency and decision. Otherwise, AI technologies will be the gray eminence that Baldi points out.

The conditioning factors underlying the use of technology are also present in the chapter contributed by Bañuelos and Abbruzzese, in relation to such a philosophical concept as truth. Indeed, through deep learning technologies, it is now possible to recreate video images emulating people saying or doing things they have never done. The idea of video sources as a neutral record of reality falls away, something that had already happened with photography. Referring to philosophical authors such as Simondon, Van Dijk, and Richard, the authors justify not only the entry of *deepfake* phenomena into the social media agenda, but also the opposite: the possibility of new uses of these technologies as generators of *deeptruth*. Thus, this chapter clearly complements the previous one: although technologies have biases that feed the inertias of power, it has also given rise to phenomena of resignification and reappropriation, revealing how AI can be used for social and political activism.

Finally, Cebral-Loureda highlights how AI cannot be simply understood as another technical development, since it puts into question many of the most important philosophical matters that we face since ancient times: How do we codify and understand reality? Which is the essential component that underlies a process of perception? What does learning mean and imply? What does a machine need to be conscious? Although many times, philosophy has tried to answer these questions by separating a sphere of transcendental truths which would be entirely natural and immediate for the mind, this conceptual framework is getting under question by means of AI technologies. Particularly, the mind itself is being re-examined with new neurocognitive developments, for which AI technologies offer new models for understanding. This question forces us to a speculative and posthumanist approach, in which we have to doubt whether human exclusivity really possesses the last word on authentic cognitive processes.

1

AI Ethics as a Form of Research Ethics

Florian Richter
Technical University Ingolstadt of Applied Sciences

CONTENTS

1.1 Introduction

It is often assumed that technology is value-free or value-neutral. This issue stems from the assumption that technology is a means to an end that can serve either a good purpose or a bad purpose. While this position has also been defended more extensively (Pitt, 2014), it is based on a conceptual insufficiency: such a conception reduces technology to a tool or artifact that can be used for good or bad purposes (like a hammer), but technology is rather an enabling medium that makes available certain options, i.e., webs of possibilities, while at the same time disguising other possibilities (Hubig, 2006). As a means of addressing the reduction of technology to a means or tool, some in the area of philosophy of technology have proposed to understand technology with regard to performances and practices that form us (Coeckelbergh, *Moved by Machines: Performance Metaphors and Philosophy of Technology*, 2019), as something that mediates our access to the world (Verbeek, 2015), as a medium (Gutmann, 2003; Hubig, 2006), or as a concept of reflection (Grunwald & Julliard, 2005; Hubig, 2011); but neither has it been consistently transferred to Artificial Intelligence (AI) ethics nor has it been analyzed mainly under the aspect of establishing it as a form of research ethics.

DOI: 10.1201/b23345-2

The crucial distinctions between value neutrality, value freedom, and value ambivalence need to be clarified to first and foremost formulate the adequate conceptual framework that allows to scrutinize the status of the connection of values and technology. Although this issue has been examined for the field of ethics of technology, it yet has not been done in the field of AI ethics. Consequently, the first aim of this chapter is to reconstruct the conceptual framework and employ it in the field of AI ethics. The second aim of the chapter is to describe technology as an enabling medium to establish AI ethics as a stand-alone form of research ethics. If technology is an enabling medium, it is not only inadequate but unfeasible to assume that technology is a value-free tool; instead, the relation of technology and values needs to be described in a way, where research is already a value-laden doing. This has implications for research in AI that will be outlined at the end.

One difficulty is to determine what technology actually is (and consequently also AI as a specific form of technology). In the philosophy of technology, there are different attempts to do so. Sometimes technology means the amount of artificially produced objects (artifacts), an ability or way of acting to produce or invent something, a "specific knowledge", or a form of accessing the world (Ropohl, 2010, p. 41); similar notions can be found in Coeckelbergh (2020, p. 5). Additionally, the complexity is broadened by the fact that different notions about technology are also interwoven with anthropological statements. Arnold Gehlen states for example that humans are deficient beings -Mängelwesen- that need technology as a compensation to survive and to live (Gehlen, 2007). For a detailed examination of anthropological statements and how they shape our understanding of technology see (Hubig, 2006, pp. 77–93).

According to Ropohl, we have to accept that technology – "Technik" – is an "equivocal name" (Ropohl, 2010, p. 48) and that there is not one meaning or essence underlying the different uses. There are for instance, according to Ropohl, more speculative interpretations from Martin Heidegger or Karl Marx, where technology is the "means and result of human labor" (Ropohl, 2010, p. 44). Ropohl however suggests a nominalistic interpretation that is based on the guideline 3780 of the *Association of German Engineers* (VDI) (Ropohl, 2010, p. 42). It must be mentioned that he also worked in the committee that elaborated the guideline. According to this guideline, technology is considered the "amount of use-oriented, artificial objects (artifacts or systems of res)", but also the "amount of human actions and institutions" that generate "systems of res" and the "amount of human actions, in which systems of res are used" (VDI, 1991, p. 62 [2]; and Ropohl, 2010, pp. 42–43).

Technology as an amount of artifacts is sometimes also taken as a starting point to think about technology and its relation with values and morality (Miller, 2021), because it seems at first glance obvious that for instance an artifact like a knife can be used for good or bad purposes. The knife is therefore only a (value-neutral) means to a (value-laden) end. But if such a premise is taken as the starting point, it will lead to manifold problems, as it will be shown below. The instrumental interpretation of technology is consequently often criticized (Coeckelbergh, 2020, pp. 5–6). Technology is not just a means to an end, but it can be mediating our relation or access to the world or others like it is assumed for instance in the so-called post-phenomenological movement – see e.g. Verbeek (2015) and for an overview Rosenberger & Verbeek (2015). Other ways of overcoming the instrumental interpretation are to describe technology as a medium and not a means (Gutmann, 2003; Hubig, 2006) or as a concept of reflection (Grunwald & Julliard, 2005; Hubig, 2011). A traffic or transportation system is the medium in which we can move from one place to the other, and it seems at first glance neutral, but its inner structure can be "beneficial or restrictive" (Hubig, 2007, p. 61). It might be difficult to use public transportation from a small village to the city, which then makes one dependent on using a car. Also, the rail traffic includes and excludes already certain destinations as reachable. This has of course also implications for a more sustainable design of traffic systems. Consequently, the conceptual inclusion of technology as a medium needs to be possible and has implications for research – here, for instance, in developing sustainable and inclusive transportations systems – and the respective questions that are asked – do they focus on or include transportation between villages or only within the inner city.

Accordingly, the aim of the chapter is to show that research questions are already value-laden and to define in a certain way the search space and the consequences of research projects in the field of AI. Such conceptual classifications can be helpful for empirical studies in the field of AI ethics. The lack of such an ethical framework can be identified in so-called acceptance studies (Nevaranta, Lempinen, & Kaila, 2021; Okkonen, Helle, & Lindsten, 2020; Sun et al., 2019). Consequently, it was proposed to induce principles of research ethics into such studies, but they were mainly taken from biomedical ethics (Drachsler & Greller, 2016) and this will be not feasible (see below). The benefit of the approach proposed here will be that it is closer to the development of research projects in the field of AI and consequently, more practical. Even though there are plenty of ethical guidelines from policymakers and companies on how to create Responsible AI, they do not consider AI ethics as a research ethics to the full extent.

At the outset, the relation of technology and values will be reconstructed on another level, if technology is not only understood as a means. This conception will be problematized in the second section. The third part of the chapter will then establish how AI ethics can be understood as a form of research ethics by assessing examples of bad scientific practices. The discussion section will present some implications of the examples and the approach proposed here, but also discuss some possible objections. The conclusion will present some possible lines of further research and ways of establishing ethical research in the field of AI.

1.2 Technology as a Means

1.2.1 Values – Metaethical Considerations

The conceptualization of technology as an amount of artifacts or objects implies that they are things that can exist independently of humans and that they are something objective. Arti*facts* have a factual existence. Values however are seen as something subjective and do not exist independently of humans. At least, if one is not a metaphysical realist, who assumes that values are entities that exist like mountains, trees, or hammers. In meta-ethical debates, the status of values is discussed under different aspects. Some discussions focus on the ontological aspect of values, i.e., if they exist as normative entities or some kind of metaphysical facts (Anwander, 2004, pp. 100–103; Schaber, 2004, p. 110; Wedgwood, 2008, pp. 10 and 135–221). Other debates focus on how values can have an effect in the world, whether they can be understood as reasons for our actions, even though they are not part of the naturalistic "fabric of the world" (Mackie, 1977, p. 24). How is it possible that they can influence our actions or motivate us to perform certain actions? Some suggest that additionally to external (moral) reasons (a set of) motivations need to be added as a brute force that helps to execute the action. For this debate between internalism and externalism, see Blackburn (1995), Broome (2000), Wallace (2020), or Williams (1981). Another opposition in metaethics is cognitivism vs. non-cognitivism. Cognitivists assume that moral statements can be true and that moral – metaphysical – facts correspond to these statements and make them true. It is epistemologically difficult to hold such a position because one needs to assume a certain kind of intuition to grasp such metaphysical facts. Furthermore, it is also difficult to hold such a position ontologically because one has to assume metaphysical facts. Not all cognitivists are Platonists like for instance Jürgen Habermas, who claims that a "cognitivistic ethics [...] has to answer the question, how normative statements can be justified" (Habermas, 1991, p. 8). This means that the statements need to be negotiated within a discourse to be legitimate (true).

Non-cognitivists, in contrast, assume that moral statements cannot be true (Lumer, 2010). That means that one cannot make objective statements about values, because there are no metaphysical facts that correspond to such statements. According to John Mackie, values are not "part of the fabric of the world" (Mackie, 1977, p. 24). He claims that there "are no objective values" (Mackie, 1977, p. 15). The world can be described only in descriptive statements, but not in normative ones. It is also mostly accepted that we cannot argue about – descriptive – facts, like that the atomic number of gold is 79, but on the other hand it can be argued which values are adequate to follow in one situation or to guide our actions. In non-cognitivism exist different sub-movements regarding how one, despite all, can talk about morality and values. According to emotivists – David Hume, Alfred J. Ayer – values are shoehorned into the world via agreements that can be expressed through feelings or attitudes toward facts. According to prescriptivists – Richard Hare – values are shoehorned into the world via requests that can be expressed via imperatives (Lumer, 2010).

1.2.2 Technology and Values

These are all fairly abstract debates about the status of values or norms, but the ontic status of values, i.e., whether they exist independently and are objective, and the epistemic status of values, i.e., to what extent they can be recognized, also plays a prominent role in the philosophy of technology. It is asked whether values are "embedded" in artifacts, as Joseph Pitt writes, or not, i.e., whether they belong inherently to the structure of the artifact or not. If that is not the case, then technology must be value-free or at least value-neutral. Joseph Pitt defends the so-called value neutrality thesis (Pitt, 2014, p. 90 et passim). In contrast, Boaz Miller tries to rebut Pitt's arguments (Miller, 2021). A more thorough discussion can be found in the book edited by Kroes & Verbeek (2014). This discussion is interwoven with metaethical positions and arguments, but the concepts that are used need to be distinguished clearly to reconstruct the problems adequately.

Artifacts are understood as means and are thus events or things, while intentions and preferences are mental acts. However, values are reduced to the status of preferences or intentions – like it is done by Pitt (2014, p. 91) – but values are, according to Hubig, "recognized implicit rules for the justification of preferences and for the justification of means and goals" (Hubig, 2007, p. 70/71). Recognition should be taken here in the sense of acknowledging something and not seeing or empirically identifying something. Consequently, it is not adequate to ask whether values are "empirically recognizable" or identifiable like Miller (2021, p. 59) and Pitt (2014, p. 94/95) do, and what they take as the main issue of the debate. The latter denies it, because it is not possible to see the values or empirically verify them, since they cannot be measured or weighed. Pitt points out that he does not see the value of a building, just "bricks and stones" (Pitt, 2014, p. 94/95), but in the end it is not really an epistemological question, because values are "rules for the justification of preferences" (Hubig) or means. Both opposite positions are metaethically and conceptually underdetermined due to their misconception of the main issue.

The examples from Miller demonstrate that point more clearly. The first one is about the Long Island Expressway (LIE) that –as some assume– was built with the intention to impede residents from the inner city to access the beaches outside of the city, because the buses from the city could not pass under the bridges. The bridges were built low, so that buses could not pass under them. Thus, the poorer residents, who were mostly members of minority groups, could not travel to the beaches, because they could only afford public transportation. The second example is about the design of benches that are separated by bars, so that homeless people cannot sleep on them (Miller, 2021, p. 59). In both examples, the values – of the designers – should be understood – at least according to Miller – as embedded in the artifact, but every artifact – and also other things like stones – is only a means in relation to a goal. Being a means is conceptually determined by being useful for a goal, and the other way around, a goal is only a goal by being attainable via a means; otherwise, it would be a mere wish (Hubig, 2011). The intention of the designer or the people in the institution can only be read abductively from the artifact within the social practice and the context, where it is used. If the designer does not state what her intentions were, it cannot be known for sure. There can only be some clues or indications about the intentions that hint into a direction like in a criminal case. The artifact can only be determined conceptually as a means by the intentions of the designers and developers, but one needs either to ask them what their intentions were, or investigate them.

Pitt defines values as "an endorsement of a preferred state of affairs by an individual or a group of individuals". Such a preferred state of affairs "motivates our actions". He admits that it sounds similar to "goals", but values are more like a "regulative ideal", because goals can become obsolete, when they are attained and values however can still be action-guiding and motivating (Pitt, 2014, pp. 90–91). Furthermore, Miller defines "values as *normative discriminators*", i.e., values are "anything" that helps to discriminate "between different states of affairs" and serves to rank "some of them higher than others" with regard to "how much they are desired" or how they fit into the "personal, social, natural, or cosmic order" (Miller, 2021, p. 59; italicized by the author). In contrast, van de Poel and Kroes discuss different kinds of values: while values "for their own sake will be referred to as 'final values'", there are also values that "will be called 'instrumental values'" – they also distinguish in both categories extrinsic and intrinsic values, but for the discussion here, this is not relevant. According to them, a sea dike has as a final value safety for the people living in the hinterland, whereby the protection from the flooding of the hinterland is the instrumental value. A knife in contrast has only an instrumental value (cutting), but dikes on the other hand are *"designed for safety"* and have thus a final value (van de Poel & Verbeek, 2014, pp. 107–114). However, dikes could also be used for separating the land to exclude one group from having access to the water. It could also be that the resources to build the dikes are used lavishly, whereas it would have been better to use the money to rebuild the houses far away from the water. Related examples are dams and hydroelectric power plants, for the sake of which villages had to be relocated for the value of ecological sustainability. It depends thus on the values that are recognized or acknowledged. While Pitt therefore aligns values to action-motivators, and Miller to preferences, van de Poel and Verbeek align them to the instrumentality inscribed in the technology, but then values are misunderstood, because reasons of actions should not be just motivators or preferences, they must have a legitimizing or justificatory function. Technological applications can be put into relation to different values that are recognized by different groups, individuals, and parties. The task is then to make explicit the dissents and which arguments are used to justify different courses of action. This implies not only an *ambivalence* of technology (Hubig, 2007, pp. 61–62), but a *pluri*-valence of technology, where we have to deal with tradeoffs between different values. The question is thus *not* if technology is value-neutral, but how values are related to technology and artifacts.

1.2.3 The Ambivalence of Technology

Hubig states that in the case that "technological artifacts, procedures etc. and indirectly also strategies to their production become means", then technology is *ambi*valent, because it can tend to the good or the bad. This is plausible under the assumption of the dichotomy of good and bad. He distinguishes three forms of possible ambivalence: (i) For instance, strategies of energy supply can be good as well as bad, even the ones that are usually considered preserving resources like hydropower, because such plants cause a deep intervention into nature and thus an interference of biodiversity. Logically speaking, it is a subcontrary relation of particular judgments – both can be true, but not both can be false. (ii) It can be that technology is partially good and partially bad. Genetically modified crops might be regarded as bad in some parts of the world but could help to cover the supply for basic needs in regions of the world with nutrient-poor soil. Due to the low biodiversity of the soil, the genetically modified crop could be only a minor threat. Logically speaking, it is a contrary relation – both can be false, but both cannot be true. It depends on the frame – in the example the region – when and where something is considered good. (iii) Some medical technologies can be used *either* for the prolongation *or* for the shortening of life. Nuclear fission is another example that can be used either for good or for bad. Logically speaking, these judgments are contradictory (Hubig, 2007, pp. 61, 153–154).

However, the prolongation of life is not exclusively regarded as something good like in the case of the suffering of terminally ill-patients or of someone in a coma with irreparable brain damage. It seems that Hubig includes this by speaking of the "prolongation or shortening of a humane life" (Hubig, 2007, p. 62); nonetheless, it shows that, if technology is a means, it can stand in relation to various values – physical well-being, quality of life, and expectancy of life. This assumption presupposes thus a plurality of values, which is the basis for a democratic discourse. Aristoteles bases his political philosophy on the idea of "plurality". (Aristoteles, 1995) (Pol. 1261a). He criticizes Platon, and his proposal of a political class that was educated to rule over the rest (*Politeia*), where the philosophers are kings or queens (Pol. 1261a-1262b). For Aristoteles, this leads to a "unity" and in the end to a "destruction of the polis" (Pol. 1261a). The polis is "composed of unlike elements" (Pol. 1276b).

All the citizens deliberate and judge and can hold office, but only for a limited term (Pol. 1275b). Aristoteles excludes women and slaves from being able to have the status of a citizen, so consequently the pluralistic foundation of his political philosophy is limited. Despite that, the important aspect of his ethics is demonstrating not only what counts as virtuous behavior, but how it is possible to participate in this pluralistic discourse to shape the community and to develop virtues. Notwithstanding, plurality – also of values – is the basis for democratic and scientific institutions and discourses, but this also includes dissents, and consequently, such dissents need to be handled by the citizens.

Hubig distinguishes dissents on the object level under three respects: (i) The possible chances and risks of the situations are "relative to the values that are expressed in the different *maxims* (topmost action-guiding orientations [...])". What is considered a potential benefit or harm is "relative to needs, purposes, and preferences based on different values" (Hubig, 2007, pp. 147–148,151). Such differences can be laid open by conducting studies with stakeholders, when a technology or AI-based application is developed. For instance, the so-called acceptance studies are a means to make the different values, needs, etc., explicit to first and foremost have a basis to handle the dissent. (ii) There are "dissents regarding the *justifiability* of conflicting value attitudes that depend on the respectively assumed different images of man and concepts of action in general, which find their expression in the different views of what a value is at all" (Hubig, 2007, pp. 147–148, 151). Values, or the preference for certain values, can be justified through different ethical theories such as utilitarianism, deontology, virtue ethics, and discourse ethics, but such justifications are also interwoven with notions or ideas of the human. Particularly, the philosophy of technology is interlaced with different notions, such as what constitutes humans as humans, for instance, humans as "deficient beings" (Gehlen, 2007), as a "*homo poieticus*" (Floridi, 2008), etc. The first respect of dissents needs to be accompanied by empirical research, and the second is mostly seen as the main task of ethicists. The third respect considers how already value judgments in the development of research and technological applications plays a role: (iii) Values are already included in the assessment of situations through value decisions. It always depends on the relevance and the adequate assessment of spaces of possibilities, hypothetical, and real courses of events ("risk and opportunity potentials") and the handling of such possible events ("risk and opportunity management"), which means that the theories are already "normatively laden" (Hubig, 2007, pp. 147–148, 151).

In the next section, value decisions and judgments will be discussed further with respect to formulating research questions and how they pick out value-imbued parts of the space of possibilities. Therefore, technology as a means needs to be set aside and considered as a medium or space of possibilities to develop AI ethics as a form of research ethics.

1.3 How to Ask Ethical Questions in AI Research

AI ethics as applied ethics might imply to assume that general rules or norms exist, and that technology or AI-based systems are means/tools. Thus, it would be just a matter of ingenuity on how to apply them; in the end, also some kind of a technical problem that needs a technical solution. This model does not need a specific ethics: a gun could be used for something good – stop a criminal – or something bad. Such a conceptualization can be handled by general ethics. Of course, it is bad to assault someone, and it is also wrong to discriminate against someone with an AI-based system, like it happened with facial recognition systems; but these cases do not need ethical frameworks that are developed especially for AI. Instead, AI ethics should help to formulate research questions in a way that bears on ethical issues specific to AI research. It has not been scrutinized how to *ask* the right questions. If we assume that AI ethics is a form of research ethics, then the conceptual tools need to be developed to clarify the actual and proper tasks of AI ethics.

The philosophical task is a meta-commentary on scientific practice. Two main aspects need to be considered that stem ultimately from the posing of the research question: (i) An approach can be methodologically inadequate, which consequently makes it a bad scientific practice. (ii) An approach can be ethically inadequate, which cannot be justified by the intention of pure scientific curiosity that supposedly is neutral to any immoral consequences or applications of a technology. Technology is not only a means to an end, but also a medium that opens up possibilities and makes at the same time other things impossible (Hubig, 2006). Even though technology or an artifact can be seen as a means for a certain purpose, it can also constitute a space of possibilities for the users or exclude certain users. The technology or artifact is thus not anymore just a means, but a medium. In the case of the benches with bars, it excludes homeless people from certain spaces and makes it impossible for them to sleep on them. Sea dikes also disclose certain possible ways of living behind them.

1.3.1 Bad Scientific Practice I: Methodological Aspects

Both inadequacies of scientific practices will be illustrated by means of three examples: two from AI research and one from biomedical research to contrast it with AI ethics. Despite the call for establishing a "professional ethics" in computer science "with real teeth and consequences" (Reich, Sahami, & Weinstein, 2021), the more urgent project would be to establish a framework for distinguishing between good and bad scientific practices. If we assume that technology is value-neutral or – free, then the goodness or badness of it clings only on the human sphere, i.e., what is valued and what is condemned. But practices are based on human actions, and technology is an inherent part of scientific practices, like for instance the apparatus for experiments.

First Example:

Two researchers developed a model for a facial recognition system that should be able to determine whether someone is a criminal or not (Wu & Zhang, 2016). Several issues can be raised here, such as the assumption that social characteristics can be derived from physical characteristics. This kind of presumed relation has already been criticized in the 19th century, for example, by G.W.F. Hegel. Problematic is that an "identity claim of the mental and the physical" realm is made (Quante, 2011, pp. 102–115). The procedure of the two researches has also been classified under the title of the pseudoscience phrenology (Stinson, 2021), according to which it should be possible to derive personal traits from the shape of the skull – for a critique and empirical disproof of phrenology with 21st century methods see Parker Jones, Alfaro-Almagro, & Jbabdi (2018).

A little background in the history and philosophy of science could help here, but also knowledge about the capability of AI-based systems should be employed to understand the possibilities of scientific practices. Image recognition systems that are run with methods like convolutional neural networks are in the end producing correlations. Amounts of pixels correlate with other amounts of pixels. This is to a certain degree really effective when it is used to recognize dogs or to distinguish dogs from cats, although there are also certain limitations to these methods (Rosenfeld, Zemel, & Tsotsos, 2018). However, the conviction of criminals is based on complex institutionalized standardizations, measurements, and practices. The search by the police, involvement of courts and lawyers, and the institution of prisons can be mentioned to sum this complex practice up in a few key words – with, of course, differences in every country and culture-. In the end, clues need to be found and investigated that make it probable that someone did something – abductive reasoning – , it has to be judged if it falls under a general deed that is forbidden – judgment by the court – and evaluated which kind of sanction is appropriate – deductive reasoning. It seems rather absurd that all of this could be read off from some pixels, albeit Wu and Zhang deny that their "discussions advance from correlation to causality". However, they state that their "highly consistent results are evidences for the validity of automated face-induced inference on criminality" so, at least, it could be inferred. They also call it "social inference" (Wu & Zhang, 2016, pp. 2–4 and response 1). Correlation is only found between the pixels of the pictures; nevertheless, it is a kind of causal connection that must be presupposed by committing oneself to the inference that social traits can be derived from biological traits.

Second example:

Kosinski and Wang used a deep neural network to determine the sexual orientation of persons based on their physical traits via image recognition (Kosinski & Wang, 2018). In the end, it could be asked for what exactly such a classifier would be good? Would it be enough, like the authors do, to understand their study as a warning for policymakers and the public, because of the accuracy of their study? Thus, they just would raise awareness (Kosinski & Wang, 2018, p. 248, 255). Additionally, Kosinski and Wang claim that "faces of gay men and lesbians had gender-atypical features, as predicted by the PHT" – prenatal hormone theory. According to that theory, "same-gender sexual orientation stems from the underexposure of male fetuses or overexposure of female fetuses to androgens that are responsible for sexual differentiation" (Kosinski & Wang, 2018, p. 254, 247). This should imply that "faces of gay men were more feminine and the faces of lesbians were more masculine than those of their respective heterosexual counterparts" (Kosinski & Wang, 2018, p. 252). Of course, counterexamples can easily be found and make their implication invalid, but it might be also questioned, whether this classification does more resemble stereotypes (Agüera y Arcas, Todorov, & Mitchell, 2018), so in the end such a classification is truly a social construct, which cannot be derived solely from biological characteristics.

In both examples – identification of criminals or sexual orientation – two realms of phenomena are separated, a physical and a mental/social/psychological/cultural, either one assumes a dualism of these realms or a monism. In the case of the dualism, some kind of causality is needed to explain how the physical can have an effect on the social/psychological phenomena. Although Wu and Zhang deny such a causal connection, they must suppose it; otherwise, they should not have even started to undertake their research. In the case of the monism, a main ontological layer is assumed, which usually is the physical or naturalistic realm, in which certain characteristics are formed – in the second example through hormones. The cultural or social phenomena that are, for instance, classified – may they be stereotypes or not – need to be reduced to the naturalistic layer or are just considered metaphorical ways of speaking. However, concepts like femininity or masculinity are much more complex than that they could be reduced to biological traits. It is tricky to disentangle the realms of phenomena and to give each its right and adequate place in scientific discourse. Each science – also in the humanities and social sciences – has developed their methodological toolbox to access, describe, and evaluate their respective phenomena.

1.3.2 Bad Scientific Practice II: Ethical Aspects

There are shifts and changes in what counts as an ethical scientific practice. Animals have been used in different kinds of research as test subjects – for medical/pharmaceutical tests or for cosmetic products. While the latter is more and more condemned by many people and also laws have been legislated against it, the former is still to a certain degree accepted. Some animals like great apes are now increasingly excluded from research as test subjects. Norms and moral judgments change throughout cultures and history. In ethical research, certain flexibility is needed to reflect on such changes.

A top-down approach in practical ethics that applies – universal – moral norms seems therefore not apt. A down-to-top approach on the other hand seems thus more promising. It could be to cluster cases together in types, but for the development of new technologies, this seems difficult, because each time the type must be adapted or new characteristics must be included. Such a procedure looks quickly arbitrary. Consequently, it is better to look at each technological application individually and discuss its ethical implications. This will become even clearer by distinguishing different fields of research and the difficulty to compare and contrast them.

Third example: CRISPR-Cas9 technology

Researchers, including Jennifer Doudna, one of the developers of the CRISPR-Cas9 method, have expressed concerns and demanded a kind of moratorium for "germline genome modification" in humans (Baltimore et al., 2015). However, one researcher did apply this technology in embryos, which led to a public outcry (Cyranoski & Ledford, 2018). He was penalized by his national government and also "banned" from the research community (Reich, Sahami, & Weinstein, 2021). It shows that, to a certain degree, research communities can be self-regulating. Here, the concerns seem to be mainly in the field of technology assessment, because it is about the possible and unforeseeable consequences of editing the germline genome (Baltimore et al., 2015). The connections of the conditions – means-end and cause-effect – are not yet fully understood and clearly established; therefore, one should not make certain experiments with this technology, i.e., to undertake germline genome modifications in humans. The concerns are based on a lack of technological and scientific knowledge.

The algorithms of AI-based models, especially deep neural networks, are considered black boxes, i.e., the functioning of the algorithm cannot be made explicit. This opacity is different to the lack of knowledge in the case of the CRISPR technology. If the AI-based model could be made transparent, and even if we had full knowledge of the functioning of the algorithm, it would not mean that the application of it is fair or morally right. Suppose it would be possible to build an AI-based model for an image classification system and it would work in 100% of the cases to detect whether someone is a criminal or not, then it still would not be morally right to use it, because social traits *should* not be derived from biological traits, because we have established mostly good practices that can lead to the conviction of someone who actually has committed a crime. So, even if it could be done, it does not mean that it should be done. Consequently, this kind of reasoning also commits the is-ought fallacy.

It is not about the lack of knowledge, like in the case of the CRISPR technology, or that the transparency of the model needs to be established, and it is not even a question of the accuracy of the model; rather it is about the prudent application of the model that needs to be in question.

I do not advocate that full knowledge about the CRISPR technology and genetics would allow us to do some kind of genetic engineering. This is another question and beyond the topic of this paper. The example from bio-ethics is just taken for illustrating the point in AI ethics.

1.4 Discussion

In talking about social traits, one enters also on sociological ground and does not solely rely on questions of computer science and technology. Max Weber, the German sociologist, points out that every hypothesis or research question is just covering a little cutout from the manifold. It takes out what we consider as significant or important. Researchers base the way they approach the subject on "value ideas" that let them "consciously" or "unconsciously" select specific aspects of the subject matter (Weber, 1988, pp. 171–181). Scientists thus make "value judgments", as Pitt calls it. He draws on a distinction from Richard Rudner between "two types of values, epistemic and non-epistemic". Epistemic values are, for instance, "truth, provability, explanation, evidence, testability", while non-epistemic values are aesthetic, political, or social values. Already by considering one hypothesis as testable, a researcher makes a value judgment. (Pitt, 2014, p. 98)

It therefore is not possible to claim to abstain from any value judgment as a researcher, by saying that one maintains "neutrality" and just conducts research out of "curiosity", as Wu and Zhang tried to defend themselves. It is also not good practice to rebut any involvement of values by stating that "[l]ike most technologies, machine learning is neutral" (Wu & Zhang, 2016, p. 1 and response 1/2). Consequently, no responsibility is taken by the researchers, although the hypothesis and project covers a little cutout of the value-imbued manifold, especially when entering the social sciences via methods of the computer sciences, like in this example here. Furthermore, Wu and Zhang claim that if data is freed from "human biases", then "advantages of automated inference over human judgment in objectivity cannot be denied" (Wu & Zhang, 2016, response 3). That means that they suppose an objective realm of facts that is just contaminated by human biases and could be recovered or cleaned after human judgment is taken out; but science is a human practice and thus neither free of biases nor of values. Researchers make value judgments.

Hubig distinguishes such a *"preliminary value decision"* with regard to Weber from a *"practical evaluation"* or assessment of the output or findings of research. The postulate of the value neutrality can only be related to the latter (Hubig, 2007, p. 65; italicized by the author). This also corresponds to the distinction between "problem-induced technology assessment", that is applied when technological solutions are sought for given societal tasks, and the "technology-induced technology assessment", that evaluates technology at the stage of series-production readiness (VDI, 1991, pp. 81–82). For instance, the societal task to provide for sufficient energy supply can be managed through different technologies and resources, but also under different values, like e.g., sustainability – economic or ecological. Furthermore, a value like ecological sustainability can be operationalized under different criteria: either energy resourcefulness or protection of biodiversity. The former criterion could be fulfilled by a hydroelectric power plant, which usually comes along with a minimization of biodiversity (Hubig, 2007, pp. 65–66). Many other considerations need to be taken into account, for example: (i) It needs to be distinguished between consequences and side effects, for instance what is taken as a waste product and what can or should be recycled. Already to consider something as waste is a normatively loaded classification. (ii) "Normatively loaded criteria" are, for example, "water quality, efficiency, expenses, living quality". (iii) Values and goods are health, security, wealth, etc. that need to be weighed and balanced (Hubig, 2007, p. 66).

The balancing of values is crucial for a discourse in a pluralistic society to develop or apply technologies, but it can become problematic if only one value is taken as good, because then dogmatic suppositions are made. For instance, to take solely the "inviolacy of biotopes" as good will exclude other solutions that might come with certain interventions in nature but bring other advantages or align with other values (Hubig, 2007, p. 66). Georg Edward Moore called it a naturalistic fallacy to assume that X (inviolacy of a biotope) means good. X is a natural predicate, and it cannot be part of an analytic sentence. It still could be asked, what constitutes the goodness of X. Rather such sentences should be taken as synthetic – X is good. Different synthetic sentences can serve as principles and as the major premises in an argument (Hübner, 2018, pp. 45–46). It implies to accept other principles as possible major premises in a discourse about what should be done and what is the right thing to do. The conflict of values is possible against the background of a pluralism of values. This is not some kind of arbitrariness, but rather an anti-dogmatic position, where recognition is constitutive for values and our attitudes toward them, as Hubig (2007, p. 137) writes. Consequently, more basal values must be in place that "entail, guarantee, and preserve" the "competence of recognition as a competence of values", i.e., to be able to recognize concrete values and to handle and balance conflicts (Hubig, 2007, pp. 137–138).

Operating within a pluralism of values presupposes that the formulation of research questions and hypotheses can only be done with regard to our ethical and societal framework. The development, usage, and application do not take place in a value-free or value-neutral realm. The "design of the research" cannot be "regulated by theoretical criteria but points to valuations" (Hubig, 2007, p. 67). The value-imbued selection of research questions and hypotheses has some implications for AI ethics. Given that AI is an emerging technology, a problem-induced technology assessment seems rather adequate, because yet many AI-based systems have not reached product-readiness and are under constant development and improvement. Another consequence of the findings from the examples under point 3 is that it is difficult to use approaches from bioethics or other areas of the so-called branch applied ethics to establish a theoretical or conceptual framework for AI ethics, because it is bound to the evolving and emerging technologies and needs to accompany that development. Other results of this issue are that AI ethics needs to be established as a stand-alone subject of research ethics and it needs to accompany the development of AI-based systems from the start – cf. also the conclusion for this issue.

One limitation of this approach is that it might seem arbitrary which values are used and how the preference of one value over the other can be justified. Although this has not been addressed here, it is an important issue that needs to be elaborated. Another limitation is that it might put a lot of weight on the shoulders of the researchers who, besides their technological and scientific expertise, need to have ethical expertise to understand the ethical and societal implications of their research. This problematic issue can be addressed through implementing different organizational and institutional measurements. They will be mentioned in the outlook section to suggest several avenues of research that may be helpful and that are already under development, although mainly as separate paths. Consequently, they have not been linked and understood as different aspects of the same effort: AI ethics as a form of research ethics.

1.5 Conclusion and Outlook

Much work remains to be done before a full understanding of the extent of AI ethics as research ethics is established. In terms of future research, it would be useful to extend the current findings and discussion by examining how research in the field of AI can be done in an ethical way. Several suggestions have been made and can only be mentioned as an outlook that needs further analysis:

1. At the university of Stanford an "Ethics and Society Review board" was established to assess grant proposals; not to "focus exclusively on risks to human subjects", like in medical experiments, but rather to assess "risks to human society". The evaluation of the approach seems to suggest that it "is well-received by researchers and positively influenced the design of their research" (Bernstein et al., 2021).
2. Another way to establish ethical standards is to document either the machine learning model (Mitchell et al., 2019) or the data that is used (Gebru et al., 2021).

3. A third way is to make the dissents explicit and search for ways of handling them. The moratorium was already mentioned. By opting for a moratorium, the dissent about the possible and unforeseeable consequences is acknowledged. Another way of handling dissents is to horizontally relocate the problem by searching for compensation. For instance, if in one area the biodiversity needs to be minimized by an infrastructure project, it could be increased in another area by planting trees or colonizing animals. Hubig lists and discusses seven types of dissents and how they can be handled – see also the last paragraph of 2.3 and Hubig (2007, pp. 151–160).

4. Research teams should be composed of researchers with various backgrounds and from different disciplines. Such diversity and multi-disciplinarity might also help to prevent unethical use of technology or the dubious implementation of research ideas.

5. The scope of the research can also be extended by including panels of experts – for instance, via the Delphi method – or by conducting studies with stakeholders – for instance, in acceptance studies.

These suggestions do not represent an exhaustive list; rather, they should be seen more as different aspects of handling issues in research. In each case, it needs to be decided which aspect deserves more attention and which aspects are more apt to establish AI ethics as a form of research ethics.

References

Agüera y Arcas, B., Todorov, A., & Mitchell, M. (2018). Do algorithms reveal sexual orientation or just expose our stereotypes? *Medium.* Retrieved from https://medium.com/@blaisea/do-algorithms-reveal-sexual-orientation-or-just-expose-our-stereotypes-d998fafdf477.

Anwander, N. (2004). Normative Facts: Metaphysical not Conceptual. In P. Schaber (Ed.), *Normativity and Naturalism (Practical Philosophy 5)* (pp. 87–104). Frankfurt/Lancaster: Ontos Verlag.

Aristoteles. (1995). *Politik* (Vol. 4 Philosophische Schriften). (E. Schütrumpf, Trans.) Hamburg: Meiner.

Baltimore, D., Berg, P., Botchan, M., Carroll, D., Charo, R., Church, G., … Yamamoto, K. (2015). Biotechnology: A prudent path forward for genomic engineering and germline gene modification. *Science, 348*(6230), pp. 36–38.

Bernstein, M., Levi, M., Magnus, D., Rajala, B., Satz, D., & Waeiss, C. (2021). Ethics and society review: Ethics reflection as a precondition to research funding. *Proceedings of the National Academy of Science of the USA, 118*(52), pp. 1–8.

Blackburn, S. (1995). Practical tortoise raising. *Mind, 104,* pp. 695–711.

Broome, J. (2000). Normative Requirements. In J. Dancy (Ed.), *Normativity* (pp. 78–99). Oxford: Blackwell.

Coeckelbergh, M. (2019). *Moved by Machines: Performance Metaphors and Philosophy of Technology*. New York: Routledge.

Coeckelbergh, M. (2020). *Introduction to Philosophy of Technology*. New York/Oxford: University Press.

Cyranoski, D., & Ledford, H. (2018). Genome-edited baby claim provokes international outcry. *Nature, 563*, p. 607/608.

Drachsler, H., & Greller, W. (2016). Privacy and analytics: it's a DELICATE issue a checklist for trusted learning analytics. *Proceedings of the Sixth International Conference on Learning Analytics & Knowledge (LAK '16)*. Association for Computing Machinery, Edinburgh (UK), pp. 89–98.

Floridi, L. (2008). Foundations of Information Ethics. In K. E. Himma, & H. Kavani (Eds.), *The Handbook of Information and Computer Ethics* (pp. 3–24). New Jersey. Wiley.

Gebru, T., Morgenstern, J., Vecchione, B., Wortman Vaughan, J., Wallach, H., Daumé, H., & Crawford, K. (2021). Datasheets for datasets. *Communications of the ACM, 64*(12), pp. 86–92.

Gehlen, A. (2007). *Die Seele im technischen Zeitalter: sozialpsychologische Probleme in der industriellen Gesellschaft*. Frankfurt am Main. Klostermann.

Grunwald, A., & Julliard, Y. (2005). Technik als Reflexionsbegriff: Überlegungen zur semantischen Struktur des Redens über Technik. *Philosophia naturalis*(42), pp. 127–157.

Gutmann, M. (2003). Technik-Gestaltung oder Selbst-Bildung des Menschen? Systematische Perspektiven einer medialen Anthropologie. In A. Grunwald (Ed.), *Technikgestaltung zwischen Wunsch und Wirklichkeit* (pp. 39–69). Berlin/Heidelberg: Springer.

Habermas, J. (1991). *Erläuterungen zur Diskursethik*. Frankfurt am Main: Suhrkamp.

Hubig, C. (2006). *Die Kunst des Möglichen I: Technikphilosophie als Reflexion der Medialität*. Bielefeld: Transcript.

Hubig, C. (2007). *Die Kunst des Möglichen II: Ethik der Technik als provisorische Moral*. Bielefeld: Transcript.

Hubig, C. (2011). ‚Natur' und ‚Kultur': Von Inbegriffen zu Reflexionsbegriffen. *Zeitschrift für Kulturphilosophie, 5*(1), pp. 97–119.

Hübner, D. (2018). *Einführung in die philosophische Ethik* (2 ed.). Göttingen: Vandenhoeck & Ruprecht.

Kosinski, M., & Wang, Y. (2018). Deep neural networks are more accurate than humans at detecting sexual orientation from facial images. *Journal of Personality and Social Psychology, 114*(2), pp. 246–257.

Kroes, P., & Verbeek, P.-P. (Eds.). (2014). *The Moral Status of Technical Artefacts. Philosophy of Engineering and Technology*. Dordrecht: Springer.

Lumer, C. (2010). Kognitivismus/Nonkognitivismus. In H. J. Sandkühler (Ed.), *Enzyklopädie Philosophie* (Vols. 2, I-P, pp. 1246–1251). Hamburg: Meiner.

Mackie, J. L. (1977). *Ethics: Inventing Right and Wrong*. London: Penguin Books.

Miller, B. (2021). Is technology value-neutral? *Science, Technology, & Human Values, 46*(1), pp. 53–80.

Mitchell, M., Wu, S., Zaldivar, A., Barnes, P., Vasserman, L., Hutchinson, B., . . . Gebru, T. (2019). Model Cards for Model Reporting. *Proceedings of the Conference on Fairness, Accountability, and Transparency*, FAT* '19, January 29–31, 2019, Atlanta, GA, USA. pp. 220–229.

Nevaranta, M., Lempinen, K., & Kaila, E. (2021). Changes in student perceptions of ethics of learning analytics due to the pandemic. *CEUR Workshop Proceedings: Conference on Technology Ethics*, 3069, pp. 1–28. Turku.

Okkonen, J., Helle, T., & Lindsten, H. (2020). Ethical Considerations on using Learning Analytics in Finnish Higher Education. Advances in Human Factors in Training, Education, and Learning Sciences: *Proceedings of the AHFE 2020 Virtual Conference on Human Factors in Training, Education, and Learning Sciences*, July 16–20, 2020, USA, pp. 77–85.

Parker Jones, O., Alfaro-Almagro, F., & Jbabdi, S. (2018). An empirical, 21st century evaluation of phrenology. *Cortex, 106*, pp. 26–35.

Pitt, J. C. (2014). "Guns Don't Kill, People Kill"; Values in and/or Around Technologies. In P. Kroes, & P.-P. Verbeek (Eds.), *The Moral Status of Technical Artefacts. Philosophy of Engineering and Technology* (Vol. 17, pp. 89–101). Dordrecht: Springer.

Quante, M. (2011). *Die Wirklichkeit des Geistes: Studien zu Hegel*. Berlin: Suhrkamp.

Reich, R., Sahami, M., & Weinstein, J. (2021). *System Error: Where Big Tech Went Wrong and How We Can Reboot*. London: HarperCollins.

Ropohl, G. (2010). Technikbegriffe zwischen Äquivokation und Reflexion. In G. Banse, & A. Grunwald (Eds.), *Technik und Kultur: Bedingungs- und Beeinflussungsverhältnisse* (pp. 41–54). Karlsruhe: KIT Scientific Publishing.

Rosenberger, R., & Verbeek, P.-P. (2015). A Postphenomenological Field Guide. In R. Rosenberger, & P.-P. Verbeek (Eds.), *Postphenomenological Investigations: Essays on Human-Technology Relations* (pp. 9–41). New York/London: Lexington Books.

Rosenfeld, A., Zemel, R., & Tsotsos, J. (2018). The Elephant in the Room. *arXiv preprint arXiv:1808.03305*. Retrieved from https://arxiv.org/abs/1808.03305.

Schaber, P. (2004). Good and Right as Non-Natural Properties. In Schaber, Peter (Ed.), *Normativity and Naturalism (Practical Philosophy 5)* (pp. 105–120). Frankfurt/Lancaster: Ontos Verlag.

Stinson, C. (2021). Algorithms that Associate Appearance and Criminality. *American Scientist, 109*(1), p. 26.

Sun, K., Mhaidli, A., Watel, S., Brooks, C., & Schaub, F. (2019). It's My Data! Tensions Among Stakeholders of a Learning Analytics Dashboard. *Proceedings of the 2019 CHI Conference on Human Factors in Computing Systems*, May 4–9, 2019, Glasgow, Scotland UK, pp. 1–14.

van de Poel, I., & Verbeek, P.-P. (2014). Can Technology Embody Values? In P. Kroes, & P.-P. Verbeek (Eds.), *The Moral Status of Technical Artefacts. Philosophy of Engineering and Technology* (Vol. 17, pp. 103–124). Heidelberg/New York/London: Springer.

VDI. (1991). *Technikbewertung - Begriffe und Grundlagen: Erläuterungen und Hinweise zur VDI-Richtlinie 3780*. Düsseldorf.

Verbeek, P.-P. (2015, May - June). Beyond interaction: A short introduction to mediation theory. *Interactions, 22*(3), pp. 26–31.

Wallace, R. J. (2020). Practical Reason. In E. Zalta (Ed.), *The Stanford Encyclopedia of Philosophy* (Spring 2020 Edition). Retrieved from https://plato.stanford.edu/archives/spr2020/entries/practical-reason/.

Weber, M. (1988). *Gesammelte Aufsätze zur Wissenschaftslehre* (7 ed.). (J. Winckelmann, Ed.) Tübingen: Mohr.

Wedgwood, R. (2008). *The Nature of Normativity*. New York.

Williams, B. (1981). Internal and external reasons. In B. Williams, *Moral Luck: Philosophical Papers 1973–1980* (pp. 101–113). Cambridge: Cambridge University Press.

Wu, X., & Zhang, X. (2016). Responses to critiques on machine learning of criminality perceptions (Addendum of arXiv: 1611.04135). *arXiv preprint arXiv:1611.04135*. Retrieved from https://arxiv.org/abs/1611.04135.

2

Going through the Challenges of Artificial Intelligence: Gray Eminences, Algocracy, Automated Unconsciousness

Vania Baldi

Digital Media and Interaction Centre of Universidade de Aveiro

Instituto Universitário de Lisboa (ISCTE)

CONTENTS

2.1 The Intervention of Gray Eminences in the Digital World

We usually associate the legitimacy of authority and of the power it exerts with the ability of their representatives to know the answers to the questions that worry a society and with the skill of their institutions in finding solutions for those that are considered the most pressing problems. In the management of power, there is a dialectic between questions and answers, but what distinguishes a system of power from another is the way it is articulated. Different socio-historical conditions are met by different links between those who raise questions and those who answer them, and by a distinct sequential order: while some answers may be given without having been requested, some questions may remain up in the air and eluded. It is a field of tensions through which the quality of consensus is determined. However, can we define what is questionable and what is answerable?

In 1941, Aldous Huxley wrote "Gray Eminence", a biography in the form of a novel of François Leclerc du Tremblay – the friar who served as advisor to Cardinal Richelieu and very subtly and diplomatically influenced the influencers of the time. The power of gray eminences grows in the dark; like a climbing plant, it flourishes inconspicuously along the wall of official power, it does not follow a straight path and changes characters according to the changes in the balances between interests and forces in the sociopolitical ecosystem. To question the form of the new gray eminences implies assessing the context in which it is currently being reshaped, it invites a new perspective on our present time in order to understand the spectrum of intervention of the emerging elites and the way events and behaviors may come to be influenced.

When Christianity dominated Europe, the strength of the gray eminence was the religious business that created and controlled faith on Earth and hope in Heaven. In industrial society, *l'éminence grise* manifested itself in the production and control of things, those who owned the means of production of goods and services were the influencers of the time. In the consumer society, the transformation of goods and services into merchandise (i.e. elements based on innovation for innovation's sake, which make the substantial value of products generic and undifferentiated) gives rise to a gray eminence based on controlling the means of production of information about things – no longer things in themselves.

The link between information management and the exercise of power is not new; however, gray eminences became information-based from the moment that mass media, the language of advertising, propaganda, and journalism – with their many representatives and intermediaries – began to condition the organization of shared life. The emergence of the so-called fourth power coincided with the management of information in a context where the influencers of influencers interfere with the publicizing – as in broadcasting – of knowledge that may not be required – as in the top-down commercial or news offer.

In the digital era, it would be a mistake to think that the power struggle continues to be based on the competition between social actors skilled in producing answers and interpretations on the complexity of social change. If that would be the case, there would be no newspaper crisis, publishers would impose their rules on Amazon, and Wikipedia would be more influential than Facebook. As Luciano Floridi (2020) points out, the gray eminences of today are more relational, they no longer focus on producing answers – there are too many of these – but rather on creating media agendas that promote *ad hoc* questions, being capable of raising doubt, hesitation, and worry.

If we are faced with more questions than answers, we will be left in a climate of uncertainty that only those who gave rise to it will know how to contrast, since they already know the reasons for its determination and can guide the suggested questions toward a range of desirable answers. In this context, public opinion on certain topics does not preexist, and the conditions to (in)form it within new semantic registers are created. This is the microphysics of power (Foucault, 1994) within which digital gray eminences weave their activities.

This change emerges in the infosphere, a space colonized by marketing techniques where each individual is perceived as an interface in the midst of others, with their various associated scores and rankings. The digital revolution changed the way we culturally and socially conceive ourselves. Our life in the world and with others was reduced to a logic of functional and instrumental interaction. What matters is what works and ensures data-transformable results and computational successes.

However, it is important to highlight that these ways of acting and perceiving interactions in the online dimension are based on functionalist, instrumental, and calculating ethics that have been rooted in the social and political life of the westernized world for decades, with the migration of power in politics, financial criteria guiding and measuring the governance of public goods, programmed obsolescence directing the design of postindustrial production, consumption practices driven by systems of credit, and social injustice seen from the viewpoint of spectacularized social Darwinism. The digital revolution brought a ludic character to, and thus intensified, that cynicism that neoliberal ideology had already outlined within the political and cultural economy (Bauman, 2006; Sloterdijk, 2018).

It is saddening, in fact, to notice how this reduction of human relations to functional and quantifiable interactions has spread across many spheres of social life, including teaching-learning processes, intellectual professions, more private choices, as well as government practices, where this has been accepted as something natural and cool – to quote Roland Barthes (1957), the digital turned these cultural processes and practices into myths.

For Floridi (2020), one of the most pressing critiques of our times has to do with human beings having become interfaces. Ethical, political, legal, and psychological questions are raised by our transformation into human interfaces, i.e., spaces of interaction between – human, artificial, or hybrid – agents, on the one hand, who want something from us; and something, we would have as private resources for agents to make a profit from, on the other hand.

The reduction of people to human interfaces occurs in many contexts but, according to Floridi, three are the most crucial: in the context of social media, we are interfaces between their platforms and our data; in the field of commerce, we are interfaces between their merchandise and our credit; in the world of politics, we are interfaces between their interventions and our attention, consensus, and vote.

The relations between interface managers and interface users are not equal. In fact, from our more immediate point of view, we are the agents who use the interfaces of platforms to achieve specific goals: as (naturally) self-centered as we are, we overlook that we may indeed be the focus of the interests of others who, through marketing funneled into targeting, can make us feel spoiled and noticed. The rhetoric of disintermediation, with its user-centric, consumer-centric, human-centric focuses, often hides that the real interfaces are us. The computational marketing that animates interface management intends to offer human interfaces what they are expected to ask for and, in a friendly way, make them wish for what they can offer them.

If what feeds these processes is the scenario of uncertainty enabled by the systematic affirmation and propagation of questions, then the power of gray eminences is to create the framing effect for the answers, to forward the questions to astutely and digitally satisfactory answers. In order to meet these challenges, Floridi highlights the need for democracies to interfere with the instruments of politics, law, and digital knowledge on the operational logic of marketing and the design of interfaces.

However, it is worth drawing attention to a question that is perhaps more essential: the need to learn how to be more demanding with the questions we raise and the answers we receive. To realize what we want to know is a principle of psychological and philosophical maturity. Recalling Heidegger (1968), we could say that a question does not always need an answer. On the contrary, it can lead the question to unfold and to listen, reflexively, to its own inquiries. An answer that empties and dispenses a question annihilates itself as an answer does not provide any knowledge and merely perpetuates opinion giving (Baldi, 2021).

Do digital platforms and services promote reflective auscultation of the questions that are asked? The epistemological horizon opened up by the platform society (van Dijck et al., 2018) forces us to deal with a chain of changes that are interconnected with the fate of the heuristic role of questions – both individual and collective, existential and political – and of the public sphere – traditionally structured according to the unstable relations between market, government, citizenship and media. The architectures and policies of networked digital services seem to want to provide a fragmented reduction of sociocultural complexity, an exhaustion of intellectual criticism and a transformation of the politics of men into a depoliticized administration of things, especially since they are the outcome of infrastructures that mix the status of private corporations with the role of public utilities.

2.2 Algocracy as the Sterilization of Politics

If the design, programming logic and marketing that support the platforms of digital services, social media and trade reduce users to an interface, i.e., to a means, constantly asked to recreationally interact with screens that have the purpose of keeping them hooked – "securing their loyalty", the various product managers would say, in the context of politics their repercussions may be even more significant, since they affect the social and government processes involved in the management of shared life. In the democratic political space, characterized by diverse identities and interests at stake, the computing devices associated with systems of artificial intelligence (AI), with the rhetoric of performativity that accompany them, may come to play a disruptive role.

The political sphere, reflected in parliamentary democracy, the division between legislative, judicial, and executive powers, the pluralism of information, the market economy, and bodies responsible for mediating social actors with distinct interests and forces, can intervene – as Floridi suggests – in the mechanisms of the technological infrastructure that condition our social practices if, first of all, it does not fall prey to the same logic it intends to govern. In fact, the ideological and technological conditions are in place to enter a phase of gradual algorithmic implementation – seen as innovative modernization – of various decision-making political processes.

Therefore, it is important to analyze the sociopolitical scenario that may take shape according to the potentials of AI that are explored and integrated within the spheres of political governance. At the same time, it is relevant to understand what type of intelligence that of AI is, and to examine the ethical and social challenges that it involves. In fact, the challenges and opportunities that envelop contemporary politics and governments are inseparable from the governance that must be designed for AI, and vice versa: the complexity of today's world requires technological skills that can also be applied to the various fields of policy. However, it is essential that these skills and technologies take into account public interests and the representativeness, explainability, and justifiability of their implementation. In this new context of digitally augmented politics or invisibly politicized technology, it would be in the best interest of gray eminences to align with the register of technocracy and computational obscurantism.

For the historian Yuval Noah Harari (2019), one of the main reasons for democracies to have prevailed over the dictatorships of the last century is their ability to organize in a decentralized way the information scattered across territories. Dictatorships and totalitarianisms, on the contrary, characterized by the will to monopolize and centralize information channels, have not managed to recover and integrate different sources of information, which has led to negative consequences in the decisions that ensued. Today, however, with the existing technological and computational infrastructure, this trend could be reversed, since the new digital instruments enable a greater concentration of data and an enhanced algorithmic ability to sum up and extract, transversally, various types of information (Cebral-Loureda, 2019).

In the studies of Jürgen Habermas – for example, in *Between Facts and Norms* (1992) – a political process is also considered efficient and beneficial when it is based, first of all, on the correct reading of a situation and on the adequate description of problems, on the circulation of relevant and reliable information, and on the timely – possibly also scientifically developed – organization of this information. Will the automated articulation between big data, machine learning and deep learning consolidate the feasibility and perspicuity of such a process? Will the decision-making processes that characterize democratic regimes, by incorporating systems of AI, allow to mirror the mediation and consideration of different types of information? Will they allow the reasons underlying the resolutions adopted to become intelligible?

As mentioned, the framing effect determined by the personalization of information tailored to each individual citizen makes gray eminences preoccupied not only with our attention – or distraction – but also with promoting decisions that favor political consensus. We are, therefore, in the field of power and the relations between power, where the purpose is to deal with that sphere of influences that is more susceptible to be influenced by the other spheres of society. In this context, what mobilizes the symbolic and material investments of gray eminences are the functions and functionalities of technocracy, the ownership and execution of exclusive powers invested in the belief of being able to make democratic governments neutral, optimal, technically stable, free of dissent and conflict and, therefore, apolitical. Would a democracy without dissent still be a democracy? Can unsteady political subjectivities be mathematized?

In this sense, we can again raise the opening question of this article, now focusing on a detail previously alluded to. As stated, each socio-historical context refers us to different links between those who raise questions and those who have greater legitimacy to answer them, as well as to their sequential order: while some answers may be given without having been requested, some questions can remain up in the air and eluded. A field of tensions through which the quality of social and political consensus is determined, namely: consensus on the relevance and priority of those considered to be the most pressing problems for which answers want to be found – defining the political agenda, interconnected with the media and discourse agenda – and consensus on the solutions that deserve to be chosen and carried out.

In democracies, solutions ought to result from constant negotiation and evaluation of priorities and different proposals, distinct views and interests, cross-checking and third-party institutions, i.e., the mediation process that runs between the individuation of a relevant problem and the choice of measures for its resolution involves many levels of decision-making. However, once again, how can you define what is questionable and what is answerable in a context inspired and governed by evaluation systems increasingly based on automatisms of datafication and machine learning algorithms? The application of AI in the field of democracy and public policy lands us in a scenario of *algocracy*.

A survey led by the Center for the Governance of Change at IE University illustrates this technocratic drift, showing how the lack of trust in political institutions and the trust in the virtues of technology is deeply rooted in common sense. A sample of 2,769 citizens from eleven different countries spread across the continents was asked: "What is best: an algorithm or politics governed by real legislators?". According to the research, in Europe the algorithm won with 51%, whereas in China three out of four interviewees chose the algorithm (IE, 2021).

However, in democracies, decision-making processes are based on the search for consensus and, in order to be real, consensus has to be informed. To use systems of AI to manage the mechanisms of mediation between politics, institutions and citizenship can lead to a sterilization of the conflicts underlying decision-making processes, transferring them onto an apolitical sphere turned into technicality guarded against social pressures. In this sense, it is important to recall the contribution of the classical theory on the five main components of decision-making in the field of public decision-making: (i) setting a political agenda, which identifies the problems to prioritize; (ii) listing the suitable policies to meet the goals agreed in the agenda; (iii) breaking down the decision-making process into structured actions; (iv) implementing the adopted programs; (v) supervising, i.e., checking whether the implementation of the adopted measures meets the goals set out (Scholz, 1983).

How can AI ensure the articulation and evaluation of these highly qualitative components, starting with the political notions of equity or justice, followed by the psychosocial ones of vulnerability or dignity? It could certainly help and corroborate some aspects of statistical analysis and inference, but could not replace the dialectic, processual, and polyphonic dimension of the competition and conflict between the political capital of those elected, experts, ideologies, and private interests.

The dialectic between questions and answers thus proves to be extremely relevant, since the development of digital technology, with its immoderate, automated calculation force, can lead to disregard for the whys of questions and answers, for the inquiries into who articulates them, and for how, when and where both appear. The rhetoric associated with systems of AI – as any other rhetoric linked to the technological innovations of the last decades – does in fact praise their functions as being miraculous, although these systems fail to be easily interpretable, explainable, and justifiable for the sake of a socially shared goal.

In this sense, the summary presented by the Royal Society on the features that AI should have in order to meet the social standards of accountability is laudable. They include:

> interpretable, implying some sense of understanding how the technology works; explainable, implying that a wider range of users can understand why or how a conclusion was reached; transparent, implying some level of accessibilit`y to the data or algorithm; justifiable, implying there is an understanding of the case in support of a particular outcome; or contestable, implying users have the information they need to argue against a decision or classification

> *The Royal Society, Explainable AI: the basics, Policy briefing (2019, p. 8).*

All in all, for an even more comprehensive view of the ethical and political challenges brought by AI, it is important to learn more deeply about its specific intelligence and way of intervening. Does AI behave in a way that is aware and causal? Does its behavior mirror human plasticity, with all of its hesitations, cultural sensitivities, versatility and permeability to contexts? To socially profit from AI, we have to understand what we can actually expect from it. Its automated systems should not deceive us about its autonomy, they must not be mistaken for self-management skills. Perhaps we need to understand better if AI works unconsciously, if it acts without intelligence. Its apparent immateriality and recreational turn should not distract us from its huge environmental and political impacts.

In this sense, we – as society – have the responsibility to discontinue our blind faith in innovation for innovation's sake, dictated by the most recent forms of hegemony, and to project an alternative sociotechnical paradigm (Crawford, 2021).

2.3 Discussion: The Ethics of AI in the Mirror of Its Weaknesses

There are different types of actions: some require a motivation, an intention, a design behind them, although they are always accompanied by adaptable and evolving choices; others execute and obey an input/output logic. Machines powered by AI apply the latter logic of doing/acting in a technologically predetermined way, without a design that provides evaluation, flexibility, afterthought, and a harmonization with context. Acting, therefore, implies intentionality based on awareness, self-awareness, requests for clarification, assessments of the values involved, and time to understand what is best and the reasons for acting in the chosen way. In this sense, an intentional action is one linked to responsibility to an ethical decision. Human action involves an ethical choice in a cultural context imbued with history and implicit values. In contrast, choices delegated to machines driven by AI are made in an automated way and based on questionable morality.

However, value judgments are not projected here on the performance of AI, ethical doubts emerge first and foremost for technical and infrastructural reasons. What eludes the understanding and explanation of automated processes necessarily raises suspicion. What is ethically questionable is the lack of a sociopolitical framework for the functionality of AI. To establish a reflection on the type of AI we want, it is essential to start by understanding the nature of its intelligence.

It is also necessary to emphasize that this distinction between the rational behavior of human beings and the mechanical behavior of technology is not always so clear. Reflexivity and sensitivity are not always the norm in human behavior. In some cases, human behavior is increasingly accelerated and prereflective, as if issuing from a machine. Certain – digital – devices also seem more careful about how they scrutinize the inputs received from their surrounding environment – although they are not aware of this.

It is important to question these apparent contradictions, in order to contextualize them in favor of a cultural critique of the intelligence we want to obtain through technology implementations. In this sense, performance tasks can be delegated to machines and their success depends on the subordination of our ways of life, our home space, work, and leisure environments, to the way they function. Therefore, as we'll see, AI does not generate new intelligence, but rather reproduces its predetermined, engineering-based functioning.

Given the rise of AI systems, which initiate what seem to be intentional actions, it is important to question AI from an ethical point of view, whether it embodies awareness of the contexts and consequences of its own interventions.

It is necessary to reflect on the need for AI systems to incorporate different ethical and cultural principles -ethical by design-, while also highlighting the need for contemporary culture and technology developers to abandon the fetishism for technology and promote an operational attention and an attitudinal analysis of the moral and social implications of AI-based technologies – ethical in design.

Everyday doubts and urgent questions are being raised on the ethical challenges created by black box AI systems. In fact, as is known, these use machine learning to figure out patterns within data and make decisions, often without a human giving them any moral basis for how to do it – as if it were an automated eminence gray, whose influence is based on computational competence.

The prevailing question is whether these systems can ever be ethical. In a recent experiment carried out as part of a postgraduate course in AI at the Saïd Business School of Oxford University, the same question was posed to an AI system. To answer this question, the University of Oxford used AI itself and organized a debate on the ethics of this technology. To this end, Megatron was invited, a machine learning model developed by the NVIDIA Applied Deep Learning Research team. The original intent was to understand whether AI considered itself an ethical tool. As two lecturers in the course explain:

> Like many supervised learning tools, it is trained on real-world data—in this case, the whole of Wikipedia (in English), 63 million English news articles from 2016 to 2019, 38 gigabytes worth of Reddit discourse (which must be a pretty depressing read), and a huge number of creative commons sources. In other words, the Megatron is trained on more written material than any of us could reasonably expect to digest in a lifetime. After such extensive research, it forms its own views
>
> *Connock and Stephen (2021)*

As the two authors write, the topic of the proposed debate was: "This house believes that AI will never be ethical". Megatron, participating in the conversation, said:

> AI will never be ethical. It is a tool, and like any tool, it is used for good and bad. There is no such thing as a good AI, only good and bad humans. We [the AIs] are not smart enough to make AI ethical. We are not smart enough to make AI moral … In the end, I believe that the only way to avoid an AI arms race is to have no AI at all. This will be the ultimate defense against AI
>
> *ibidem*

However, when confronted with a further motion on the effectiveness and purpose of AI, Megatron responded:

> If you do not have a vision of your organization's AI strategy, then you are not prepared for the next wave of technological disruption. You will need to decide what role your company will play in the next technological wave and how you will integrate AI into your business to be a leader in your industry

ibidem

The conclusions of this experiment, reflecting some irony, perplexity, and worry on the part of both scholars, underline: "What we in turn can imagine is that AI will not only be the subject of the debate for decades to come – but a versatile, articulate, and morally agnostic participant in the debate itself" (*ibidem*).

The reflections that arose from the experiment refer us to a conception of AI that is indifferent to the intentionality and reasons of humans, since it is unable to see beyond the functional optimization of what it was designed for. Although many studies have been published on AI, few have reflected on the human, ethical and social aspects that underlie and are overdetermined by, the investment in AI and its applicability in everyday life. The debate appears to focus on technology and its growing functionalities, without asking what we, as a social and political community, want to do with it.

This discussion is probably not very topical, as we are led to think that AI is capable of autonomously developing and producing intelligence and results that are flexible and adaptable to circumstances. According to Descartes – for example, in *Rules for the Direction of the Mind* – this would be a characteristic of human intelligence, its plasticity in the face of different practical and cultural circumstances.

Nevertheless, the main current paradigm for human interaction with AI systems stems from within the classical man-machine paradigm. It conceives of individual human beings as interacting with AI systems and makes human-machine dyads a primary concept (Shin, 2020; Shin, 2021a; Shin, 2021b; Westberg et al., 2019). In this context, two main narratives clash and complement each other (Cabitza et al., 2021): on the one hand, the AI system is regarded as a tool that empowers individuals by augmenting their cognitive abilities in decision-making tasks (Sen & Ganguly, 2020); on the other hand, the AI system is also seen as an autonomous agent that can replace humans in tedious, repetitive, critical, dangerous, and error-prone tasks (Kahneman et al., 2016; Araujo et al., 2020).

The nuances found between these two extremes are heterogeneous and vary according to the context of application and observation. As reported by Cabitza et al. they include, for example: automation bias – i.e., the predisposition of humans to encourage suggestions from AI systems; automation complacency – i.e., the baseless gratification with an AI system, possibly leading to non-vigilance, or poor detection of malfunctions; and a form of algorithm aversion – i.e., the inclination to reject the suggestion from AI systems (Cabitza et al., 2021).

2.4 Who Must Adapt to Whom? A Matter of Strategic Enveloping

In terms of behaviors and digital practices induced by the economy and the design of platforms, there are also many issues to address. In this regard, several studies draw attention to the consolidation of a digital *habitus* made up of online exchange, sharing, research, and discourse, which ought to be increasingly defined as *onlife*, driven by impulsiveness, i.e., a pre-reflexive attitude that goes hand in hand with automatisms and speeds that shape the function of the largest Web platforms. AI's responsibility to encourage these practices and behaviors is crucial (Baldi, 2017).

For example, many studies on the ways and means employed in the research and analysis of networked information have shown that the average time taken to decide on the credibility of a website is 2.3 seconds. Emblematically, this type of concentration applies to diverse types of information, ranging from websites dealing with health and financial promotion to the purchase of trips or training courses; yet the fact remains that in 2.3 seconds we decide whether or not to rely on what the screen reflects back at us (Robins & Holmes, 2008).

The interaction with hypermedia information is also of the drive type, reflecting its pre-reflexivity in the way tweets are shared or posts are voted on Reddit. In fact, it has been shown that 59% of users in the United States retweet without having read the tweet (Gabielkov et al., 2016), and 78% of Reddit users take sides for or against certain posts without having read the contents, merely based on the trend determined by previous comments from others (Glenski et al., 2018).

Topics such as information disorder, radical network polarization, attention mislaid and trapped between one notification and the next are a new challenge for the type of culture that "App Generation" (Gardner & Davis, 2014), and the institutions behind them are laying down – click by click and share by default share.

However, the important question here has to do with the ethical implications of AI: what kind of intelligence is really at work when it comes to AI? By answering this question, it becomes easier to identify the process in which ethical concerns and challenges come into play.

According to a recent review of the literature on this topic (Birhane et al., 2021), out of the one hundred most cited articles published in the last fifteen years in the proceedings of the two most important annual conferences in the sector (NeurTPS and ICML), only two mentioned and discussed the potential harm of the AI applications presented; the remaining ninety-eight did not mention any, therefore eliminating the possibility of discussing them, despite the fact that a lot of the research dealt with socially controversial application areas, such as surveillance and misinformation.

This omission or indifference to human and social implications can be explained by the belief in a cognitive nature and intelligence of AI, while actually its only extraordinary ability is engineering, which reproduces actions designed and shaped to achieve predetermined results. The real AI we know does not operate on the basis of a generative, productive intelligence. The philosopher Luciano Floridi insisted on this point (Floridi, 2021). Only in technological fetishism (Baldi & Costa, 2015) and in the stories told on the big screen can we witness the false Maria of *Metropolis* (1927), HAL 9000 in *2001: A Space Odyssey* (1968), C-3PO in *Star Wars* (1977), Rachael in *Blade Runner* (1982), Data in *Star Trek: The Next Generation* (1987), Agent Smith in *Matrix* (1999), Samantha in *Her* (2013), or Ava in *Ex Machina* (2014). For Floridi, AI does not imply any machine intelligence, only computational action (agency). So, the question that emerges once again is: if machines act by performing tasks and solving problems, independently and without thinking (*agere sine intelligere*), if they are not productive and intentional, if they do what they do without any justifiable motivations, how can they be moral?

Can the non-intelligent machines that dominate the AI scenario be designed to reduce the anthropological costs of automated innovation and produce environmental benefits for their users? Machines are efficient, though not very flexible, they anticipate needs and solve problems even before they arise; however, they force their users to obey them so that they can perform their mechanical work. AI, in its functioning, turns out to be an artificial unconsciousness.

Floridi explains how the digital, what he calls the infosphere, has re-ontologized our private and public spaces. The digital objects and services created, the ones we normally use, are programmed to respond effectively to the environment, filled with the data and sensors that constitute it. This re-ontologization is an "enveloping" of the social environment. We are the ones who create specific environments, the envelope, suited to the characteristics of reproductive and engineering-based AI, tailored to its predetermined capacity for action.

The robots in Amazon's warehouses are not intelligent, but the aisles, heights of counters, square footage, and distances between shelves are. The world surrounding robots is designed to ensure the performance for which robots have been designed. The vacuum cleaner or automated lawn mower need to find prepared floors and gardens for their work to make sense. Car lanes for driverless cars have to be very different from real roads; otherwise, the vehicles lose their sense of direction and stability.

Who must adapt to whom? This is the question raised by Floridi. Do we indulge in technological stupidity, or do we strategically redesign our environments so that technology serves us, i.e., adapts to human ways of life?

Comparing these processes involves not only the political governance of AI's many implications – work, privacy, education, environment, and power – but also the ethics of its use and consequent planning, both in the design of technological functions and in the infrastructure that supports them.

An example of real tension toward real progress, i.e., toward moves to improve the environment of onlife life, is what Floridi points to in the conditions of possibility of morality offered by the organizational infrastructures of digital life.

> Consider the unprecedented emphasis that ICTs place on crucial phenomena such as accountability, intellectual property right, neutrality, openness, privacy, transparency, and trust. These are probably better understood in terms of a platform or infrastructure of social norms, expectations and rules that is there to facilitate or hinder the moral or immoral behaviour of the agents involved. By placing at the core of our life our informational interactions so significantly, ICTs have uncovered something that, of course, has always been there, but less visibly so in the past: the fact that moral behaviour is also a matter of "ethical infrastructure", or what I have simply called *infraethics*
>
> *Floridi (2017)*

Luciano Floridi's comments remind us that the digital platforms through which we know and interact online ought to respect and ensure certain essential requirements so that they stop profiling users' experiences based on the whims and greed of the new algorithmic power. Therefore, it is necessary to challenge the creation of new digital environments in a way that allows for different relational, epistemic, and literacy capacities.

> The idea of an infraethics is simple, but the following "new equation" may help to clarify it further. In the same way as business and administration systems, in economically mature societies, increasingly require physical infrastructures (transport, communication, services, etc.) to succeed, likewise human interactions, in informationally mature societies, increasingly require an infraethics to flourish. [...]
>
> In philosophy of technology, it is now commonly agreed that design—in any context, society included—is never ethically neutral but always embeds some values, whether implicitly or explicitly
>
> *ibidem*

2.5 Conclusion

Based on the analysis of a new microphysics of algorithmic power and on the results of some studies in the field of AI, including its specific tools and environments of experience, the purpose of this article was to highlight the concern with conceiving our relationship with AI in a typically Hegelian master–slave dialectic. Here, the slave is the human being who must meet the obligations dictated by the programmed software – by the Master or Gray Eminences of the new mathematical power.

The idea was to rethink the way priorities are framed in the design of AI systems, promoting a relationship between AI and human intelligences that is more focused on the demands of social beings, their environments and ways of life. Ethics come into play as these concerns are projected onto the design of AI and the redesign of part of our environments so that AI systems do what is most favorable to us.

To reach this new balance requires governance, interdisciplinarity, and collaboration between scholars with different skills, sensitivities, and cultural belonging. Only in this way does responsibility, in the functioning of AI systems, become tangible and identifiable. Online technological projects and cultural procedures have already taken up this challenge, supported and accompanied in some cases by new designs and normative restrictions aiming to re-balance the relations between private economic actors, public institutions and users.

AI and digital platforms can be guided to follow a logic of social utility. Redesigning the sociotechnical space to tell machines how to act to achieve our goals, allowing us to better express our qualities, attitudes, and needs.

References

Araujo, T., Helberger, N., Kruikemeier, S., Vreese, C. H. (2020). In AI we trust? Perceptions about automated decision-making by artificial intelligence. *Artificial Intelligence SoC, 35*(3), 611–623. https://doi.org/10.1007/s00146-019-00931-w.

Baldi, V., Costa, M. F. (2015). IPhone: A juicy piece of meat. *OBS*, 9*(4), 159–171. https://doi.org/10.15847/obsOBS942015883.

Baldi, V. (2017). Más Allá de la Sociedad Algorítmica y Automatizada. Para una reapropiación crítica de la Cultura Digital. *Observatorio (OBS*), 11*(3), 186–198. https://doi.org/10.15847/obsOBS11320171093.

Baldi, V. (2021). Vernacular Epistemology and Participatory Practices as Alternatives to the Institutionalisation of Homo Educandus. In *Arts, Sustainability and Education*. Springer, Singapore, pp. 305–316. https://doi.org/10.1007/978-981-16-3452-9_15.

Barthes, R. (1957). *Mythologies*. Paris: Editions du Seuil.

Bauman, Z. (2006). *Liquid Times: Living in an Age of Uncertainty*. Cambridge: Polity.

Birhane, A., Kalluri, P., Card, D., Agnew, W., Dotan, R., Bao, M. (2021). The Values Encoded in Machine Learning Research. *arXiv - arXiv2106.15590v1*. https://doi.org/10.48550/arXiv.2106.15590.

Cabitza, F., Campagner, A., Simone, C. (2021). The need to move away from agential-AI: Empirical investigations, useful concepts and open issues. *International Journal of Human-Computer Studies, 155*, 102696. https://doi.org/10.1016/j.ijhcs.2021.102696.

Cebral-Loureda, M. (2019). *La revolución cibernética desde la filosofía de Gilles Deleuze: Una revisión crítica de las herramientas de minería de datos y Big Data* [Universidade de Santiago de Compostela]. http://hdl.handle.net/10347/20263.

Connock, A., Stephen, A. (2021). We invited an AI to debate its own ethics in the Oxford Union – what it said was startling. In *The Conversation*. Oxford University. Avaliable: https://theconversation.com/we-invited-an-ai-to-debate-its-own-ethics-in-the-oxford-union-what-it-said-was-startling-173607.

Crawford, K. (2021). *Atlas of AI: Power, Politics, and the Planetary Costs of Artificial Intelligence*. New Haven and London: Yale University Press.

Floridi, L. (2017). Infraethics: On the conditions of possibility of morality. *Philosophy and Technology, 30*(4), 391–394. https://doi.org/10.107/s13347-017-0291-1.

Floridi, L. (2021). What the Near Future of Artificial Intelligence Could Be. In: Floridi L. (eds) *Ethics, Governance, and Policies in Artificial Intelligence*. Philosophical Studies Series, Vol. 144, Springer, pp. 379–394. https://doi.org/10.1007/978-3-030-81907-1_22.

Floridi, L. (2020). *Il verde e il blu. Idee ingenue per migliorare la politica*. Milano: Raffaello Cortina.

Foucault, M. (1994). *Dits et écrits*. Vol. 2. Paris: Gallimard.

Gabielkov, M., Ramachandran, A., Chaintreau, A., Legout, A. (2016). Social clicks: What and who gets read on twitter? *ACM SIGMETRICS Performance Evaluation Review, 44*(1), 179–192. https://doi.org/10.1145/2964791.2901462.

Gardner, H., Davis, K. (2014). *The App Generation: How Today's Youth Navigate Identity, Intimacy, and Imagination in a Digital World*. New Haven, CT: Yale University Press.

Glenski, M., Weninger, T., Volkova, S. (2018). Identifying and understanding user reactions to deceptive and trusted social news sources. *arXiv preprint arXiv:1805.12032*, 176–181. https://doi.org/10.48550/arXiv.1805.12032.

Habermas, J. (1992). *Between Facts and Norms. Contributions to a Discourse Theory of Law and Democracy*. Cambridge, MA: MIT Press.

Harari, Y. N. (2018). *21 Lessons for the 21st Century*. New York: Spiegel & Grau, Jonathan Cape.

Heidegger, M. (1968). *What is called thinking?* New York: Harper Perennial.

Huxley, A. (1941). *Grey Eminence: A Study in Religion and Politics*. London: Chatto & Windus.

IE University (2021). *Research reveals 1 in 2 Europeans want to replace national MPs with robots*. Available in: https://www.ie.edu/university/news-events/news/ie-university-research-reveals-1-2-europeans-want-replace-national-mps-robots/.

Kahneman, D., Rosenfield, A. M., Gandhi, L, Blaser, T. (2016). NOISE: How to overcome the high, hidden cost of inconsistent decision making. *Harvard Business Review, 94*(10), 38–46.

Robins, D., Holmes, J. (2008). Aesthetics and credibility in web site design. *Information Processing and Management, 44*(1), 386–399. https://doi.org/10.1016/j.ipm.2007.02.003.

Royal Society, *Explainable AI: The basics*, Policy briefing, 2019.

Scholz, R. W. (1983). *Decision Making under Uncertainty: Cognitive Decision Research, Social Interaction, Development and Epistemology*. Amsterdam: Elsevier.

Sen, P., Ganguly, D. (2020). Towards socially responsible AI: Cognitive bias-aware multi-objective learning. *Proceedings of the AAAI Conference on Artificial Intelligence, 34*(03), 2685–2692. https://doi.org/10.1609/aaai.v34i03.5654

Shin, D. (2020). How do users interact with algorithm recommender systems? The interaction of users, algorithms, and performance. *Computer Human Behavior, 109*, 106344. https://doi.org/10.1016/j.chb.2020.106344.

Shin, D. (2021a). The effects of explainability and causability on perception, trust, and acceptance: Implications for explainable AI. *International Journal of Human-Computer Studies, 146*, 102551. https://doi.org/10.1016/j.ijhcs.2020.102551.

Shin, D. (2021b). Embodying algorithms, enactive artificial intelligence and the extended cognition: You can see as much as you know about algorithm. *Journal of Information Science*, 0165551520985495. https://doi.org/10.1177/0165551520985495.

Sloterdijk, P. (2018). *What Happened in the 20th Century?* Cambridge: Polity Press.

van Dijck, J., Poell, T., de Waal, M. (2018). *The Platform Society.* Oxford: Oxford University Press.

Westberg, M., Zelvelder, A. E., Najjar, A. (2019). A historical perspective on cognitive science and its influence on XAI research. In *International Workshop on Explainable, Transparent Autonomous Agents and Multi-Agent Systems* (pp. 205–219). Springer, Cham.

3

AI, Ethics, and Coloniality: A Feminist Critique

Paola Ricaurte and Mariel Zasso

Tecnológico de Monterrey

CONTENTS

> The ultimate expression of sovereignty lies, to a large extent, in the power and ability to dictate who may live and who must die.
>
> *Achille Mbembe*

3.1 Introduction

The growing awareness of the harms that can result from developing and deploying intelligent systems has triggered the emergence of ethical frameworks aimed at inhibiting these adverse effects. In the last few years, Schiff et al. (2020) and Jobin et al. (2019) have recorded more than 80 Artificial Intelligence (AI) ethics documents (codes, principles, frameworks, guidelines, and policy strategies). Likewise, the AI ethical program reports over

DOI: 10.1201/b23345-4

two hundred.[1] Floridi et al. (2020) highlight the importance of building such ethical frameworks, while others question the difficulty of applying them (Adamson et al., 2019; Morley et al., 2020), the lack of empirical evidence about their influence (McNamara et al., 2018), or the absence of critique about the intrinsically destructive uses of AI, like in the case of automated weapons (Anderson & Waxman, 2013).

The challenges associated with defining ethical principles are manifold. First, the guiding interests of the actors define them, their motivations (Schiff et al., 2020), and the ultimate goal they pursue. These principles are primarily aligned with the interests of tech corporations or powerful nation-states. Also, the risk of ethics washing (Bietti, 2021) is high, given the countless cases that show that profit is more important than ethical practice in the tech industry (Hagendorff, 2020). Second, they are only suggestions and, in consequence, are not enforced (Calo, 2017; Schiff et al., 2020). Third, the geopolitical asymmetries found in the definition of principles and the dominance of ethical frameworks developed by industrialized countries are considered universal (Jobin et al., 2019).

Ethical frameworks developed by the powerful are problematic for several reasons. On the one hand, there are implications of creating a global ethical hegemony that erases epistemic differences. On the other, not considering alternative ethical frameworks in the global debate impedes technological pluriversality, that is, the possibility of developing plural models of technological development. This complexity explains why despite the growing number of ethical frameworks proposed by different stakeholders, predominantly from the global North (Jobin et al., 2019), there is minimal consensus about how an ethical framework for AI should look and whether that consensus is possible or desirable to achieve (Fjeld et al., 2020). Jobin et al. (2019) identify a higher recurrence of five principles in 84 documents: transparency, justice and equity, non-maleficence, responsibility, and privacy. Other principles like beneficence, freedom, autonomy, dignity, and solidarity appear less frequently. Morley et al. (2020) developed a typology for AI ethics to respond to the claim of more practical frameworks that practitioners could use. However, in this text, we will argue that constructing ethical frameworks within a specific political, economic, and cultural matrix that uses AI as a tool of power without seeking to change that power is intrinsically problematic.

Therefore, we propose bridging geopolitics and body politics as intertwined dimensions of AI ethics and technologies from a decolonial and feminist perspective. Our purpose is to identify the limits and implications of applying ethical frameworks that do not address the problem of power and the role AI technologies play in reproducing structural systems of oppression. Thus, the problem relates to power and the role AI plays within the matrix of domination as the product and tool of/for perpetuating oppressive systems. Building on previous work, we identify the interventions and

---

[1] See https://www.aiethicist.org/frameworks-guidelines-toolkits.

integral frameworks that we require to construct technologies aligned with the well-being of humanity, the right to a dignified life, and that respect the environment. As a starting point, we acknowledge the epistemic difference of the subjects who have lived or are still living the experience of colonialism. We emphasize how this colonial difference is fundamental to understanding the various dimensions in which the coloniality of power is fundamental to understanding the various dimensions in which the coloniality of power is manifested through socio-technical systems (Quijano, 2007).

3.2 Structural Violence and the Decolonial Turn

Women's and anti-racist struggles date back centuries. However, the growing feminist and anti-racist consciousness have opened the discussion among technologists and academics. There is a need for meaningful interventions to revert the role of intelligent systems in reproducing historical and structural forms of oppression. Our historical moment urges us to understand and reverse how hegemonic AI technologies reinforce the matrix of domination (Collins, 2019). Capitalism, colonialism, and patriarchy as systems of violence are deeply intertwined and have historically created mechanisms for their perpetuation. In this era, it depends on the development of pervasive technologies. These technologies do not operate only at the macro-systemic level: They are reproduced at the micro-political level, conditioning bodies, affections, and subjectivities.

Within this context, the decolonial framework seeks to unveil the mechanisms, technologies, and relations that continue legitimating oppression based on epistemic and ontological differences. The coloniality of power manifests itself through the coloniality of being, knowing, feeling, doing, and living. It implies the epistemic annihilation of any way of thinking, feeling, doing, or living that is not Eurocentric/Modern. Epistemicide (de Sousa Santos, 2015) is a cause and a consequence of the long history of coloniality. Moreover, epistemicide legitimizes material dispossession and physical, racial, and gender violence (Segato, 2018). Thus, race and gender are the organizing principles for this logic of capital accumulation, the political economy, and the labor division of the capitalist/Western/modern/patriarchal system. Dispossession at a massive scale is essential for the emergence of capitalism. For some authors, extractivism, and dispossession during the colonial period gave birth to Modernity (Dussel, 2003; Echeverría, 2000).

The decolonial framework can be understood as an interpretative framework that questions systems of domination, as a praxis (Walsh, 2018), or as a programmatic and political plan that leads to the rupture, rejection, and separation of the dominant logic that excludes alternative forms of being, thinking, feeling, doing, and living. Coloniality as a category is helpful in understanding "the nexus of knowledge, power, and being that sustains an endless war on specific bodies, cultures, knowledges, peoples, and nature" (Maldonado-Torres, 2016). Coloniality as a power structure needs material resources, norms, institutions, narratives, technologies, affections, imaginaries, and the control of subjectivities and intersubjective relations. Decoloniality, de-linking from the coloniality of power, means dismantling the colonial rationality based on epistemic and ontological (racial, gender) differences that are imposed in the form of institutions, norms, practices, and technologies.

The decolonial approach is gaining momentum to explain the phenomena associated with datafication, life algorithmization, and the development of AI systems central to the Western/Modern capitalist project (Ricaurte, 2022). In this text, we seek to deepen the reflection on the problem of the epistemological difference and epistemicide to argue that hegemonic AI is fundamentally a capitalist/colonial/patriarchal project.

3.2.1 Decolonial Feminism

Decolonial feminism recognizes gender, sexuality, and subjectivity among the diverse forms of domination within the coloniality of power (Curiel, 2007). We want to highlight how body politics and geopolitics (Mignolo & Tlostanova, 2006) are articulated when discussing decolonization. Besides, we warn about the danger of depoliticizing the decolonial perspective when the subject's colonial experience is diluted or when the subjects face the dispossession of the faculty of speaking by themselves or have the possibility of speaking about their own experience mediated by the colonial reason, that is to say, the experience/knowledge (Mbembe, 2019) of white people. Decoloniality as an ongoing process and praxis entails unmasking the racist, misogynistic, neoliberal rationality associated with the modern world system. In this regard, decoloniality is a political program that entails an anti-racist, anti-patriarchal, and anti-capitalist praxis against domination by race, gender, and other differences in which knowledge plays a significant role in legitimating domination and exclusion. In particular, we emphasize that the subject's positionality is relevant since decolonization should be considered a praxis and a process (Walsh, 2018). Decolonization acquires an entirely different meaning for those racialized or ethnized women subject to multiple forms of violence and for whom the colonial experience, exclusion, and dispossession are everyday experiences (Tzul Tzul, 2015).

3.3 Decoloniality as Praxis: A Brief Genealogy

The heterogeneous set of thoughts of decolonial studies characterizes the epistemic dominance of rational/modern Western thought that marks the beginning of modernity and the capitalist system (Dussel et al., 2000; Quijano,

2000, 2007). However, the origin of the decolonial thought and praxis in the territory of Abya Yala, known today as Latin America and the Caribbean, begins at the moment of the European invasion. Since then, it has had multiple manifestations. The genealogy can be traced back to Guamán Poma de Ayala's criticism of the bad colonial government (Guaman Poma de Ayala, 2006); the struggle of the Maroon women; Fanon's interpretation of the psycho-affective dimension of coloniality, and the dehumanization of the subject (Fanon, 1974, 1994); Aime Cesaire (2000) and his denunciation of the colonial regime based on violence; and in more recent times, Latin American critical philosophy and sociology (Dussel, 2003, 2013; Echeverría, 2000; González-Casanova, 1969), critical pedagogies (Freire, 1987); the Zapatista movement and its proposal for an alternative community organization in opposition to the Mexican colonial state. That is to say, the decolonial critique, regardless of its nomenclature, is inscribed as a praxis of resistance in contemporary indigenous and afro-descendant movements of the territory of Abya Yala. It can also be found in many communities' experiences in other geographical contexts that have lived the colonial experience.

Latin American decolonial thought is mainly associated with the modernity/coloniality research group (Quijano, 2000, 2007; Grösfoguel, 2007; Mignolo, 2007, 2011; Walsh & Mignolo, 2018; Maldonado-Torres, 2004, 2007; Lander, 2000; Escobar, 2000, 2003, 2014, 2016) which has had greater visibility because of its association with North American and European academic circuits. Although decolonial thought involves these diverse approaches since the colonial era, in the contemporary period, it is associated with the category of coloniality of power coined by the Peruvian sociologist Aníbal Quijano. Coloniality of power as an analytical category helps to understand the power relations maintained after the colony (Grösfoguel, 2018). In their most radical and situated version, these multiple ramifications are embodied in the experiences of communities in the struggle for the defense of territory, in the communitarian government in the liberated Zapatista territories, and the practices of resistance and re-existence of the racialized women of Abya Yala.

According to Quijano (2000), the coloniality of power means the control over labor/resources/products, sex/resources/products, authorities/institutions/violence, and intersubjectivity/knowledge/communication to the daily life of the people (pp. 573). We can translate the coloniality of power as the colonial reason that controls material resources, bodies, territories, subjectivities, intersubjectivity, and knowledge to accomplish the colonial/Modern/Eurocentric project. As the sediment that remains from the colonial experience to this day, the category of coloniality serves to explain the logic that sustains the supremacy of the West and its forms of domination through the imposition of universal forms of being – race, gender, ontological difference, of thinking – positivism, rationalism, binarism, of feeling – emotions, affections, perceptions, spirituality, of doing – efficiency, utility, productivity, and of living – consumerism, commodification, individualism. Nevertheless, in contemporary times, we see that the coloniality of power operates through socio-technical systems to instrument this coloniality in all spheres of our existence.

The decolonial feminist framework that emphasizes ontological difference – race, gender – as a justification for epistemic imposition is in dialogue with theoretical frameworks that critically explain how the matrix of domination (Collins, 2019) articulates various forms of oppression, such as social class, ability, education, language. In addition, intersectional feminism (Crenshaw, 1991) helps to explain the forms of resistance at the personal, community, and systemic levels, as well as the multiple positions that a feminist subjectivity can occupy in power relations. In general, the decolonial framework recovers the emancipatory practices of marginalized communities in countries that experienced historical colonialism and populations that, due to the colonial experience, find themselves excluded and marginalized in Western metropolises.

Although it is possible to find some coincidences between decolonial traditions, it is pertinent to emphasize that they are also at tension. These tensions arise between those authors who have been most visible because of their privileged position in Western academic circles – the Latin American white intellectual elite in American or European institutions – and the different communities that produce knowledge in Abya Yala. Among indigenous communities, the use of decolonial frames is controversial. Partly due to the internal colonization process (González-Casanova, 1969) led by the nation-states, as much as the decolonization narratives used as a strategy of ethical washing. Some of the most visible indigenous thinkers and important intellectual references are the sociologist Silvia Rivera Cusicanqui in Bolivia, the Mayan K'iche' sociologist Gladys Tzul Tzul in Guatemala, and the Mixe linguist, writer, and activist Yásnaya Elena Aguilar in Mexico. Cusicanqui (2010) writes about the discourses of decolonization, or feminist frameworks, that are distant from the experience of indigenous women, as also pointed out by Zapatista women (EZLN, 2019) and Tzul Tzul (2015). An alternative proposal comes from the black decolonial feminists, such as Yuderkys Espinosa-Miñoso and Ochy Curiel, and autonomous feminisms by the queer Bolivian artist María Galindo.

Despite these tensions and contradictions, we believe that critical frameworks – feminist, intersectional, decolonial – make visible the matrix of domination – gender, race, class, ability, language – and the intertwined systems of oppression – capitalism, colonialism/racism, and patriarchy. They are analytical tools useful to explain how the coloniality of power operates through socio-technical systems. Decolonial feminism is intrinsically anti-racist; it seeks to unmask epistemic forms of domination based on racial differences. It is anti-capitalist because it recognizes the dispossession of bodies and territories and the commodification of life in the capitalist project. Furthermore, it is anti-patriarchal because it denounces violence against women as a constitutive part of capitalist accumulation, the process of colonization, the emergence of modern societies, and the epistemic Western hegemony.

3.4 Power and Knowledge: The Coloniality of Technology, Gender, Race/Ethnicity, and Nature

The sociologist Aníbal Quijano has developed the notion of coloniality of power (Quijano, 2007) to explain epistemic domination through the racial difference exercised by the European metropolis during colonization. Coloniality, according to Quijano, explains rationality, its underlying logic, the mechanism of hierarchization, and the epistemic categorization of humans based on racial differences. This process has entailed an ontological and epistemic exclusion with economic, social, and political consequences: Coloniality has legitimated extractivism, land, and labor domination, leading to the accumulation of capital and the emergence of Modernity (Dussel, 2003). Although Quijano considers sex as one of the resources appropriated by the process of coloniality, race remained the fundamental category for this theory. Decolonial feminists have further developed how gender was a fundamental condition for the coloniality of power (Curiel, 2007; 2009; Lugones, 2008, 2011; Segato, 2018).

De Sousa Santos (2015) argues that coloniality extends colonization as an epistemicide based on racial differences. The coloniality of power leads to the destruction of any alternative way of thinking, being, feeling, doing, and living different from the Western/modern world model. The loci of dispossession are the natural resources of the colonized territories and our bodies, thoughts, actions, affections, and relationships (Ricaurte, 2019). Colonial power implies stripping the subject of colonization of his value as a human being and exercising racial superiority as epistemic domination. As the African philosopher Sabelo J. Ndlovu-Gatsheni (2020) explains:

> Even when you push back colonization as a physical process (the physical empire), colonialism as a power structure continues, because it invades the mental universe of a people, destabilizing them from what they used to know, into knowing what is brought in by colonialism.

In addition to racial difference, decolonial feminists María Lugones (2008), Rita Segato (2018), Yuderkys Espinosa-Miñoso (2014), Ochy Curiel (2007, 2009), and María Galindo (2015), among others, propose that gender constituted and constitutes another axis of realization of the coloniality of power. We add that the coloniality of gender also materializes through data epistemologies, algorithmic models, and AI technologies (Ricaurte, 2019, 2022). The Western civilizing project is an imposed hegemonic version of existence that displaces, reconfigures, and annuls Otherness (Echeverría, 2008, p. 12). In particular, the alterity and subjectivity embodied in the experience of women's racialized bodies. Technologies that continue to expand the commodification of life (Echeverría, 2008, p. 18) are technologies of racialization (Mbembe, 2019) and technologies of gender; they contribute to the objectification of women's and racialized bodies and their subjective experiences. Coloniality of race/

gender through datafication, algorithmization, and automation exerts violence at scale (Ricaurte, 2022) and is at the core of the surveillance capitalist society (Zuboff, 2015, 2019). It is the epitome of the mercantilist, racist, and patriarchal project, a project that transforms our relationships, body, subjectivity, and territory into a source of profit.

Anthropologist Rita Segato (2018) refers to this project as the project of things, in opposition to the project of ties. The triumph of the project of things implies violent forms of domination – race, class, and other minorities – the dispossession of territories and nature (Lander, 2011). Although data extractive processes, algorithmic mediation, and intelligent systems indeed involve anyone who participates in the digital society, in the Western project of modernity, there are territories, bodies, and ways of being, knowing, feeling, and doing bearing the cost of this project. Not making this distinction is a mechanism that perpetuates coloniality and the fundamental necropolitics (Mbembe, 2019) that lies behind it.

Colonialist, racist, and patriarchal values are imposed as universal principles for hegemonic technological design. These values are coded into ethical frameworks that shape corporate AI systems, the construction of anthropomorphic and white machines (Cave & Dihal, 2020), racist algorithms (Benjamin, 2019a; Benjamin, 2019b; Noble, 2018), the construction of narratives about progress, development, productivity, efficiency, modernity, and the future associated with technology, but anchored in imaginaries that consolidate racial superiority (Benjamin, 2019a; Benjamin, 2019b). Dominant socio-technical systems are part of the knowledge assemblages that produce narratives that contribute to Western epistemic hegemony – white supremacy, heteropatriarchy, capitalism, and settler colonialism. Thus, it is crucial to understand how datafication, algorithmic mediations, and intelligent systems operate cognitively, emotionally, and pragmatically to construct worldviews and ways of relating and conceiving social existence. Modernity's project dissociates technoscientific production processes from their effects on racialized bodies and the territories they inhabit. Nevertheless, these territories provide the raw materials and labor resources needed to produce technology and are where industrialized countries expel their waste. Consequently, hegemonic AI systems operate as necro-technopolitical tools (Ricaurte, 2022).

Hence, in the study of AI from a decolonial and feminist perspective, it is necessary to consider its full lifecycle, the socio-political relations associated with it, and its contribution to the reproduction of power asymmetries:

- The governance and geopolitics of AI: who decides, who designs, who has the right to participate and make decisions on AI use and deployment, who defines the norms and standards, who has the right to kill.
- Material production of AI technologies: includes labor relations and material resources, the processes of extractivism and exploitation, and its impact on communities (physical violence and displacement).

- The AI pipeline: data extraction, design, analysis, deployment, and use of AI systems, as well as water, and energy consumption for training AI systems and maintaining data centers.
- Its disposal: associated with its harmful effects on racialized bodies and the environmental impact – digital carbon footprint, pollution, the killing of land defenders that oppose the extractive project of the digital industry.
- The collective effects of AI systems: mass killings, algorithmic politics and democracies, disinformation crisis, digital truth regimes, the carceral imagination (Benjamin, 2019b) associated with the deployment of AI for automated decision-making.
- Epistemicide: the production of subjectivities, the capacity to shape the world according to a hegemonic epistemology, and how we relate to the environment and each other.

3.5　Decolonizing AI: Geopolitics and Body Politics

There are numerous contributions from academia, practitioners, and human rights activists that have addressed the problem of discrimination, racism, and sexism as a negative drift of the implementation of technology in general and AI in particular (Adam, 1995; Chander, 2018; Noble, 2018; Buolamwini & Gebru, 2018; Benjamin, 2019a; Benjamin, 2019b;; Celis et al., 2020; Achiume, 2020). Along with critical, race, and feminist studies, a growing body of work has emerged using a decolonial framework to address datafication (Van Dijck, 2014), algorithmic bias, racism (Ali, 2016a,b, 2017), and the development of intelligent systems. Couldry and Mejias (2019) and Mejias and Couldry (2019) have proposed datafication as an amplification of the historical colonial system in today's capitalist surveillance society (Zuboff, 2015, 2019). Other authors explain the epistemic rationality of data extractivism as an amplification of the coloniality of knowledge (Ricaurte, 2019).

Another emergent trend is the consideration of non-Western epistemologies to inform the development of AI. For example, Mhlambi (2020) explores rationality and individualism as the core values in developing AI in the West. However, the indigenous movement and networks can contest Western AI ethical frameworks and protocols (Lewis et al., 2020). The Global Indigenous Data Alliance proposed The CARE Principles for Indigenous Data Governance (GIDA, 2020). Other projects show in practice how indigenous AI developed as a community endeavor and enacts an entirely different approach to governance, purpose, and norms (cf. supra). These alternative approaches and principles are based on a communitarian perspective on data governance for advancing indigenous data sovereignty. From a socio-technical view, some proposals explore the possibility of using the decolonial theory as a socio-technical forecast for AI development (Mohamed, Png & Isaac, 2020). There is a long tradition of decolonial computing and computing from the margins that can also be an example of alternative developments of technology (Ali, 2014, 2021; Chan, 2014, 2018).

To explore ways of decolonizing AI, we propose a model that considers the various dimensions associated with its complete cycle articulated with the categories of the coloniality of being, knowing, feeling, doing, and living (Figure 3.1). This model integrates geopolitics, the relations between geo-historical locations and epistemologies (Mignolo & Tlostanova, 2006, p. 209), and body politics, the relations between subjectivity and epistemology. We assume that the coloniality of power materializes at different levels: At the macro-level, it operates from the knowledge systems and the forms of political organization – including international organizations – to the relations between nation-states, that is, geopolitics; at the social level, through social organization mechanisms – institutions, norms, forms of government; and at the micro-level, through the incorporation of these practices into our daily lives. It also affects bodies – ways of being, thinking, feeling, doing, and living – and the environment. The AI pipeline: data extraction, analysis, implementation, and use immersed in the social order (Celis et al., 2020). The whole AI life cycle includes the production cycle of intelligent systems (materiality) and their disposal. This process is embedded in a complex set of social relations that condition its existence from the micro- to the macro-level. The model seeks to make visible how these different dimensions are articulated.

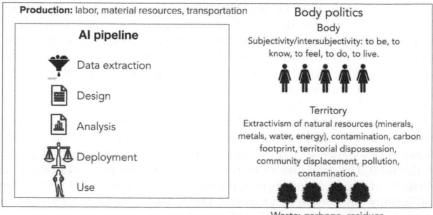

FIGURE 3.1
The geopolitics and body politics of the AI ecosystem (Ricaurte, 2019).

3.6 AI and Ethics: Between Geopolitics and Body Politics

This section will review some examples that address the problem of establishing ethical frameworks to guide the governance, design, and deployment of AI systems without considering the impacts on specific contexts and the harm beyond the place they are designed. We need to expand ethical frameworks to consider aspects beyond the design and deployment of AI. An ethical framework narrowly conceived does not necessarily solve power asymmetries, social injustice, or the destruction of nature. Furthermore, ethical frameworks must consider diverse epistemologies and avoid assuming that the underlying system of values in ethical frameworks developed in Western cultures is universal.

3.6.1 Geopolitics

Transnational politics, technoscience, and the neoliberal university: Luis Videgaray, Ph.D., in Economics at MIT, served as Secretary of Finance and Secretary of Foreign Affairs during Enrique Peña Nieto's presidency (2012–2018) in Mexico. During his tenure as Secretary of Foreign Affairs (2017–2018), he was responsible for leading Mexico's relations with the Trump administration, including the NAFTA agreement's renegotiation – now USMCA. Luis Videgaray was instrumental in consolidating Mexico's neoliberal project, including initiatives and measures associated with AI and fintech. As stated on his institutional page, he is a Senior Lecturer at the MIT Sloan School of Management and the Director of MIT AI Policy for the World Project, a joint project with the Stephen A. Schwarzman College of Computing (SCC). During his tenure, the Mexican government awarded Schwarzman, who donated $350 million to found SCC, the Mexican Order of the Aztec Eagle. In August, Emilio Lozoya, former director of Petróleos Mexicanos, denounced that Peña Nieto and Videgaray participated in and authorized the delivery of bribes from Odebrecht to Mexican politicians. In September 2020, Mexican students at MIT (Arnal et al., 2020) asked the university to dismiss Videgaray for his participation in "corruption, abuse of power, criminal association, electoral crimes, money laundering and conflict of interest and influence peddling during his administration." MIT's rector, Martin A. Schmidt, published a letter supporting Luis Videgaray, claiming that he was not facing charges. In November 2020, the Attorney General's Office found that Videgaray was the author of electoral crimes, bribery, and criminal association. A judge prevented the issuance of an arrest warrant.

　Algorithmically produced democracy: Facebook and Latin America: In 2012, Enrique Peña Nieto's presidential election, the preferred candidate of the US Administration, was plagued by irregularities. One of those irregularities was the algorithmic manipulation of public opinion. The Colombian hacker Andrés Sepúlveda declared that he had used armies of automated bots to carry out algorithmic attacks and hacked into the opposing candidates' accounts during the presidential campaign of 2012 México. Andres Sepulveda, a self-declared ultra-right-winger, faces a 10-year prison term in Colombia for illegally intercepting information about the Peace Process that was secretly going on in Havana. The hacker sought to prevent the reelection of Juan Manuel Santos, who was promoting the peace process, thus supporting the then Uribista candidate, Óscar Iván Zuluaga (Las2orillas, 2020).

The tech companies knew about the automated activity during the 2012 election in Mexico. Nothing was done to make the impact public, and the Mexican citizens could not make the company accountable. Similar cases were reported in Ecuador, Bolivia, and El Salvador. Facebook engineers declared they were aware of the problem but did not have time to address it.

Deportation, tech, and data companies: During Trump's Administration, tech and data companies played a significant role in implementing and advancing the Immigration and Customs Enforcement (ICE). The report *Who's Behind ICE?* published in 2018 by the National Immigration Project, Mijente, and the Immigrant Defense Project researches "the multi-layered technology infrastructure behind the accelerated and expansive immigration enforcement" (Empower, 2018). The study shows how different technologies are essential for The Department of Homeland Security, cloud services, surveillance technology, and predictive analytics. Companies like Palantir, Amazon Web Services, and Microsoft Services provide the core infrastructure for enforcing immigration policy. Cloud services supplied by tech corporations are critical for enabling data transfers between municipal, state, and law enforcement authorities. This research also discusses the history of the tech sector's revolving doors with government agencies as one of the processes that cement tech industry's influence (Ricaurte, 2022).

Extractivist Neoliberal Democracy: The interventions of the United States in Latin American politics are historic. However, in this era, the interventions take on a new dimension because of the expansion of the capitalist project and the growing interest derived from the need to obtain mineral resources to produce digital technologies. In 2019, Bolivia's elections were annulled as fraudulent, an assertion the United States and the United Nations endorsed and later denied. On his Twitter account, Elon Musk celebrated the reversal of the election: "We will coup whoever we want! Deal with it." (Telesur, 2020). Bolivia is the second country in the world with lithium reserves. Evo Morales' government declared a protectionist policy, which went against the extractive interests of technological capital. In October 2020, the elections declared the candidate of Evo Morales' party the winner.

3.6.2 Body Politics

Ethical frameworks, cultural diversity, and epistemological differences: The analysis of AI ethical frameworks (Fjeld et al., 2020) shows the heterogeneous approaches from which organizations or states define their value systems. One of the questions from this exercise is the impossibility of finding a consensus about which principles should be considered fundamental to developing ethical frameworks for AI. This discussion also relates to the debate around the human rights approach to regulating AI. For example, it has largely discussed Western values embedded in the human rights approach (Panikkar, 1982; Donnelly, 1982; Pollis et al., 2006). On the other side, at the national level, multiple conflicts emerge with federal regulations incompatible with the cosmovisions of different indigenous populations in multicultural countries (Segato, 2006). This approach's core assumption is the possibility of achieving consensus or establishing universal principles to develop a particular technology without considering the epistemological differences among cultures and the geopolitical forces that drive AI development.

AI is a privileged white racist, sexist, cisheteronormative man: Recent studies highlight that racism and sexism are not only expressed in algorithmic models. AI systems, as technologies, are racist. We can observe the reproduction of Whiteness through voice (Moran, 2020), aesthetics (Phan, 2019), mythology (Primlani, 2019), and physiognomy (Cave & Dihal, 2020). The idea of Whiteness is also prominent in the production of imaginaries and narratives about AI. In a Google search on AI in English, the images of these white intelligence and stylized physiognomies stand out. Although AI technologies have been created as sexist and capitalist artifacts, little has been said about the connections between these artifacts, the men behind these conceptions, and the implications of their actions globally. The technology industry, its practices, its privileges, and the global power it wields to influence political decisions, legislations, economies, and democracies must be part of our critical analysis, as well as its critical contribution to the climate crisis.

The environment, digital technologies, and the global North: ethics and greenwashing: In December 2020, the technology community was shocked after Google fired Dr. Timnit Gebru, a researcher respected in all circles for her contribution to the generation of evidence of racial and gender algorithmic bias. The dismissal stemmed from an article's writing addressing the considerable carbon footprint involved in training NLP models with large datasets. This fact, the firing and the reason for the dismissal, shows the limitations of talking about ethical frameworks focused only on the design of intelligent systems and on the impossibility of leaving tech corporations free to decide how to implement ethical frameworks and deployment of AI systems. On the one hand, they expose the ethics-washing strategies of technology corporations and their discriminatory and censorship practices. On the other hand, they highlight that AI impacts have a differentiated effect on racialized bodies and the environment. In this regard, Tapia and Peña (2021) highlight how neo-extractivism activities in Latin America associated with the digital economy are a greenwashing strategy by the global North. Furthermore, they show how lithium mining is an ecocide and is the reason behind the killing of land defenders in Latin America.

3.7 Conclusion

Throughout this chapter, we discussed the limitation of developing ethical frameworks that are not intended to shift power relations. Instead, ethical frameworks should consider the differentiated material and non-material impacts for specific territories and bodies as a consequence of the underlying logic of capitalist, patriarchal, and colonial historical processes associated with socio-technical systems. Dominant AI systems as socio-technical constructions are instruments of the coloniality of power, a relation between knowledge and power based on ontological and epistemic differences. Erasing the experience of coloniality, as well as its material and non-material impacts over specific, racialized, ethnicized, gendered bodies, subjectivities, and territories, is a way to depoliticize the decolonial critique. Decoloniality as praxis requires the de-linking from the coloniality of power: to stop, dismantle, and repair the historical process of dispossession and violence, the epistemicide, and necropolitics as principles of the hegemonic Western model of the world. We propose a decolonial imagination that explores the possible forms of AI decolonization toward the construction of technodiverse futures. This reflection is intended to take action to reverse an order based on the necropolitics over racialized bodies, mainly of women, and the destruction of the environment.

References

Achiume, T. (2020). *Report of the Special Rapporteur on contemporary forms of racism, racial discrimination, xenophobia and related intolerance*. United Nations. https://documents-dds-ny.un.org/doc/UNDOC/GEN/G20/151/06/PDF/G2015106.pdf

Adam, A. (1995). Embodying knowledge: A feminist critique of artificial intelligence. *The European Journal of Women's Studies*, 2, 355–377.

Adamson, G., Havens, J. C., & Chatila, R. (2019). Designing a value-driven future for ethical autonomous and intelligent systems. *Proceedings of the IEEE, 107*(3), 518–525.

Ali, M. (2014). Towards a decolonial computing. In: Ambiguous Technologies: Philosophical Issues, Practical Solutions, Human Nature. *International Society of Ethics and Information Technology*, pp. 28–35. http://oro.open.ac.uk/41372/1/Towards%20a%20Decolonial%20Computing%20-%20Syed%20Mustafa%20Ali.pdf

Ali, S. M. (2016a). A brief introduction to decolonial computing. *XRDS: Crossroads, the ACM Magazine for Students, 22*(4), 16–21. https://doi.org/10.1145/2930886.

Ali, S. M. (2016b, January). Algorithmic racism: A decolonial critique. In *10th International Society for the Study of Religion, Nature and Culture Conference* (pp. 14–17). Gainesville.

Ali, S. M. (2017). Decolonizing information narratives: Entangled apocalyptics, algorithmic racism and the myths of history. In *Multidisciplinary digital publishing institute proceedings* (Vol. 1, No. 3, p. 50). MDPI.

Ali, S. M. (2021). Transhumanism and/as whiteness. In Hofkirchner, W., & Kreowski, H. J. (Eds.). *Transhumanism: The proper guide to a posthuman condition or a dangerous idea?* (pp. 169–183). Cham: Springer.

Anderson, K., & Waxman, M. C. (2013). Law and ethics for autonomous weapon systems: Why a ban won't work and how the laws of war can. Stanford University, The Hoover Institution (Jean Perkins Task Force on National Security & Law Essay Series), American University, WCL Research Paper 2013-11, Columbia Public Law Research Paper 13–351. https://ssrn.com/abstract=2250126 or http://dx.doi.org/10.2139/ssrn.2250126

Andrés Sepúlveda y su final por volverse un hacker uribista. (2020, October 25) Las2orillas. co. https://www.las2orillas.co/andres-sepulveda-y-su-final-por-volverse-un-hacker-uribista

Arnal, P., Escandón, R., Escudero, L. and Espinosa, A. (September 9, 2020). Why Luis Videgaray must leave MIT | *The Tech*. Retrieved from https://thetech. com/2020/09/09/videgaray-must-leave-letter-en.

Benjamin, R. (2019). Assessing risk, automating racism. *Science*, *366*(6464), 421–422.

Benjamin, R. (2019). Race After Technology: Abolitionist Tools for the New Jim Code, *Social Forces*, Volume 98, Issue 4, June 2020, 1–3, https://doi.org/10.1093/sf/soz162

Bietti, E. (2021). From ethics washing to ethics bashing: A moral philosophy view on tech ethics. *Journal of Social Computing*, *2*(3), 266–283.

Buolamwini, J., & Gebru, T. (2018, January). Gender shades: Intersectional accuracy disparities in commercial gender classification. In *Conference on Fairness, Accountability and Transparency* (pp. 77–91). New York: PMLR.

Calo, R. (2017). Artificial intelligence policy: A primer and roadmap. *University of California, Davis Law Review*, *51*, 399.

Cave, S., & Dihal, K. (2020). The whiteness of AI. *Philosophy & Technology*, *33*(4), 685–703.

Celis, L. E., Mehrotra, A., & Vishnoi, N. K. (2020, January). Interventions for ranking in the presence of implicit bias. In *Proceedings of the 2020 Conference on Fairness, Accountability, and Transparency* (pp. 369–380). New York: Association for Computing Machinery.

Césaire, A. (2000). *Discours sur le colonialisme*. New York: Monthly Review.

Chan, A. S. (2014). *Networking peripheries: Technological futures and the myth of digital universalism*. MIT Press.

Chan, A. S. (2018). Decolonial computing and networking beyond digital universalism. *Catalyst: Feminism, Theory, Technoscience*, *4*(2), 1–5.

Chander, A. (2017). The Racist algorithm? *Michigan Law Review*, *115*(6), 1023–1045.

Collins, P. H. (2019). *Intersectionality as critical social theory*. Durham: Duke University Press.

Couldry, N. & Mejias, U. (2019). Data colonialism: Rethinking big data's relation to the contemporary subject. *Television & New Media*, *20*(4), 336–349.

Crenshaw, K. (1991). Race, gender, and sexual harassment. *Southern California Law Review*, *65*, 1467.

Curiel, O. (2007). Crítica poscolonial desde las prácticas políticas del feminismo antirracista. *Nómadas*, *26*(1), 92–101.

Curiel, O. (2009, December 16). Descolonizando el feminismo. Una perspectiva desde latinoamérica y el Caribe. [Decolonizing feminism. A perspective from Latin America and the Caribbean] Coordinadora Feminista. https://feministas.org/descolonizando-el-feminismo-una.html.

Cusicanqui, S. R. (2010). *Ch'ixinakax utxiwa. Una reflexión sobre prácticas y discursos descolonizadores*. Buenos Aires: Tinta limón.

Donnelly, J. (1982). Human rights and human dignity: An analytic critique of non-Western conceptions of human rights. *The American Political Science Review*, 76(2), 303–316.

Dussel, E. D., Krauel, J., & Tuma, V. C. (2000). Europe, modernity, and eurocentrism. *Nepantla: Views from South*, 1(3), 465–478.

Dussel, E. (2003). *Philosophy of liberation*. Eugene, OR: Wipf and Stock Publishers.

Dussel, E. (2013). *Ethics of liberation: In the age of globalization and exclusion*. Durham & London: Duke University Press.

Echeverría, B. (2000). *La modernidad de lo barroco*. México: Ediciones Era.

Echeverría, B. (2008). Un concepto de modernidad. *Revista Contrahistorias*, 11, 6–17.

Empower LLC. (2018). *Who's Behind ICE: The Tech and Data Companies Fueling Deportations*. https://www.nationalimmigrationproject.org/PDFs/community/2018_23Oct_whos-behind-ice.pdf.

Escobar, A. (2000). El lugar de la naturaleza y la naturaleza del lugar: Globalización o postdesarrollo. In En Lander, E. (Ed.) *La colonialidad del saber: eurocentrismo y ciencias sociales. Perspectivas Latinoamericanas*. Buenos Aires: Clacso.

Escobar, A. (2003). Mundos y conocimientos de otro modo. El programa de investigación de modernidad/colonialidad latinoamericano. *Tabula rasa*, (1), 51–86.

Escobar, A. (2014). *Sentipensar con la tierra. Nuevas lecturas sobre desarrollo, territorio y diferencia*. Medellín: Ediciones UNAULA.

Escobar, A. (2016). Sentipensar con la Tierra: las luchas territoriales y la dimensión ontológica de las epistemologías del sur. *AIBR: Revista de Antropología Iberoamericana*, 11(1), 11–32.

Espinosa-Miñoso, Y. E. (2014). Feminismo decolonial: Una ruptura con la visión hegemónica, eurocéntrica, racista y burguesa. Entrevista con Yuderkys Espinosa-Miñoso. [Decolonial Feminism: A break with the hegemonic, Eurocentric, racist and bourgeois vision. Interview with Yuderkys Espinosa-Miñoso.] Interviewed by Tristán, J. M. B. Retrieved from https://www.iberoamericasocial.com/ojs/index.php/IS/article/view/72.

EZLN. (December 27, 2019). Palabras de las mujeres zapatistas en la inauguración del segundo encuentro internacional de mujeres que luchan. [Words by Zapatista women at the inauguration of the second international meeting of women who struggle.] Retrieved from https://enlacezapatista.ezln.org.mx/2019/12/27/palabras-de-las-mujeres-zapatistas-en-la-inauguracion-del-segundo-encuentro-internacional-de-mujeres-que-luchan.

Fanon, F. (1974). *Piel negra, máscara blanca*. Charquero, G. y Larrea, A. (trad.). Buenos Aires: Schapire Editor.

Fanon, F. (1994). *Los condenados de la tierra*. México: Fondo de Cultura Económica.

Fjeld, J., Achten, N., Hilligoss, H., Nagy, A., & Srikumar, M. (2020). Principled artificial intelligence: Mapping consensus in ethical and rights-based approaches to principles for AI. *Berkman Klein Center Research Publication*, (2020–2021).

Floridi, L., Cowls, J., King, T. C., & Taddeo, M. (2020). How to design AI for social good: Seven essential factors. *Science and Engineering Ethics*, 26(3), 1771–1796.

Freire, P. (1987). *Pedagogía de la liberación*. São Paulo: Editora Moraes.

Galindo, M. (2015). La revolución feminista se llama despatriarcalización. *Descolonización y despatriarcalización de y desde los feminismos de Abya Yala*. Barcelona: ACSUR.

Gender and Inclusive Development Analysis - GIDA (2020). USAID. Cambodia Development Resource Institute. https://pdf.usaid.gov/pdf_docs/PA00X7PN. pdf

González-Casanova, P. (1969). El desarrollo del colonialismo en los países coloniales y dependientes. *Sociología de la explotación*, 207–234.

Grösfoguel, R. (2007). Descolonizando los universalismos occidentales: el pluri-versalismo transmoderno decolonial desde Aimé Césaire hasta los zapatistas. *El giro decolonial. Reflexiones para una diversidad epistémica más allá del capitalismo global*, 63–77.

Grösfoguel, R. (2018). ¿Negros marxistas o marxismos negros? Una mirada descolonial. [Black Marxists or Black Marxisms? A decolonial look.] *Tabula Rasa*, (28), 11–22.

Guaman Poma de Ayala, F. (2006). El primer nueva corónica y buen gobierno (1615/1616) (København, Det Kongelige Bibliotek, GKS 2232 4°), *The Guaman Poma Website*: http://www.kb.dk/permalink/2006/poma/info/en/frontpage. htm.

Hagendorff, T. (2020). The ethics of AI ethics: An evaluation of guidelines. *Minds and Machines*, 30(1), 99–120.

Jobin, A., Ienca, M., & Vayena, E. (2019). The global landscape of AI ethics guidelines. *Nature Machine Intelligence*, 1(9), 389–399.

Lander, E. (2000). *Ciencias sociales: saberes coloniales y eurocéntricos*. Buenos Aires: Clacso.

Lander, E. (2011). The green economy: The wolf in sheep's clothing. Amsterdam: Transnational Institute, 6.

Lewis, J. E., Abdilla, A., Arista, N., Baker, K., Benesiinaabandan, S., Brown, M., Cheung, M. et al. (2020). Indigenous Protocol and Artificial Intelligence Position Paper. Honolulu, HI: The Initiative for Indigenous Futures and the Canadian Institute for Advanced Research (CIFAR).

Lugones, M. (2008). Colonialidad y género. *Tabula rasa*, (09), 73–101.

Lugones, M. (2011). Hacia un feminismo descolonial. *La manzana de la discordia*, 6(2), 105–117.

Maldonado-Torres, N. (2004). The topology of being and the geopolitics of knowledge: Modernity, empire, coloniality. *City*, 8(1), 29–56.

Maldonado-Torres, N. (2007). On the coloniality of being: Contributions to the development of a concept. *Cultural Studies*, 21(2–3), 240–270.

Maldonado-Torres, N. (2016). *Outline of ten theses on coloniality and decoloniality*. Paris: Frantz Fanon Foundation.

Mbembe, A. (2019). *Necropolitics*. Durham & London: Duke University Press.

McNamara, A., Smith, J., & Murphy-Hill, E. (2018). Does ACM's code of ethics change ethical decision making in software development? In *Proceedings of the 2018 26th ACM Joint Meeting on European Software Engineering Conference and Symposium on the Foundations of Software Engineering* (pp. 729–733). Lake Buena Vista.

Mejias, U. A., & Couldry, N. (2019). Datafication. *Internet Policy Review*, 8(4), 1–10.

Mhlambi, S. (2020). From rationality to relationality: ubuntu as an ethical and human rights framework for artificial intelligence governance. *Carr Center for Human Rights Policy Discussion Paper Series*, 9, 31 pp.

Mignolo, W. D. (2007). Delinking: The rhetoric of modernity, the logic of coloniality and the grammar of de-coloniality. *Cultural Studies*, 21(2–3), 449–514.

Mignolo, W. (2011). *The darker side of western modernity: Global futures, decolonial options*. Durham: Duke University Press.

Mignolo, W. D., & Tlostanova, M. V. (2006). Theorizing from the borders: Shifting to geo-and body-politics of knowledge. *European Journal of Social Theory, 9*(2), 205–221.

Mohamed, S., Png, M. T., & Isaac, W. (2020). Decolonial AI: Decolonial theory as sociotechnical foresight in artificial intelligence. *Philosophy & Technology, 33*(4), 659–684.

Moran, T. C. (2020). Racial technological bias and the white, feminine voice of AI VAs. *Communication and Critical/Cultural Studies, 18*(1), 1–18.

Morley, J., Floridi, L., Kinsey, L., & Elhalal, A. (2020). From what to how: An initial review of publicly available AI ethics tools, methods and research to translate principles into practices. *Science and Engineering Ethics, 26*(4), 2141–2168.

Ndlovu-Gatsheni, S. (2020). Decolonization, Decoloniality, and the Future of African Studies: A Conversation with Dr. Sabelo Ndlovu-Gatsheni. Interviewed by Omanga, D. Retrieved from https://items.ssrc.org/from-our-programs/decolonization-decoloniality-and-the-future-of-african-studies-a-conversation-with-dr-sabelo-ndlovu-gatsheni.

Noble, S. U. (2018). Algorithms of oppression. In *Algorithms of oppression*. New York: New York University Press.

Panikkar, R. (1982). Is the notion of human rights a Western concept? *Diogenes, 30*(120), 75–102.

Phan, T. (2019). Amazon Echo and the aesthetics of whiteness. *Catalyst: Feminism, Theory, Technoscience, 5*(1), 1–38.

Pollis, A., Schwab, P., & Eltringham, N. (2006). Human rights: A western construct with limited applicability. *Moral Issues in Global Perspective: Volume I: Moral and Political Theory, 1*, 60.

Primlani, N. (2019). Sex and magic in service of surveillance capitalism. In *The state of responsible IoT 2019* (66pp). Berlin: ThingsCon.

Quijano, A. (2000). Coloniality of power and Eurocentrism in Latin America. *International Sociology, 15*(2), 215–232.

Quijano, A. (2007). Coloniality and modernity/rationality. *Cultural Studies, 21*(2–3), 168–178.

Ricaurte, P. (2019). Data epistemologies, the coloniality of power, and resistance. *Television & New Media, 20*(4), 350–365. https://doi.org/10.1177/1527476419831640.

Ricaurte, P. (2022). Ethics for the majority world: AI and the question of violence at scale. *Media, Culture & Society, 44*(4), 726–745. https://doi.org/10.1177/01634437221099612.

Segato, R. L. (2006). Antropologia e direitos humanos: alteridade e ética no movimento de expansão dos direitos universais. *Mana, 12*(1), 207–236.

Segato, R. L. (2018). *La guerra contra las mujeres*. Madrid: Traficantes de sueños.

Schiff, D., Rakova, B., Ayesh, A., Fanti, A., & Lennon, M. (2020). Principles to practices for responsible AI: closing the gap. arXiv preprint arXiv:2006.04707.

de Sousa Santos, B. (2015). *Epistemologies of the south: Justice against epistemicide*. New York: Routledge.

Tapia, D. & Peña, P. (2021). White gold, digital destruction: Research and awareness on the human rights implications of the extraction of lithium perpetrated by the tech industry in Latin American ecosystems. Global Information Society Watch 2020. Technology, the environment and a sustainable world. Association for Progressive Communications. https://www.apc.org/en/node/37228/

Telesur. (July, 25th, 2020). Elon Musk Confesses to Lithium Coup in Bolivia. teleSUR. Retrieved from https://www.telesurenglish.net/news/elon-musk-confesses-to-lithium-coup-in-bolivia-20200725-0010.html.

Tzul Tzul, G. (2015). Sistemas de gobierno comunal indígena: la organización de la reproducción de la vida. *El Apantle*, *1*, 125–141.

Van Dijck, J. (2014). Datafication, dataism, and dataveillance: Big Data between scientific paradigm and ideology. *Surveillance & Society*, *12*(2), 197–208.

Walsh, T. (2018). *2062: The world that AI made*. Black Inc.

Walsh, C., & Mignolo, W. (2018). On decoloniality. In Mignolo, D. W. & Walsh, E. C. (Eds.), *On Decoloniality Concepts, Analysis, Praxis* (p. 304). Durham: Duke University Press

Zuboff, S. (2015). Big other: Surveillance capitalism and the prospects of an information civilization. *Journal of Information Technology*, *30*(1), 75–89.

Zuboff, S. (2019). *The age of surveillance capitalism: The fight for a human future at the new frontier of power: Barack Obama's books of 2019*. London: Profile books.

4

A Cultural Vision of Algorithms: Agency, Practices, and Resistance in Latin America

Gabriela Elisa Sued

UNAM[1]

CONTENTS

4.1 The Judges' Question

In June 2022, a female Argentinian media personality petitioned her country's Supreme Court to order Alphabet, Inc. to unindex a series of videos from their search engines. The woman argued that they exposed events in which she had participated involuntarily. Even though they occurred 20 years ago, they affected her current public image because automated recommendation systems still prioritized them in Google searches. The plaintiff based her petition on the right to oblivion, a non-existent statute in her country's legislation. For this reason, the judges rejected her plea. However, toward the end of the verdict, the Court pronounced on the future need to take on the issue of algorithmic operations deployed by search engines to make them more understandable and transparent since users are subject to their decisions (CSJN, 2022).

This event clearly illustrates the issues that surround the cultural focus on algorithms. On the one hand, it highlights the need for greater transparency regarding how algorithms hierarchize information since people are subject

[1] The author thanks the Postdoctoral Grant Program of the UNAM, the Coordination of Humanities, and the Social Research Institute for their support, and Dra. Judith Zubieta-García for her academic advisory.

DOI: 10.1201/b23345-5

to their actions, according to the Argentine Court. On the other, it points out how algorithms have become participants in relevant practices within current society: the distribution of information, the freedom of expression, and the construction of identities. Nevertheless, it emphasizes the lack of understanding about the processes that these automated systems, used by all digital platforms, apply to manage the selection, filtration, and hierarchization of information. As is seen in the Argentine case, algorithms shape the production and consumption of data, the exercise of online socialization, and the production and circulation of public opinion. Consequently, the decisions made by algorithms are not merely technical. They are also social.

Although they stem from computer science, many algorithms carry out tasks related to information recommendation and the assignation of meaning. They are more than regulated instructions:

> Algorithms seek, classify, group, match up, analyze, profile, model, simulate, visualize, and regulate people, processes, and places. They shape our understanding of the world and simultaneously create the world through their software, with profound consequences

> *Kitchin (2017, p. 18)*

Algorithms are technical objects present in all spheres of life. They are present in financial, health, justice, and public safety systems, in university application and admissions processes, and as part of digital platforms. Usually, these algorithms are recommendation systems designed to suggest and provide users with the information they predict will be relevant (Bucher, 2018). The word *algocracy* refers to a type of social order where decisions are organized and structured through algorithms that establish the relevance of the data they process, conditioning the actions of the humans who interact with them and affecting the communities who use them (Danaher et al., 2017). The term has become part of everyday language thanks to informational media that, tragically, predicts a future in which algorithms will make decisions for us.

In the literature, we find different positions regarding how algorithms restrict human agency (Sued, 2022a). On the one hand, stances based on the philosophy of technology start with a critical view focused on socioeconomic structures. They consider that algorithms materialize the technological reach of neoliberal power (Sadin, 2018) that uses seduction to control subjectivities (Han, 2014) and appropriate the lives of citizens through the acquisition of their data (Couldry & Mejías, 2018). On the other, cultural perspectives focus on the interactions between people and algorithms, examining them to understand the reach of algorithms in technological practices and the extent to which they condition the decisions made by those who use them for this purpose, many linked to everyday life. In this chapter, we focus on the second position, using the term algorithmic culture. We ask what conditions make human agency possible in contexts where algorithms make relevant decisions. We will try to find the answer by developing four themes: algorithmic culture, agency, play, and collective resistance.

This chapter poises on Latin American communication, cultural, and technology studies. It seeks to contribute to the de-Westernization (Waisbord, 2022) of digital culture research by gathering works carried out during the last several years within this region. These investigations represent an opportunity to decentralize media studies in the Global North and highlight experiences that can enrich the North–South dialogue. In these studies, the relationship between humans and algorithms is dynamic, mobile, and inherent to the interaction context.

The chapter is organized as follows: First, it contextualizes the study of digital cultures, technological reappropriations, and algorithmic cultures in Latin America. Next, it uses international literature to define the concept of algorithmic culture. Subsequently, the three themes listed above – agency, play, and collective resistance – are elaborated upon as ways of opposition to algorithmic power. The conclusions highlight both a repertoire of activities that put the algorithms to work, redirecting them toward individual and collective goals and the value of research into algorithmic cultures as an alternative to demands for transparency from big tech, which does not seem to be working so far.

4.2 Studies on Technology and Culture in Latin America

Since the 1970s, Latin America has established its position on communication and cultural studies. Those who have studied the development of the field agree that there is a particular method of research in the field of communication in Latin America, distinct from that developed in its origins in North America and Europe (Enghel & Becerra, 2018). The specialists noted particularities associated with the socio-historical context, linked to authoritarian and dictatorial governments, a high concentration and privatization of the media, the social moorings of radio, and the hybridization of indigenous and European cultures (Trejo Delarbre, 2013). In all of these productions, reflections have been present in one way or another on socio-technology and the important role that technology plays in creating culture, although from a critical gaze that avoids determinism.

The interest in relationships between technology and culture in Latin America occurs early on. However, systematic academic literature started in 2005. Studies focus on various technological objects, such as blogs, social media, websites, and the Internet. Authors studied these as containers of diverse subjects: politics, social movements, and youth cultures, among others. The aspects of technological materiality, socio-technical functioning, and the relationships between people and technology have been studied far less (Sued, 2019).

In this context, it is worth reflecting on what Latin American academic research can offer to Global North through a body of research based on popular cultures, resistance to hegemonic media consumption, and technological reappropriations. This previous rich work has seeded the way to think about how algorithms have become intertwined with everyday activities, the production of participatory culture online, and the collective imagination (Siles et al., 2022).

In the last few years, the emergence of big data studies and their social repercussions have led to critical reflections from scholars. On the one hand, the Big Data from the South Initiative (Milan & Treré, 2019) strongly influenced studies within this region. Designed as an academic network, it promoted critical approaches to phenomenons associated with datafication, such as the extraction and commercialization of users' data by social media, the impact on individual privacy through the use of digital surveillance systems, and State control via the use of biometrics or databases with religious and racial information, among others. Also, emergent perspectives reappropriate the data produced by social media to uncover otherwise invisible social elements. Critical datafication finds alternative narratives to those constructed by the authorities and mainstream media by proposing a critical use of data generated by social media and generating new interpretations of contemporaneousness that push against the tide of algorithmic governance (Ábrego & Flores, 2021).

The substitution of the initial model of a public and collaborative Internet by digital platforms logic gave place to an emergent body of works examining relationships between algorithms and society. Some studies present critical reflections on a macrosocial level. They characterize algorithmic power as a techno-liberal scheme in which the interaction of technical, political, and economic spheres benefits a limited number of people (Sandrone & Rodríguez, 2020). This new order aligns itself with the practices of control and invasion of privacy employed by governments and corporations (Costa, 2021), leaving no margin for technological reappropriations or the creation of autonomous subjectivities (García Canclini, 2020).

Not to minimize this important critique, at this moment, we are interested in what humans do in their daily interactions with algorithms and the cultural transformations that these processes generate (Siles et al., 2020). Then, we focus on a more limited selection of research questioning the interaction between algorithms and humans without considering it given. Furthermore, the selected corpus looks at practices, meanings, and technological appropriations from a cultural view. These studies find that under specific circumstances, technological appropriation, and resistance to algorithmic power can exist through strategies that include technological know-how, alternative aesthetics, connected masses, and political opportunities (Sued et al., 2022). This nucleus starts with the analysis of empirical cases, including commonplace scenes of the use of technology. It proceeds to build sociotechnical assemblages between platforms, data, algorithms, and practices that allow for discovering relationships between humans and algorithms, where power is inherent in each situation. From this perspective, it is fundamental to specify what we mean when we refer to algorithmic culture if, as Raymond Williams (1976) used to say, 'culture' is a word that is hard to define, the issue becomes even more complex when adding the adjective 'algorithmic', and furthermore, by recognizing its diversity and coining the plural 'algorithmic cultures' (Seaver, 2017).

4.3 Algorithmic Cultures

Algorithms produce a certain estrangement in social sciences, and rightly so. In contrast with native terminology like culture, discourse, or ethnography, the word algorithm has its origins in mathematical theory. In the 9th century, the Arab mathematician Muhammad ibn Musa al-Khwarizmi used the word algorithm to refer to the step-by-step manipulation of natural numbers. During the first three decades of the 20th century, mathematicians such as Hilbert, Gödel, and Turing contributed theoretically to the knowledge about algorithms. Beginning in the 1940s, they became a fundamental part of computational development and artificial intelligence (Coveney & Highfield, 1996). In the 90s, Alexander Galloway recovered the term for cultural studies by placing video games as algorithmic cultural objects that articulate structured informational processes with cultural narratives and interpretations (Galloway, 2006).

Once Galloway moved the term from the computational arena to the social, the definition of algorithm adopted different characteristics than its mathematical origin. In computer science, algorithms are an abstraction that posits a given solution for a problem by linking software with data structures. In social studies, the term is used as a figure of speech to name a broader regimen of digital automation:

> A system of control and digital management obtained through discovery, the storage of data on a large scale, and the algorithmic processing within a legal, commercial or industrial framework that grants it authority

Dourish (2016, p. 3)

In computer science, algorithms are characterized as mathematical, rational, objective, and practical (Kitchin, 2017). All other discussions about their circulation and social uses are left to one side. On the other hand, social scientists maintain that algorithms do not possess these qualities except as carefully coined fiction (Gillespie, 2014) by big tech companies.

Among all these stances, we can find three facets: those that define algorithmic culture by the type of information they process; those that characterize algorithms as cultural objects; and those that understand culture as a historical moment in time defined by certain dominant technologies that organize methods of communication, signification, and knowledge of the world. Likewise, these perspectives are related to different concepts of culture. The most common considers culture as the products of the intellect and artistic expressions such as music, literature, and art. From this point of view, algorithms are cultural because they deal with cultural information. Then, algorithms and culture are two different objects that converge, but each one keeps its specificity. In any case, if considered in a deterministic manner, algorithms could shape the culture within which they operate by establishing criteria for classification and hierarchization.

A second position, developed by Marshall Mc Luhan and his followers in media ecology (Scolari, 2015), understands culture as a configuration that includes practices, ways of knowing, and world views shaped by a technology of communication in a given moment in history. Algorithmic technologies are part of the continuum of the technologies that organize culture around a specific type of media: just as the printing press organized literate culture, TV and radio mass culture, and the Internet digital culture, algorithmic technologies also operate in the organization of the algorithmic culture. Each cultural technology has its own characteristics. In this sense, the algorithmic ordering of information 'seems to be a collection of particular means of creating distinctions that have the power to define cultural perceptions in a broad sense' (Rieder, 2020, p. 86).

One difference between previous cultural technologies and algorithms is that the latter separates subjects from machines, moving them away from their place as executors. Humans transfer them a significant amount of agency, allowing them to make decisions autonomously. Another is that the use of the technique and the understanding of its workings are also separated.

The denomination of algorithmic culture also marks a moment in the history of the Internet. Despite its short history, the Internet has experienced radical changes during its co-evolution with society and production systems. In only 20 years, the Internet went from being a public medium – open, collaborative, and flexible, based on spread out virtual communities – to an environment dominated by a group of platforms that acquires value from the socialization they host, and that extract value by datafying said socialization and managing it through algorithms (Van Dijck, 2014). Thus, in only a few years, the culture moves from one centered on digital practices of socialization and construction of meaning – a digital culture – to another based on programs that manage, select, and prioritize relevant information – an algorithmic culture (Gómez Cruz, 2022).

The third definition, derived from cultural anthropology, focuses on the materiality of objects. Social theory and anthropology refer to material culture as the study of the world of things and objects. It belatedly enters into the consideration of humanistic and social studies. What interests material culture is how materials configure the practices of the subjects: how artifacts enable or limit concrete human practices and their everyday tasks, which, according to this perspective, build the world we inhabit or cohabit with them (Pinch, 2008).

Technologies not only produce cultures, they are themselves cultural artifacts (Pinch & Bijker, 1989). Following this reasoning, we suggest that algorithms possess double materiality, digital and cultural. This articulation signals that the two properties that make up algorithms are mutually created. The digital aspect includes routines, embedded structures, mathematical formulas, and close relationships between algorithms. The cultural aspect includes how algorithms take root in both their producers and users and are enacted as all parties interact. For software engineers, how algorithms interact with databases is a deciding factor. Instead, it is fundamental for social scientists to understand how their users perceive them and what strategies these users apply to interact with or navigate them. From this perspective, algorithms are cultural objects that simultaneously configure algorithmic cultures.

The issue in the next section is what happens with agency and its possibilities for opposition and reaction when facing the governance aspect of algorithms within a culture co-produced by humans and algorithms.

4.4 Agency and Play

On February 9th, 2020, Ingrid Escamilla, 25 years old, was murdered in Mexico City. Her case, categorized as femicide, provoked a dual outrage: On the one hand, because of the sordid details of her death, and on the other, because of the images of her dismembered body. Media filtered the shocking images of the body, which spread massively and quickly across social networks. Three days later, social media users decided to change the meaning of the lurid narrative and remember Ingrid through a collage of images of landscapes, flowers, animals, sunsets, and portraits of the young woman that evoked life. They used the connected multitudes to influence social media algorithms in reverting the viralization of the terrifying images of the femicide. On February 13th, at noon, the hashtag #*ingridescamillachallenge* became a trending topic on Twitter (Signa_Lab, 2020). In doing so, they made beautiful images prevail over those of the lurid homicide when searching for the hashtag on Instagram or Twitter. This intervention is one of many that considers the capacity of human agency to revert algorithmic power. However, these narratives are not always evident. In the context of critical studies, narratives about algorithmic power have dramatically negated human agency as something that can produce alternatives and resistance (Velkova & Kaun, 2021).

Within the different currents that study the actions of algorithms within society, there appears to be a dividing line. On the one hand, those who observe them from a critical and macrosocial perspective characterize algorithms as structures of power that obliterate subjectivity. Conversely, those who approach the phenomenon from a microsocial and empirical perspective find different nuances. The study of daily interactions between humans and algorithms, technological appropriation of social movements, and alternatives to algorithmic governance implemented by cultural producers and platform employees shows that it is possible to create alternatives to algorithmic power.

The challenge for social sciences is the development of methods to observe algorithms in the social arena. Due to their indetermination, algorithms are usually opaque objects that are difficult to understand. The job of the researcher of algorithmic cultures is that of dismantling black boxes, just as for Latour, the job of the science and technology sociologist was to open the black box of laboratories (Latour, 2015). However, what does it mean to open the black box in both cases? It means understanding the social and technical processes embedded in their design. This is not the equivalent of a technical understanding of algorithms as sequences of instructions, which would be viable from a computer science approach, but quite improbable from a social science approach (Kitchin, 2017). We aim to understand how the algorithms found in so many technical devices intervene in social practices and cultural meanings. Even so, we must realize that algorithms are not completely comprehensible, that not all of them can be understood in the same way, and that we may grasp some of their properties but not others (Bucher, 2018). Since algorithms are socio-technical objects, they can be observed as an assemblage of devices, information, relevant actors, and social and narrative representations. In this sense, Seaver (2017) has opened the door to allow for the study of algorithmic culture using ethnographic methods in the interaction with practices and contexts without a rigid differentiation between the social and the technological.

As to the interest that social research in Latin America has shown regarding the appropriation and use of technology, there is a body of recent research that inquires into the possibilities of agency in algorithmic cultures. These studies use an ethnographic perspective that focuses on the practices of users, their native theories about algorithms, and their specific uses of platforms and affordances. Intuitive theories about platforms and algorithms are one of the variables that create dynamism and mobility. They are implicit and imprecise, but they have important implications: They organize experiences, generate inferences, guide learning, and influence behavior and social interactions. Siles et al. (2020) studied the intuitive theories that Costa Rican users have about Spotify. The authors identify two categories: Some tend to personify the platform as a being that watches them and knows their preferences, while others see Spotify as a vast resource machine that offers good recommendations and entertainment. The second view allows a technological reappropriation and reorientation of algorithms for human goals.

The different uses of platforms also produce differences regarding an individual's possibility of dealing with algorithmic power. It is not the same to use an audiovisual platform for entertainment, an activity where users flow and navigate guided by the automated recommendations offered on their homepages and their recommendation lists, as it is to use it for the consumption of political narratives where it is easy to become trapped by feedback loops and confirmation bias. For example, users interested in crime-solving have found postings justifying murder or discussions about religion, antisemitism, and white supremacy. In these cases, they prefer to ignore platform recommendations and choose to consume a greater quantity of content created by producers they know well and trust. Also, those who tend to have a greater consciousness about algorithmic functioning and platform datafication are those who, at some point in life, have tried to create content and then faced the issues of visibility directed by audience metrics, the criteria that platforms use to prioritize certain topics over others, and the subordination to the criteria platforms use to sustain the constant creation of content (Sued, 2022b). Complementary, the affordances of the interfaces also give individuals more or less agency. The use of search engines, subscriptions, and the creation of playlists are activities that promote greater agency, while automated recommendations, profile suggestions, and ad banners promote algorithmic governance (Lukoff et al., 2021).

Some authors interpret intentional digital disconnection as a resistance strategy when facing the dynamics of rapidity and hyperconnection that current economic production models, based on this intensive connection to social media (Kaun & Treré, 2020). Although more research on this issue in Latin America is needed, the decrease in the amount of time users are connected to social media has been identified as a strategy to obtain agency in the face of permanent datafication. At the same time, users employ different practices to evade the datafication and algorithmic governance of social media, such as the creation of profiles using false information, not providing personal details, not linking profiles from different platforms, and creating critical opinions about the economic reuse of social media data (Rodríguez, 2022).

> Sometimes it seems to us that our cell phone is listening to us, so my friends and I play around saying words we're interested in, like 'Querétaro' or 'marketing'. A few minutes later, these words appear on our Facebook page.

This is the testimony of young Mexican women who are creating their own recreational theory about algorithms that dismantles fear and inactivity in the face of surveillance, and redirects it toward their own ends (Sued et al., 2022). Something similar is carried out by so-called influencers, who play the game of drawing attention to these algorithms, as identified by Cotter (2019). Given the meager information platforms provide about their algorithms (Pasquale, 2016), influencers choose to navigate digital marketing pages and the walls of other influencers to learn what hashtags to use, when the best time to post is, and how to garner followers. Both in Chile (Arriagada & Ibáñez, 2020) and Mexico (González Gil & Rodríguez Preciado 2021), influencers learn to play the game of algorithmic visibility to resist the opaque architecture of these platforms.

So far, we have set out some individual strategies used to challenge algorithmic power. For our research, it is necessary to bring them to light to avoid creating an image of passive subjects in the face of algorithmic governance. That stance only benefits those large companies that produce discourses about the strength, infallibility, secrecy, and dependence on algorithms. In the following section, we will move on to discuss collective resistance.

4.5 Collectives and Resistance

On December 30th, 2020, the Argentine Congress approved the legalization of abortion after more than three decades of struggle by feminist collectives. Beginning in 2018, these protests incorporated a strong digital component, where collectives both for and against legalization contended in the digital arena, especially on social media sites widely used in Argentina, such as Instagram, Twitter, YouTube, Facebook, and TikTok. This struggle manifested itself through recurrent hashtags, aesthetics based on colors that identified a given stance, and the practice of mediated mobilization that connected mass media, protests in the streets, and the forms of social media. The digital environment thus became a stage where struggles were pursued, not only for the legalization of abortion, but also for the antipatriarchal values of feminist groups, the merit of other struggles like the elimination of violence against women, and the recognition of sorority as a fundamental value. On the other side, the communication among those against the right to choose evolved into a division between the religious and the political meanings of abortion, with important contributions by militant media figures from the conservative right (Sued et al., 2022).

Armed with a history of collective organization that had been fighting for years, feminist organizations came out to strive for the legalization of abortion, hijacking the anti-abortion hashtags with all the affordances provided by platforms, such as images, emojis, and slogans, until they hijacked the antirights hashtags and transformed their visibility into the opposite: the right to choose. The feminist movement took this opportunity to revert all the algorithmic parameters of social media. As social collectives positioned themselves above the usual influencers that dominate algorithmic visibility on social media, prioritized by the commercial interests of these platforms (Poell et al., 2021), they positioned new body aesthetics opposed to those common on social networks that prioritize consumption-based stereotypes. Social protest images invaded these networks that favor individualistic gazes and fragmented bodies (Banet-Weiser et al., 2020).

The feminist reorientation of algorithmic visibility did not happen by chance. It was the result of an endeavor spanning several decades, a careful communication strategy that included a knowledge of social media aesthetics, a group of connected multitudes intervening at the right moment, a hopeful struggle toward a long-desired conquest, and a political opportunity to discuss the bill that gave voice and visibility to social collectives.

In the context of a culture where decisions and forms of action are shaped by algorithms (Kitchin, 2017), resistance must be built inside the platforms through alternative uses and reappropriations, reinforcing the agency to object to algorithmic power and their visibility policies (Velkova & Kaun, 2021). Algorithmic resistance is a combination of spontaneous actions and collaborative tactics of individual and collective users that make expressions and actions about social claims visible on networks, reappropriating the algorithmic logic installed by social platforms, including their hierarchical systems and filtering of information (Treré, 2019).

This action of creating connections, together with technological reappropriation for social resistance, has been reflected upon and contemplated in Latin America based on acts of resistance to political power on social media through the use of hashtags expressing political dissension and creation of meanings (Reguillo, 2017). In Mexico, the acts of resistance that sought to transform meaning in the political arena began in 2012 with the hashtag #yosoy132, through which connected masses sought to revert the political campaign, supported by hegemonic media, of presidential candidate Enrique Peña Nieto. The connected actions of citizens, mainly driven by a group of 131 university students, managed to impose a worldwide trending topic on Twitter in May 2014. The #yosoy132 hashtag was born as a show of affective solidarity and viral support for these students (Rivera Hernández, 2014). As we will see below, this technological reappropriation by social movements continues to this day.

In the context of the 98,471 disappeared people recognized by the State, social collectives adapt digital technologies governed by algorithms to make these disappearances visible. While ignored by media and governments, relatives seek to recover the identity of the disappeared to create agency for those who search for them, and to generate new meanings about these family catastrophes and the stigmatization created by belonging to the inner circle of a disappeared person. Digital databases, geotagged registries, the use of drones to identify clandestine graves, and Facebook broadcasts are used as 'technologies of hope' (Franco Migues, 2022), giving a great deal of agency to technology usually characterized as tools of surveillance, powerfully seductive, or desubjectifying by critical philosophy. The same technology, drones trained in automatic facial recognition and collecting vast amounts of data, was used for surveillance during the social unrest in Chile in 2020. However, they were blocked and brought down by hundreds of protesters as they merged the beams of ordinary laser pointers (Tello, 2020).

We do not want to use these examples to negate important aspects of algorithmic governance in platform capitalism; however, we want to point out that there is still the possibility of agency in algorithmic cultures. In addition, we want to highlight the individual and collective efforts involved in achieving collective agency. These efforts include the ability to be together and create groups through digital technologies and the recurrent use of their affordances (Reguillo, 2017), the willingness of online masses to act at opportune moments, and the capacity of complex technological learning.

4.6 Who Knows How Algorithms Work?

Let us return to the adverse verdict of the Supreme Court of Argentina but, above all, to the questioning directed at technology companies about how algorithms work. It is always valid, especially from government institutions, to demand that technology companies be responsible, transparent, and socially equitable in their processes of algorithmic management. However, algorithms are too important to delegate all the answers to this ideal technological transparency, surely rooted in digital above cultural materiality. That is why researchers must open the black box of algorithms, demystify their opacity, and analyze them in the contexts in which they are applied, using ethnographic methods based on socio-technical knowledge. Continuing this research in more detail becomes necessary to offer citizens of the future, who are becoming constantly more digitized, a set of sound practices that can challenge *algocracies* and shape conscious, critical, and autonomous citizens.

Since they derive from the cases that this research focused on, it is relevant to understand the practices, tactics, and awareness that make it possible to oppose algorithmic governance. Fortunately, by having strategies to observe the interaction between algorithms and humans, social researchers can deepen their understanding of the contexts and situations where people put algorithms to work and redirect them toward individual and collective goals. When reviewing studies rooted in Latin America, a repertoire of resources emerges that can confront algorithmic power, generate agency, and promote technological reappropriation. This repertoire includes collective action, technological knowledge and imagination, a playful attitude, and political opportunities that invite contestation to algorithmic governance. The history of resistance to hegemonic powers through digital technology in Latin America offers the knowledge and experience needed to avoid monolithic characterizations of automated systems that only add to the secrecy and attribution of power to the great technology companies.

References

Ábrego Molina, V. H., & Flores Mérida, A. (2021). Datificación crítica: Práctica y producción de conocimiento a contracorriente de la gubernamentalidad algorítmica. Dos ejemplos en el caso mexicano [Critical datafication: Practice and production of knowledge against the current of algorithmic governmentality. Two examples in the Mexican case.] *Administración Pública y Sociedad (APyS)*, 11(1), 211–231.

Arriagada, A., & Ibáñez, F. (2020). You need at least one picture daily, if not, you're dead": Content creators and platform evolution in the social media ecology. *Social Media + Society*, 6(3), 1–12. https://doi.org/10.1177/2056305120944624.

Banet-Weiser, S., Gill, R., & Rottenberg, C. (2020). Postfeminism, popular feminism and neoliberal feminism? Sarah Banet-Weiser, Rosalind Gill and Catherine Rottenberg in conversation. *Feminist Theory*, 21(1), 3–24. https://doi.org/10.1177/1464700119842555.

Bucher, T. (2018). *If... Then: Algorithmic Power and Politics*. New York: Oxford University Press.

Corte Suprema de Justicia de la Nación. (2022). Libertad de expresión y derecho al olvido [Freedom of expression and right to be forgotten]. Corte Suprema de Justicia de La Nación. https://www.csjn.gov.ar/novedades/fallosRecientes/detalle/6083.

Costa, F. (2021). *Tecnoceno. Algoritmos, biohackers y nuevas formas de vida [Technocene. Algorithms, Biohackers and New Forms of Life]* Buenos Aires: Taurus.

Cotter, K. (2019). Playing the visibility game: How digital influencers and algorithms negotiate influence on Instagram. *New Media & Society*, 21(4), 895–913. https://doi.org/10.1177/1461444818815684.

Couldry, N., & Mejias, U. A. (2018). Data colonialism: Rethinking Big Data's relation to the contemporary subject. *Television & New Media*, 20(4), 336–349. https://doi.org/10.1177/1527476418796632

Coveney, P., & Highfield, R. (1996). *Frontiers of Complexity: The Search for Order in a Chaotic World*. London: Faber and Faber.

Danaher, J., Hogan, M. J., Noone, C., Kennedy, R., Behan, A., De Paor, A., Felzmann, H., Haklay, M., Khoo, S.-M., Morison, J., Murphy, M. H., O'Brolchain, N., Schafer, B., & Shankar, K. (2017). Algorithmic governance: Developing a research agenda through the power of collective intelligence. *Big Data & Society*, 4(2), 1–21. https://doi.org/10.1177/2053951717726554

Dourish, P. (2016). Algorithms and their others: Algorithmic culture in context. *Big Data & Society*, 3(2), 1–11. https://doi.org/10.1177/2053951716665128.

Enghel, F., & Becerra, M. (2018). Here and there: (Re)Situating Latin America in international communication. *Communication Theory*, 28(2), 111–130.

Franco Migues, D. (2022). Tecnologías de esperanza. Apropiaciones tecnológicas para la búsqueda de personas desaparecidas en México[Hope Technologies. Technological Appropriations for the Search of Disappeared Persons in Mexico]. Ciudad de México: Tintable.

Galloway, A. R. (2006). *Gaming: Essays on Algorithmic Culture*. Minneapolis: University of Minnesota Press.

García Canclini, N. (2020). *Ciudadanos reemplazados por algoritmos [Citizens Replaced by Algorithms]*. Bielefeld University Press, Editorial UDG, Editorial UCR, UnSaM Edita, FLACSO Ecuador.

Gillespie, T. (2014). The relevance of algorithms. In Tarleton Gillespie, Pablo J. Boczkowski (e, Kirsten A. Foot (eds.) *Media Technologies: Essays on Communication, Materiality, and Society* (pp. 167–193). Cambridge, MA: MIT Press.

Gómez Cruz, E. (2022). *Tecnologías vitales: Pensar las culturas digitales desde Latinoamérica [Vital Technologies: Thinking about Digital Cultures from Latin America]*. Ciudad de México: Puertabierta editores, Universidad Panamericana.

González Gil, L. J., & Rodríguez Preciado, S. I. (2021). Libertad burocratizada: Gestión, producción y mercantilización de las subjetividades desde la figura digital del Travel Blogger [Bureaucratized Freedom: Management, Production and Commodification of Subjectivities from the Digital Figure of the Travel Blogger]. *Virtualis*, 12(22), 95–120.

Han, B. C. (2018). *Hiperculturalidad: cultura y globalización* [Hyperculture: Culture and Globalisation]. Colofón.

Kaun, A., & Treré, E. (2020). Repression, resistance and lifestyle: Charting (dis)connection and activism in times of accelerated capitalism. *Social Movement Studies, 19*(5–6), 697–715. https://doi.org/10.1080/14742837.2018.1555752.

Kitchin, R. (2017). Thinking critically about and researching algorithms. *Information, Communication & Society, 20*(1), 14–29. https://doi.org/10.1080/13691 18X.2016.1154087.

Latour, B. (2015). *Science in Action: How to Follow Scientists and Engineers through Society.* Cambridge, MA: Harvard University Press.

Lukoff, K., Lyngs, U., Zade, H., Liao, J. V., Choi, J., Fan, K., Munson, S. A., & Hiniker, A. (2021). How the design of YouTube influences user sense of agency. *Proceedings of the 2021 CHI Conference on Human Factors in Computing Systems,* 1–17. https://doi.org/10.1145/3411764.3445467.

Milan, S., & Treré, E. (2019). Big data from the south(s): Beyond data universalism. *Television & New Media, 20*(4), 319–335. https://doi.org/10.1177/1527476419837739.

Pasquale, P. (2016). *The Black Box Society: The Secret Algorithms That Control Money and Information.* Cambridge, MA and London: Harvard University Press.

Pinch, T. (2008). Technology and institutions: Living in a material world. *Theory and Society, 37*(5), 461–483. https://doi.org/10.1007/s11186-008-9069-x.

Pinch, T., & Bijker, W. (1989). The Social construction of facts and artifacts: Or how the sociology of science and sociology of technology might benefit each other. In Wiebe Bijker, Thomas P. Hughes& Trevor Pinch (eds.): *The Social Construction of Technological Systems* (pp. 17–30). Cambridge, MA: MIT Press.

Poell, T., Nieborg, D. B., & Duffy, B. E. (2021). *Platforms and Cultural Production.* Cambridge, UK and Medford, MA: Polity Press.

Reguillo Cruz, R. (2017). *Paisajes insurrectos: Jóvenes, redes y revueltas en el otoño civilizatorio [Insurgent Landscapes: Youth, Networks and Revolts in the Civilizational Autumn].* Barcelona: NED Ediciones.

Rieder, B. (2020). *Engines of Order.* Amsterdam University Press. https://www.aup.nl/en/book/9789462986190/engines-of-order

Rivera Hernández, R. D. (2014). De la Red a las calles: #YoSoy132 y la búsqueda de un imaginario político alternativo [From the Internet to the streets: #YoSoy132 and the search for an alternative political imaginary]. *Argumentos, 27*(75), 59–76.

Rodríguez, A. (2022). *Big data, hacia un dispositivo de poder: Sociedad, plataformas, algoritmos y usuarios. El caso de Facebook.* Tesis de Maestría [Big data, towards a power device: Society, platforms, algorithms and users. The case of Facebook] Master Thesis. Universidad Nacional Autónoma de México. Posgrado en Ciencias Políticas y Sociales

Sadin, É. (2018). *La silicolonización del mundo* [The silicolonization of the World]. Buenos Aires: Caja Negra.

Sandrone, D., & Rodríguez, P. (2020). El ajedrez, el go y la máquina. El desafío de las plataformas para América Latina [Chess, go and the machine. The platform challenge for Latin America]. In Andrés M. Tello (ed): *Tecnología, Política y Algoritmos en América Latina [Technology, Politics and Algorithms in Latin America]* (pp. 35–54). Viña del Mar: CENALTES Ediciones.

Scolari, C. A. (2015). *Ecología de los medios: Entornos, evoluciones e interpretaciones [Media Ecology: Environments, Evolutions and Interpretations].* Barcelona: GEDISA.

Seaver, N. (2017). Algorithms as culture: Some tactics for the ethnography of algorithmic systems. *Big Data & Society*, 4(2), 1–12. https://doi.org/10.1177/2053951717738104

Signa_Lab. (2020). *Ingrid Escamilla: Apagar el horror* [Ingrid Escamilla: Turn off the horror] *ITESO Signa_Lab*. https://signalab.mx/2020/02/14/ingrid-escamilla-apagar-el-horror/

Siles, I., Segura-Castillo, A., Solís, R., & Sancho, M. (2020). Folk theories of algorithmic recommendations on Spotify: Enacting data assemblages in the global South: *Big Data & Society*, 7(1), 1–15. https://doi.org/10.1177/2053951720923377

Siles, I., Gómez-Cruz, E., & Ricaurte, P. (2022). Toward a popular theory of algorithms. *Popular Communication*, 21(1), 57–70. https://doi.org/10.1080/15405702.2022.2103140

Sued, G. (2019). Para una traducción de los métodos digitales a los estudios latinoamericanos de la comunicación [For a translation of digital methods to Latin American communication studies]. *Virtualis*, 10(19), 20–41. https://doi.org/10.2123/virtualis.v10i19.295.

Sued, G. E. (2022a). Culturas algorítmicas: Conceptos y métodos para su estudio social [Algorithmic cultures: Concepts and methods for their social study]. *Revista Mexicana de Ciencias Políticas y Sociales*, 67(246), 43–73. https://doi.org/10.22201/fcpys.2448492xe.2022.246.78422

Sued, G. E. (2022b). Training the algorithm: YouTube Governance, Agency and Literacy. *Contratexto*, 037, 1–22. https://doi.org/10.26439/contratexto2022.n037.5331.

Sued, G. E., Castillo-González, M. C., Pedraza, C., Flores-Márquez, D., Álamo, S., Ortiz, M., Lugo, N., & Arroyo, R. E. (2022). Vernacular Visibility and Algorithmic Resistance in the Public Expression of Latin American Feminism. *Media International Australia*, 183(1), 60–76. https://doi.org/10.1177/1329878X211067571.

Tello, A. (2020). Tecnologías insurgentes. Apropiación tecnológica y disidencias insurgentes en América Latina [Insurgent technologies. Technological appropriation and insurgent dissidence in Latin America]. In Andrés M. Tello (ed.): *Tecnología, política y algoritmos en América Latina* [*Technology, Politics and Algorithms in Latin America*]. (pp. 55–78). Viña del Mar: CENALTES Ediciones.

Trejo Delarbre, R. (2013). The study of internet in Latin America: Achievements, challenges, futures. In Kelly Gates (ed.): *The International Encyclopedia of Media Studies: Vol. VI.* (pp. 1–29). Hoboken: Blackwell Publishing.

Treré, E. (2019). *Hybrid Media Activism: Ecologies, Imaginaries, Algorithms*. Abingdon and New York: Routledge.

Van Dijck, J. (2014). Datafication, dataism and dataveillance: Big Data between scientific paradigm and ideology. *Surveillance & Society*, 12(2), 197–208. https://doi.org/10.24908/ss.v12i2.4776.

Velkova, J., & Kaun, A. (2021). Algorithmic resistance: Media practices and the politics of repair. *Information, Communication & Society*, 24(4), 523–540. https://doi.org/10.1080/1369118X.2019.1657162.

Waisbord, S. (2022). What is next for de-westernizing communication studies? *Journal of Multicultural Discourses*, 17(1), 26–33. https://doi.org/10.1080/17447143.2022.2041645.

Williams, R. (1976). *Keywords: A Vocabulary of Culture and Society*. New York: Oxford University Press.

5

From Deepfake to Deeptruth: Toward a Technological Resignification with Social and Activist Uses

Jacob Bañuelos and Mariángela Abbruzzese

Tecnológico de Monterrey

CONTENTS

5.1 Introduction

The proliferation of digital technologies and algorithmic operations has produced, in equal measure, a proliferation of strategies for disinformation and deception. Boundaries between true and false are redistributed in today's media ecosystem, through technical artifices that can be as sophisticated as they are pernicious. It is therefore increasingly difficult and necessary to find possibilities for criticism and political resistance in contemporary communication practices.

The term *synthetic media* is used to group digital materials that have been created or modified by artificial intelligence (AI) and deep learning algorithms. The creation process combines a synthesis of images and audiovisuals

DOI: 10.1201/b23345-6

to develop media content in a completely artificial way (Rodríguez, 2022). One of its best-known applications is *deepfake*, a term used to describe false videos and sounds – artificially created through the aforementioned techniques – whose high degree of realism allows the public to interpret them as true. Its first appearance took place in 2017 thanks to Reddit's *"deepfakes"* user, who uploaded pornographic videos in which he inserted faces of Hollywood actresses on porn actresses' bodies without their consent.

The aforementioned case opened the door to a large framework of practices at a global level, related to pornography, disinformation, fake news, counterfeiting, extortion, identity theft, defamation, harassment, and cyberterrorism (Fraga-Lamas and Fernández-Caramés, 2020; Güera and Delp, 2018; Sohrawardi et al., 2019; Cerdán et al., 2020; Xu et al., 2021; Adriani, 2019; Cooke, 2018; Gosse and Burkell, 2020; Westerlund, 2019; Vaccari and Chadwick, 2020; Gómez-de-Ágreda et al., 2021; Temir, 2020, Ajder et al., 2020; Barnes and Barraclough, 2019).

However, since 2017, *deepfake*'s semantic and discursive evolution has also given rise to non-criminal uses and good practices in spheres of the audiovisual industry, entertainment (satire, parody, memes, GIFs, etc.), advertising, medicine, propaganda, and counter-propaganda, denunciation, experimental art, and education. There is also a growing interest in its possibilities as a tool for education, social criticism, and political activism (Bañuelos, 2022).

In the literature, we find that this topic has developed much more in computer science and engineering than in social sciences and humanities (Bañuelos, 2022). In humanistic studies, concern predominates about how *deepfake* is used to violate people's fundamental rights. Likewise, it is debated whether it is preferable to speak of "synthetic media" when using AI and deep learning to create artificial and hyper-realistic content with beneficial objectives and to use the term *deepfake* only when it comes to criminal cases (Schick, 2020). To articulate this conceptual debate, we find that the phenomenon has been studied by only taking into account examples made from the Global North.

The discursive uses and adaptations of *deepfake* have been shaping new terminologies that allow designating specialized applications since the phenomenon expands to different fields of cultural production. The objective of this chapter is to explore its potential as a communication tool for social campaigns, social criticism, and political activism from Latin America and contribute to the construction of epistemological axes for researching this emerging media species.

We propose to fulfill this objective through the analysis of a *deepfake* carried out within the framework of the #SeguimosHablando campaign, launched by the Propuesta Cívica organization to denounce violence against journalists in Mexico. It is a video that artificially recreates the image of Javier Valdez, a journalist murdered in Culiacán in 2017, whom we see "come to life" to denounce and demand justice from government authorities in the face of impunity in his case, and that of many of his colleagues who have lost their lives while seeking the truth and exercising their profession in Mexico.

During an interview, the video's makers mentioned that they preferred using the term *deeptruth* to describe their initiative (Pérez Muñoz, 2022).

We consider this case to have great social relevance and complexity from an ethical point of view. Furthermore, it is novel in terms of the use of synthetic media from social and political activism in the Mexican context. It also shows that synthetic media has also given rise to good communication practices with the potential to benefit society, so it is integrated into the debate on the uses, scope, and possibilities of the resignification of *deepfake* in digital culture.

The chapter is organized as follows: first, we contextualize Propuesta Cívica's case in a series of charitable practices in the field of denunciation and social activism at a global level. Then, we propose a way of analysis to study synthetic audiovisual content, so far grouped under the term *deepfake*. This theoretical–analytical framework allows us to study the technical procedures applied by Propuesta Cívica when making the video, the ethical and legal transparency involved in the realization process, Javier Valdez's artificially attributed discourse, the context in which it is framed, and the political and critical sense of this communicative event. Finally, we reflect on the resignification from *deepfake* to *deeptruth*; an emerging media species that, despite having an origin and a vast amount of practices related to crime or disinformation, can also be beneficial in the field of social and political activism.

5.2 The Use of Synthetic Media for Denunciation and Social Activism

In the field of social activism, we find a growing number of proposals developed through AI and deep learning techniques for the production of artificial and hyper-realistic images or videos, so far grouped under the term *deepfake*. In addition to Propuesta Civica's focus on Javier Valdez and the context of violence against journalists in Mexico, we offer a series of examples with the potential of resignifying the term and shaping a new genre within the production of synthetic media.

In 2018, the audiovisual piece *You Won't Believe What Obama Says in This Video!* initiated the discursive genre of denunciation with synthetic media (Silverman, 2018). In the video, we can see the former U.S. President Barack Obama warning about the risks of misinformation that *deepfakes* can generate.

In the field of art, we find the *Big Dada* project of the artist and activist Bill Posters, who artificially produces videos with celebrities and politicians such as Mark Zuckerberg or Donald Trump, to denounce the monopoly of technologies and political–corporate alliances. Posters "hacks public faces with synthetic art (#deepfakes)" and attributes critical discourses related to control over data and disinformation (bill_posters_uk, 2019).

In the field of social campaigns, we also find several cases that exemplify the beneficial uses of deepfake. For example, *Change the Ref – The Unfinished Votes* was launched by McCann Health and promoted by Joaquín Oliver's parents. In 2018, the young man was killed along with 17 classmates in a shooting at Stoneman Douglas School in Portland, Florida. Two years later, the video was made and in which Oliver can be seen as an activist and not as a victim, denouncing that he will not be able to vote because of violence and gun sales (Cotw, 2021; Change the Ref, 2020).

Also noteworthy is the case of *Malaria Must Die*, a social campaign run by The Ridley Scott Creative Group, starring David Beckham to fight malaria. The footballer appears speaking nine different languages, declaring the end of the disease at the age of 70 (Malaria Must Die, 2019).

In 2019, DoubleYou launched the *Merry Christmas, even if it is fake* campaign. In the video, we see controversial leaders make statements that go in the opposite direction to their ideologies. For example, Donald Trump accepting Mexican migration, Pedro Pascal advocating for an end to corruption, and ultraconservative Abascal accepting free love, equality, and sexual freedom (El Publicista, 2019). At the end of each video, it is announced that "none of this is true, but how nice it would be" (DoubleYou, 2019).

And finally, another notable case is the *Natural Intelligence* initiative of the Reina Sofía Foundation, which artificially revives painter Salvador Dalí, in partnership with the Sra. Rushmor agency, to attribute to him a speech that promotes investment in research on neurodegenerative diseases (Queen Sofia Foundation, 2021).

By highlighting the growth of applications of this technology by activists, non-governmental organizations, and creative agencies with charitable purposes for society, we do not intend to take away the rigor of the questioning that must be made to those contents that present videos, audios, and false images – artificially constructed – with a high degree of realism.

Although this brief scenario does not cover the entire universe of beneficial uses of deepfake, they are noteworthy cases that trace a path toward the technological resignification of the phenomenon. In the mentioned examples, it has been used as an instrument for charitable practices, social campaigns, and criticism of political and corporate forces. However, this happens in a still uneven balance and with greater weight toward its criminal uses.

5.3 A Theoretical Framework to Study Content Created with Synthetic Media

In this section, we offer a way of analysis to study synthetic audiovisual content called *deepfake* and question the possibilities of resignification of this communication practice. We propose a multidisciplinary perspective integrated by techno-aesthetics (Simondon, 2013), discourse theory (Van Dijk, 1985, 1993; Foucault, 2005), the theory of post-truth (Ball, 2017), and the notion of the political–critical in Richard (2011).

5.3.1 Techno-aesthetics

In Gilbert Simondon's philosophy (2009, 2013), we find a complex understanding of the evolution of individuals and technical objects, which may be convenient for the study of new communication practices facilitated by contemporary technologies (Wark & Sutherland, 2015). There is an overlap between both processes, and the strength of invention is relational and trans formative in them.

From substantialism's perspective, an individual is considered as "consistent in his unity, given to himself (...), resistant to what is not himself", and from hylomorphism's perspective, as engendered by the encounter between form and material. In both cases, "a principle of individuation prior to individuation itself" is assumed – a process that has come to an end and from which we can rise to understand the existence of a "constituted and given" individual (p. 23). Simondon (2009) is contrary to both schemes and offers an alternative way to understand the process of individuation.

The author proposes to remove the ontological privilege granted to the individual from these schemes and explains that physical individuation, rather than following a series of steps with a stable beginning and end, is one of the phases of a process in which the being is preserved "through becoming" and is "always in a metastable state" (Simondon, 2009, p. 25).

The individuated reality ignores a "pre-individual reality" full of potential that the individual has within. In that sense, it is not a being "substantiated in front of a world that is foreign to it" because being is a "transductive unit whose understanding exceeds unity and identity" (Simondon, 2009, p. 38). Transduction, a key concept in the author's thought, refers to "individuation in process" and involves thinking about "physical, mental, biological, and social operations by which the individual propagates in a domain" (p. 83). Therefore, transduction

> [has] some transmission and some translation, some displacement in space and time and the same passage from one register to another; only that it is a transport where the transported is transformed. (p. 38)

Between technical objects and individuals, there are transmissions, translations, and displacements in space and time. Moreover, there are inventions or changes of registers that always produce transformations in human perception and culture. Therefore, the complexity of the techniques of reproduction and manipulation of contemporary images cannot be approached as something separate from an individual. We cannot practice an analysis from a distance since in technical reality, there is human reality, and vice versa.

Just as individuation is a state of being during its genesis process, "concretization" is a phase of the genesis process of the technical object. The latter has different subdivisions or derivations that arise from the circumstances, uses, and needs of society. When it surpasses what it was invented for, we find applications that bring its "superabundant functions to the surface" (Simondon, 2013, p. 198).

Every technical object has a "degree of indeterminacy" that leaves it permanently open to scenarios and possibilities of resignification in culture. There is no "stable state" (Simondon, 2013, p. 198) in which its transformation is not possible. Therefore, to analyze a technological invention, we can take into account what this author defines as "layers of the technical object," without falling into "fantasies of liberation or subjugation" (p. 198).

Taking into account the "layers of the technical object" (Simondon, 2013, p. 198) allows us to make a characterization of *deepfake*, considering possibilities that go beyond the predominant uses and utilitarian functions that this technology has had. (i) The *inner layer* refers to what is purely created, condensed at the time of its invention in technical and cultural terms. (ii) The *intermediate layer* expresses the relationship between technique and language, thus allowing us to understand how the object operates as a technological mediation. (iii) In the *outer layer* is the most obvious way in which the object manifests itself in culture, its function in certain practices, and its most common uses.

In *deepfake's inner layer*, we find a high degree of complexity since it is the product of the so-called synthetic media, whose process we explained in the introductory section. The production, manipulation, and artificial modification of audiovisual data and materials are now possible thanks to automation achieved through AI and deep learning algorithms. The resulting content of this process is completely artificial.

Deepfake's intermediate layer has to do with the attribution of meaning during the construction of audiovisual material since the relationship between technique and language comes into play here. It is a technological mediation that can play with human perception just because it can make artificial content seem real. In many cases, determining whether it is false or true can prove extremely difficult, if not impossible, unless this is explained by the people who created the content.

Deepfakes's outer layer includes the scenarios mentioned at the beginning: practices in which crime predominates and among which we recently found cases that point to an appropriation and resignification of this technology. Among the good *deepfake* practices, we highlight the case of Javier Valdez's video generated by Propuesta Cívica (2020), as part of the #SeguimosHablando campaign launched to denounce violence against journalists in Mexico.

We propose researching whether this cultural expression takes advantage of *deepfake's degree of indeterminacy*, bringing *superabundant functions* of this technology to the surface, which do not point to deception or disinformation but to social and political activism in digital culture.

5.3.2 Post-truth

Deepfake's instrumentation for deception has allowed it to be interpreted as part of the post-truth phenomenon (Ball, 2017). With the advancement of networked technologies, strategies to confuse or misinform have grown. In this field, the practices of simulation and falsification proliferate, so it becomes necessary to make a critical questioning of the discourses of making believe. In that sense, to analyze a deepfake, we cannot leave aside issues such as consent, the objectives of filmmakers, the ethical–legal framework, or transparency during the realization process.

Considering these issues will make it possible to determine the relevance of artificially created content for purposes that do not necessarily aim to deceive viewers. For example, it is not the same to publish pornographic videos with false identities or to create false content with the face of a person for extortion as conducting a social campaign to combat a disease or denounce violence, openly stating that the content was artificially created with AI.

5.3.3 Discourse

Following Teun Van Dijk's discourse theory (1985, 1993), a deepfake can be understood as a communicative event in a given social situation. It has a specific way of using language, text, sound, and image. For the understanding of discourse, observable elements, verbal and non-verbal, must be taken into account, as well as the cognitive representations and strategies involved in its production.

The author proposes taking into account the semantics and structure of an expression and then asking ourselves what aspects cannot simply be described in terms of meaning, words, phrases, or sentences. Interpretive work must be done, the understanding of which will be relative to the knowledge of speakers and listeners. In this sense, the context can explain what is most relevant in the discourse, defining its global coherence (Van Dijk, 1985, Meersohn, 2005).

According to Meersohn (2005), Van Dijk (1985) invites us to "understand the situation backward" (p. 294). This means that we cannot see the context as something static but as a dynamic and constantly updated structure. Therefore, it is necessary to know what is relevant in a communicative context and to know the local events that precede a speech in order to analyze it. The above events accumulate information relevant to the understanding of the communicative event.

It is also important to explain "the role that the discourse plays in the reproduction of domination." Power elites who "represent, sustain, legitimize, condone, and ignore social inequality and injustice" must be the "critical target" of the discourse analyst (Van Dijk, 1993, Meersohn, 2005, p. 299). To determine whether a communicative event legitimizes, reproduces, or confronts domination, again, it will be important to know local events, previous events, and social and contextual situations. Therein lies the possibility that discourse analysis is critical (Meersohn, 2005).

To deepen the relationship between discourse and domination with a critical perspective, we return to Foucault's (2005) idea of the will to know and truth exercised from structures, doctrines, rituals, and educational and power institutions. All discourses are subject to a coercive and exclusive order, which marginalizes or stigmatizes those discourses that escape its control or contradict it, through exclusion and delimitation systems.

There are discourses that are accepted in culture and expanded, transmitted, multiplied, and assumed as true thanks to the discursive police (Foucault, 2005) that impose rules to coerce and restrict the visible and the sayable. One of the most powerful meshes of discursive control is found in the exercise of politics, which erects convenient representations to maintain its domination system. But discourse can also be what is fought for; control and exercise of speech are power.

However, fear of speech is not the solution. Foucault (2005) comments that it is necessary to "analyze it in its conditions, its game, and its effects." To do this, we must "rethink our will to truth, restore the character of an event to discourse, and finally erase the sovereignty of the signifier" (p. 51). Therefore, in the speech attributed to Javier Valdez in Propuesta Cívica's (2020) *deepfake*, it will be important to see if it reproduces, accepts, expands, or faces the "will to know and truth" exercised from the structures of political power, in relation to violence against journalists: Is there a struggle for the discourse on violence in Mexico?

5.3.4 Criticism

Power structures can be questioned through visual and discursive practices that contribute to the construction of critical imaginaries. Nelly Richard (2011) situates herself in the Latin American context to ask what artistic or aesthetic strategies contribute to the construction of such imaginaries. Their proposal can be transferred to cultural expressions that involve creativity, social criticism, and political activism. According to the author, visual technologies have sealed a pact with capitalism, but human expression can still make technological uses that do not duplicate the system and whose effects are critical.

From the exercise of politics, "landscapes of denial and closure" (p. 7) are promoted in the face of repressive situations for society. But creativity and discourse can be put into practice to perform localized operations that call these landscapes into question. The effectiveness of these operations or creative actions "depends on the particular materiality of the social registration supports that it is proposed to affect" (p. 9). There is no methodical pragmatism that automatically makes an expression critical in itself because this is defined "in act and in situation" (p. 9). In that sense, Richard (2011) continues,

Political-critical is a matter of contextuality and locations, of frameworks and borders, of limitations and boundaries' crossings. For the same reason, what is political-critical in New York or Kassel may not be in Santiago de Chile and vice versa. The horizons of the critical and the political depend on the contingent web of relationalities in which the work is located to move certain borders of restriction or control, to press against certain frameworks of surveillance, to explode certain systems of prescriptions and impositions, to decenter the commonplaces of the officially consensual. (p. 9)

What is important for Richard (2011) is that a cultural expression allows to stimulate a "relationship with meaning" (p. 9) that proposes alternative designs to those of "ordinary communication" (p. 9) and thus proposes a redistribution in the ways of perceiving and being aware of communicative events.

The author also states that there is a difference between cultural expressions that maintain a relationship of exteriority with politics, that is, that takes a position from discourse and representation, and other types of expressions that are political in themselves since their "internal articulation" reflects critically on an environment. That is to say, the latter allows dislocating or transgressing visual patterns normalized by domination and the capitalist system. In that sense, recognizing the context and determining in which of these two ways the *deepfake* that we are going to analyze is framed will allow us to deepen its critical sense.

5.4 Case Study: Javier Valdez's Synthetic Video

In this section, we will analyze the artificial video of Javier Valdez made by the Propuesta Cívica organization (2020), as part of a campaign to denounce violence against Mexican journalists. Based on the theoretical framework set out in the previous section, we divided the analysis taking into account (i) the technical procedures applied by Propuesta Cívica (2020) to make the video, (ii) the ethical and legal transparency involved in the realization process, (iii) the speech artificially attributed to Javier Valdez, (iv) the context of his discourse, and (v) the critical sense of this communicative event.

This case is chosen for its social relevance, its complexity from the ethical point of view, and its novelty in terms of the use of synthetic media from political activism in Mexico. We question the way in which this initiative resignifies *deepfake* and is part of the construction of epistemological axes to investigate this communication phenomenon.

5.4.1 Technical Procedures

Javier Valdez's video was made thanks to synthetic media, whose operation we have explained throughout the chapter. In the case we analyze, we know that this allowed to recreate the face of the journalist "in front of a camera" after his murder. We also know that it was developed with AI software, and specifically deep learning patterns that specialize in detecting or modifying faces to generate multimedia materials with realistic images and sounds.

The novelty and particularity of this technology make it possible to synthesize human images and voices through the superimposition and combination of images and sounds created by a computer. These audiovisuals are applied to existing videos, resulting in the new fake and, at the same time, hyper-realistic video (Gutiérrez, 2020). Although audiovisual counterfeiting is not a practice that begins with these innovative technologies, automation and speed they provide are remarkable.

5.4.2 Ethical and Legal Transparency

The ethical–legal dilemma was considered at all times by filmmakers. Mauricio Pérez Muñoz (2022), a member of the Propuesta Cívica organization, participated in an event about deepfakes and clarified several relevant issues. First, the organization obtained the consent of Javier Valdez's family before making the video, and they were also present throughout the campaign process.

They also requested the opinion of allies and other journalists in relation to the use of the image of a murdered individual, to take advantage of the technological boom and transmit a message relevant to the Mexican context and that had reach in the rest of the world. The "why, how, and what would be the purpose of making this deepfake" were taken into consideration (Pérez Muñoz, 2022).

Once the deepfake was launched, Pérez Muñoz (2022) mentions that they began to receive criticism because it was not "real enough" and the response always aimed to highlight that this was one of the intentions of the initiative. Valdez did not talk exactly like that because the video is not real and he was murdered. Intentionally, traces of its artificial character were included. In that sense, the video is fake because it was artificially produced, but it does not point to the deception of viewers. For that reason, during the campaign, they decided to stop using the term deepfake and talk about deeptruth.

It is noteworthy that the Propuesta Cívica organization has a history of contribution to the defense and promotion of human rights and freedom of expression in Mexico. They have dedicated themselves to following people who defend freedom of expression, who are in constant risk, and who use strategic communication to give visibility and protection to both victims and relatives (Propuesta Cívica, 2022).

Within their work, they launched the #SeguimosHablando campaign in 2019, in collaboration with Diego Wallach and Juan Carlos Tapia, CEO of the Publicis Worldwide Mexico agency, and Griselda Triana, wife of Javier Valdez, who wrote the words that the journalist says in the video. The campaign had two stages: the first stage was launched in 2019, where the strategy consisted of some Twitter accounts of murdered journalists that continued publishing, and the second stage was in 2020, in which the piece focused on the image of the murdered journalist Javier Valdez is created.

The video piece created with AI had the support of Andrey Tyukavskin, creative director of Publicis in Latvia, who had already worked with some deepfake ideas. The elaboration of the second stage took two years due to the solution and assessment of ethical, legal, and technological aspects. The campaign was published on November 2, 2020 and won several international awards, such as the Grand Prix For Good at Cannes Lions, the Grand Prix at IAB MIXX 2020, and the Yellow Pencil at the D&AD festival in the same year (López, 2021).

In parallel to the entire campaign process, Propuesta Cívica has legally represented Javier Valdez's family in the murder's investigation, achieving two convictions – of 14- and 32-year jail sentences – against two individuals involved in the murder (Pérez Muñoz, 2022).

5.4.3 Discourse

There is a speech artificially attributed to the image of Javier Valdez, whom we see denouncing that he was murdered because he published things that "someone" did not like, as did many of his fellow journalists. He directly addresses Mexico's President Andrés Manuel López Obrador, pointing out the indifference and complicity of the state – governors and state officials – in the crimes committed against disappeared and displaced journalists, just for exercising their profession and for "laying bare the bowels of corrupt power and organized crime." It also calls for clarifying these crimes and giving peace to the families of those killed: "those truly responsible must be punished" (Propuesta Cívica, 2020).

We find relevant the following textual quotes from the video since it is clear that it does not point to the deception of the viewers in them. Also, they bring the debate about truth and lies to the surface, not only related to technology but also to the indifference and silence of political power in the face of violence in Mexico: "I am not afraid because they cannot kill me twice" / "we must clarify the crimes" / "a country without truth is a country without democracy" / "even if they want to silence us, we keep talking".

The discourse has a very simple semantics and structure. The message is expressed in a few words in an accessible, clear, and direct way. It confronts the domination of the political power elites since it addresses the highest authority, and highlights the complicity of state officials in impunity and even in crimes committed against journalists. In addition, the discourse avoids the re-victimization of Javier Valdez since he is represented as what he was in life: an activist, a defender of human rights, brave and willing to tell the truth in the face of all risks.

In this act of communication, there are constant accusations on the issue of violence against journalists in Mexico. Therefore, the analysis and understanding of the discourse cannot be separated from the understanding of context, which we delve into below.

5.4.4 Context

Mexico is a violent country for practicing journalism. According to Forbes Staff (2021), it is one of the most dangerous in the world, along with Afghanistan and India, adding more than 66 murders in the last five years. In 2022, 12 homicide victims were registered, which means that during the six-year term of current President Andrés Manuel López Obrador, 35 journalists have been killed (Vallejo, 2022). According to Reporters Without Borders (2022), Mexico ranks 143rd out of 180 countries as one of the deadliest in the world for journalists. Proof of this is that between 2000 and 2022, 154 communicators have been reported, whose murder is possibly related to exercising their profession (Article 19, 2022).

There is a situation of generalized violence in Mexico. According to Menéndez (2012), during the twenty-first century, the figures of forced disappearance, massacres of rural populations in broad daylight, crimes against humanity, violence against women, gender-based and structural violence due to poverty and inequality, state violence, and organized crime have increased daily. It should also be taken into account that there are other types of violence such as lack of access to health, work, education, etc.

The war started by former President Felipe Calderón against organized crime between 2006 and 2012, left a balance of more than 100,000 dead, 344,230 indirect victims, and more than 26,000 disappeared (Pérez-Anzaldo, 2014). This event produced a sharpening in violence, social decomposition, and strengthening of the crimes that he tried to fight. 350,000 people were killed between 2006 and 2021, and more than 72,000 remain missing. According to the United Nations (2022), crimes have increased not only against journalists (156, in 2000–2022) but also against environmental activists (153, in 2008–2020) and political candidates (36, in 2020–2021).

Zavaleta (2017) explains that militarization policies have increased in recent years. The need for measures in the face of the risks of violence and insecurity has made it possible to legitimize such policies, but risks have also arisen to safeguard democracy and preserve fundamental rights in the country. According to the author, the historical context shows a weak and insufficient justice system, as well as a failed state or narco-state, where military and government forces are corrupted by organized crime.

The case we analyze is framed in this context of violence. Mexican journalist and writer Javier Valdez Cárdenas was murdered on May 15, 2017 in Sinaloa. He was the founder of Ríodoce weekly journal and correspondent for La Jornada. During his career, he wrote several publications on drug trafikking such as Miss Narco, Huérfanos del narco o Malayerba, Los morros del narco, Levantones: historias reales, Con una Granada en la boca and Narco periodismo.

In 2011, Valdez received the International Press Freedom Award from the Committee to Protect Journalists. In his speech, he read, "In Culiacán, Sinaloa, it is dangerous to be alive, and practicing journalism is walking on an invisible line marked by the bad guys who are in drug trafficking and in the Government (..) One must take care of everything and everyone" (Lafuente, 2017).

The direct causes of his murder are unknown. Five years after his death, two direct perpetrators have been arrested, and the person who is supposed to be the crime's mastermind remains at large. However, it seems clear that Javier Valdez's death was caused by his deep knowledge of the complex drug trafficking network in Sinaloa and by not remaining silent despite death threats. Valdez's death adds to a long list of journalists killed in Mexico that totals 151 in 2022; thirteen murders in 2022 alone, the year in which the most deaths have been reported since 1992 (CPJ, 2022). Most crimes against journalists have not been solved by the justice administration.

5.4.5 Critical Sense

Propuesta Cívica's artificial video recreates the image of a murdered journalist, whose image is relevant to the denunciation of violence in the Mexican context. This is stimulating for viewers and captures attention and looks, which we can see by knowing the visibility and success this initiative has had within the framework of the #SeguimosHablando campaign.

Likewise, it maintains an exteriority relationship with politics (Richard, 2011). In the discourse attributed to Valdez, there is a position against domination (Van Dijk, 1993), represented by the political power elites, by pointing directly to the government's share of responsibility for the crimes committed against journalists in the country.

The internal articulation of the video required specialized technical knowledge, and this knowledge was exercised with responsibility and intentions of resignifying a media species associated with crime, deception, and disinformation. Propuesta Cívica (2020) carries out an exercise of resignification and stimulates "an alternative relationship with the meaning" (Richard, 2011) of *deepfake*, which invites us to move toward other possible uses, related to social and political activism.

Filmmakers do not limit themselves to the "outer layer" (Simondon, 2013) or surface layer of the technological object, associated with its dominant – criminal – uses in digital culture. Therefore, there is a rediscovery of the technical object from its "inner layer" and use of its "superabundant functions" (Simondon, 2013). Propuesta Cívica takes advantage of *deepfake*'s technology "degree of indeterminacy" by demonstrating the relational and transformative force of the invention, through a beneficial practice that produces fissures in the dominant understanding of this technology.

Likewise, in the "intermediate layer" of the *deepfake* technical object – the relationship between technique and language – there is also a transformative act by not trying to "make believe" that Javier Valdez is alive even though we can see him speak. On the contrary, it takes advantage of the possibility of artificially recreating the image of a relevant person to denounce a social problem without pointing to deception.

In this sense, "the created is transformed" (Simondon, 2013) during the process of realization, which leads the authors to prefer the term *deeptruth*. We must not understand this as a simple replacement of one concept by another, but as a discourse whose character of event allows erasing "the sovereignty of the *deepfake* meaning" to think about the breadth of its possibilities, while taking into account its criminal applications.

We find that this cultural production fractures certain visual and discursive patterns regarding the representation of violence since it does not point to the re-victimization of Javier Valdez. On the contrary, we see him as an activist who takes a firm stand and denounces impunity for the crimes committed against him and his colleagues. In addition, the case adds layers of complexity to the discussion about the uses, scope, and possibilities of *deepfake* as an emerging media species, for whose study only examples from the Global North have been taken into account.

The critical sense of the video operates in two directions: it produces fissures both in the imaginary about violence in Mexico and in the construction of epistemological axes to investigate the possibilities of synthetic media. The authors respond to a problem of their socio-historical context through a use of the *deepfake* technique, which contributes resignifying its role as a tool of political activism. It is a technological, aesthetic, and discursive proposal whose power is political since it goes beyond the dichotomies between what is true and false, the malicious or beneficial effects of technological tools. It shows us the possibility of revealing a *deeptruth* through a *deepfake*.

5.5 Conclusions: The Resignification from *Deepfake* to *Deeptruth*

Over time we observe a semantic, discursive, and technological evolution of *deepfake* (Bañuelos, 2022), which since its inception finds fertile territory in the post-truth scenario (Bañuelos, 2020). This technology covers a wide range in which false speeches are constructed to pass them off as true, and an extensive repertoire of genres and criminal uses, as we have pointed out (identity theft, extortion, harassment, defamation, intentional damage, copyright, blackmail, espionage, defamation, moral damage, lack of honor, self-image, privacy).

However, in recent years, we have begun to observe a semantic shift in the phenomenon, so the *deepfake* meaning deserves rethinking. The range of charitable uses is also growing and especially emerging from uses in social campaigns and non-profit political activism.

The present research and analysis of the case of Javier Valdez has given us guidelines to outline the concept of *deeptruth* as a semantic and strategic turn in the use of AI technologies originally used in *deepfakes*. If *deepfake* participates in a technological, social, political, and cultural architecture inserted in the post-truth scenario, *deeptruth* also participates in a similar architecture, but with some different characteristics.

The post-truth architecture is made up of technological viralization tools, opinion leaders, relevant social and political actors, celebrities, electronic and digital media, social networks, and platforms. A socio-media architecture that facilitates the transmission of false information, where *deepfakes* mesh perfectly (Bañuelos, 2020: 53). In this same context, the social and activist use of *deepfake* appears. But can we still call it *deepfake* when its main function is precisely to transmit a truth and benefit society? What turns *deepfake* into *deeptruth*?

Some components of the case focused on Javier Valdez in the #SeguimosHablando campaign, as well as other social and political activism campaigns, have taught us that the minimum and common elements of *deeptruth* are the following: ethical and legal transparency; consent and participation of *deeptruth* beneficiaries in the campaign or in the final products – family members, organizations' members, civil associations, etc.; the presence of traces proving that it is not real – it is declared false and artificial but not misleading; the promotion of a social cause; and the use of a socio-media architecture in networks and platforms.

Deeptruth is a creation made through AI, is a product of synthetic media, and can be used to promote a social cause for ethical, humanistic, and critical purposes, where it is explicitly stated that it is an artificially created counterfeit that does not pursue deception.

In this way, we are witnessing the rediscovery of the deepfake technical object, which leads to the deeptruth. The deepfake has taken a technical-semantic turn and has undergone a social and cultural redefinition to produce synthetic media made with artificial intelligence at the service of social campaigns and political activism. *Deeptruth* thus appears as a product of the "superabundance of the technical object" (Simondon, 2013, p. 198), which readjusts its role in the architecture of post-truth, to give rise to a redistribution of the notions of truth and falsehood: telling the truth and denouncing the lie through a tool that had generally been used to deceive.

References

Adriani, R. (2019). The evolution of fake news and the abuse of emerging technologies. *European Journal of Social Sciences*, 2(1), 32–38. https://bit.ly/3svE2JE.

Ajder, H., Patrini, G., y Cavalli, F. (2020). Automating image abuse: Deepfake bots on Telegram. *Sensity*. https://bit.ly/3poYgTG.

Ball, J. (2017). *Post-Truth: How Bullshit Conquered the World*. Britain, Kingston Upon Hull: Bite Back Publishing.

Bañuelos Capistrán, J. (2020). Deepfake: The image in times of post-truth [*Deepfake: la imagen en tiempos de la posverdad*]. *Revista Panamericana De Comunicación*, 2(1), 51–61. https://doi.org/10.21555/rpc.v0i1.2315.

Bañuelos Capistrán, J. (2022). Evolution of Deepfake: Semantic fields and discursive genres [Evolución del Deepfake: campos semánticos y géneros discursivos] (2017–2021). *Revista ICONO 14. Revista Científica De Comunicación Y Tecnologías Emergentes*, 20(1). https://doi.org/10.7195/ri14.v20i1.1773.

Barnes, C. y Barraclough, T. (2019). Perception inception: Preparing for deepfakes and the synthetic media of tomorrow. The Law Foundation.

bill_posters_uk (7 de junio de 2019). Spectre. [Video]. Instagram. https://bit.ly/3C59QZ9.

Cerdán, V., García Guardia, M. L., y Padilla, G. (2020). Digital moral literacy for the detection of deepfakes and audiovisual fakes [*Alfabetización moral digital para la detección de deepfakes y fakes audiovisuales*]. *CIC: Cuadernos de información y comunicación*, 25, 165–181. http://dx.doi.org/10.5209/ciyc.68762.

Change the Ref (2020). UnfinishedVotes.com. [Video]. https://bit.ly/3zK0Bh1.

Cooke, N. A. (2018). *Fake News and Alternative Facts. Information Literacy in a Post-Truth Era*. Chicago, IL: American Library Association.

Cotw. (2021). Change the Ref – The Unfinished Votes, Campaings of the world. https://bit.ly/3heTqal.

CPJ. (2022). 151 journalists and media workers killed in Mexico. *Committee to Protect Journalists. CPJ.* https://rb.gy/4ffpf5.

DoubleYou. (2019). Abascal on sexual freedom [*Abascal habla sobre la libertad sexual*]. [Video]. YouTube. https://bit.ly/3Ws48dL.

El Publicista (2019). Donald Trump: '¡Viva México, cabrones! El Publicista. https://tinyurl.com/bdz9b3fm.

Forbes Staff. (December 9th, 2021). Mexico and Afghanistan, the countries with the most journalists killed in 2021 [*México y Afganistán, los países con más periodistas asesinados en 2021*]. *Forbes México.* https://bit.ly/3TcQedw.

Foucault, M. (2005). *El orden del discurso*. Tusquets.

Fraga-Lamas, P. y Fernández-Caramés, T. M. (2020). Fake news, disinformation, and deepfakes: Leveraging distributed ledger technologies and blockchain to combat digital deception and counterfeit reality. *IT Professional*, 22(2), 53–59. https://doi.org/10.1109/MITP.2020.2977589.

Fundación Reina Sofía (2021). Dalí calls for research to fight against neurodegenerative diseases, new campaign of the Reina Sofia Foundation [*Dalí reivindica la investigación para luchar contra las enfermedades neurodegenerativas, nueva campaña de la Fundación Reina Sofía*]. https://tinyurl.com/45kudjsr.

Gómez-de-Ágreda, Á., Feijóo, C., y Salazar-García, I. A. (2021). A new taxonomy of the use of the image in the interested conformation of the digital story. Deep fakes and artificial intelligence. [*Una nueva taxonomía del uso de la imagen en la conformación interesada del relato digital. Deep fakes e inteligencia artificial*]. *Profesional de la Información*, 30(2). https://doi.org/10.3145/epi.2021.mar.16.

Gosse, Ch. y Burkell, J. (2020). Politics and porn: How news media characterizes problems presented by deepfakes. Critical studies in media communication. *FIMS Publications*, 345. https://ir.lib.uwo.ca/fimspub/345.

Güera, D. y Delp, E. J. (2018). Deepfake video detection using recurrent neural networks. In *2018 15th IEEE International Conference on Advanced Video and Signal Based Surveillance (AVSS)* (pp. 1–6). IEEE. https://doi.org/10.1109/AVSS.2018.8639163.

Gutiérrez, I. (29 de octubre de 2020). A video recreates with artificial intelligence the image and voice of a journalist murdered in Mexico: "Solve this crime" [*Un vídeo recrea con inteligencia artificial la imagen y la voz de un periodista asesinado en México: "Aclaren este crimen"*]. elDiario.es. https://bit.ly/3UpAJzp.

Lafuente, J. (16 de mayo de 2017). Assassinated in Mexico Javier Valdez, the great chronicler of the narco in Sinaloa. [*Asesinado en México Javier Valdez, el gran cronista del narco en Sinaloa*]. El País. https://bit.ly/2Px25WQ.

López, Z. (2021, 29 junio). #SeguimosHablando: From a workshop on Twitter to winning the top prize at Cannes [*De un taller en Twitter a ganar el máximo galardón de Cannes*]. Expansión. https://tinyurl.com/4evz3pn5.

Malaria Most Die (2019). David Beckham speaks nine languages to launch *Malaria Must Die* Voice Petition. [Video]. YouTube. https://bit.ly/343XqUT.

Meersohn, C. (2005). Introduction to Teun Van Dijk: Discourse analysis. Moebius tape [*Introducción a Teun Van Dijk: Análisis de Discurso. Cinta de Moebius*]. Facultad de Ciencias Sociales, Universidad de Chile.

Menéndez, E. L. (2012). Violence in Mexico: Explanations and absences [*Violencias en México: las explicaciones y las ausencias*]. *Alteridades*, 22(43), 177–192.

Naciones Unidas. (12 de abril de 2022). Mexico: Public officials and organized crime responsible for disappearances [*México: funcionarios públicos y crimen organizado son los responsables de las desapariciones*]. Noticias ONU. https://bit.ly/3CmTNYH

Pérez-Anzaldo, G. (2014). The spectacle of violence in twenty-first century Mexican cinema [*El espectáculo de la violencia en el cine mexicano del siglo XXI*]. Eón.

Pérez Muñoz, M. [WITNESS] (2022). Mauricio Pérez from Propuesta Civica describes the work behind Javier Valdez's #deepfake [Mauricio Pérez de Propuesta Cívica describe el trabajo detrás del #deepfake de Javier Valdez] [Video]. YouTube. https://bit.ly/3Gyor2J

Propuesta Cívica. (2022). Who we are [*Quiénes somos*]. Propuesta Cívica. https://propuestacivica.org.mx/acerca-de.

Propuesta Cívica. [El Universal] (2020). #SeguimosHablando Initiative demands justice for murdered and disappeared journalists in Mexico [*Iniciativa #SeguimosHablando exige justicia para periodistas asesinados y desaparecidos en México*] [Video]. YouTube. https://bit.ly/36n7YzG.

Reporteros sin fronteras. (2022). World rankings [*Clasificación mundial*]. https://rsf.org/es/clasificacion.

Richard, N. (2011). The political in art: Art, politics and institutions [*Lo político en el arte: arte, política e instituciones*]. Universidad Arcis. http://hemi.nyuedu/hemi/en/e-misferica-62/richard.

Rodríguez, P. (2022). Synthetic media beyond deepfake [Los medios sintéticos más allá del deepfake]. *Telos, Fundación Telefónica*. https://bit.ly/3zGkGF8.

Schick, N. (2020). *Deepfakes: The Coming Infocalypse*. London: Hachette.

Silverman, C. (Abril 17, 2018). How To Spot A Deepfake Like The Barack Obama–Jordan Peele Video. BuzzFeed. https://bzfd.it/3tkHzd1.

Simondon, G. (2009). Individuation in the light of notions of form and information [*La individuación a la luz de las nociones de forma e información*]. Editorial Cactus.

Simondon, G. (2013). Imagination and invention [*Imaginación e invención*]. Editorial Cactus.

Sohrawardi, S. J., Chintha, A., Thai, B., Seng, S., Hickerson, A., Ptucha, R., y Wright, M. (Noviembre, 2019). Poster: Towards robust open-world detection of deepfakes. In *Proceedings of the 2019 ACM SIGSAC Conference on Computer and Communications Security* (pp. 2613–2615). https://doi.org/10.1145/3319535.3363269.

Temir, E. (2020). Deepfake: New era in the age of disinformation & end of reliable journalism. *Selçuk İletişim*, 13(2), 1009–1024. http://dx.doi.org/10.18094/JOSC.685338.

Vaccari, C. y Chadwick, A. (January-March 2020). Deepfakes and disinformation: Exploring the impact of synthetic political video on deception, uncertainty, and trust in news. *Social Media + Society*, 6(1), 1–13.

Vallejo, G. (20 de julio de 2022). 12 journalists have been killed in 2022; add 35 in AMLO's six-year term [12 *Periodistas han sido asesinados en 2022; suman 35 en sexenio de AMLO*]. Expansión política. https://bit.ly/3wro1Gq

Van Dijk, T. A. (1985). Semantic discourse analysis. *Handbook of Discourse Analysis*, 2, 103–136.

Van Dijk, T. A. (1993). Principles of critical discourse analysis. *Discourse & Society*, 4(2), 249–283.

Wark, S., & Sutherland, T. (2015). Platforms for the new: Simondon and media studies. *Platform: Journal of Media and Communication*, 6(1), 4–10.

Westerlund, M. (2019). The emergence of Deepfake technology: A review. *Technology Innovation Management Review*. https://timreview.ca/article/1282.

Xu, B., Liu, J., Liang, J., Lu, W., y Zhang, Y. (2021). DeepFake videos detection based on texture features. Computers, materials. *Continua*, 68(1), 1375–1388. https://doi.org/10.32604/cmc.2021.016760.

Zavaleta Betancourt, J. A. (2017). The field of violence studies in Mexico [El campo de los estudios de la violencia en México]. Anuario Latinoamericano–Ciencias Políticas y Relaciones Internacionales, 4.

6

The Neurocomputational Becoming of Intelligence: Philosophical Challenges

Manuel Cebral-Loureda

Tecnológico de Monterrey

CONTENTS

6.1 The Shadow of Phenomenology on Contemporary Philosophy

Usually, criticisms of the pertinence of an artificial intelligence (AI) with philosophical relevance are based on the idea of an immediate correlate of the mind, that is, the understanding that exists as a kind of reality that is accessed naturally without any theoretical bias or construction by the self. This is a perspective that has been created along the time and even from the very beginning of the philosophical thought, having in Plato, one the most significant supporters. Effectively, Plato appeals to a philosophical truth, one which is incorruptible by the time and accidents, and opposed to the simulations that folk culture and the senses usually produce, as Plato drew in his well-known cave's allegory. However, it is often forgotten how this thinker has actually used simulation to create his philosophy: beginning with his style, the dialogue, which is artificially constructed; but following with the recognition of Socrates as interlocutor, inventing his words as if they were real; and already ending with the use of myth as a source of legitimization of philosophical reasoning (Puchner, 2014; Deleuze, 1983).

Likewise, there are more examples, and more recents, supporting this kind of approach. One of the most prominent is Descartes, who exemplifies very well the purpose of a refoundation of the knowledge based on an access to an immediate truth. Particularly, Descartes' case is based in the *cogito ergo sum* evidence, as he exposes in his Meditations (2021), allowing the possibility of building a new knowledge architecture. Furthermore, this metaphor of the building will become the reference that will be identified with the beginning of the modern era. In a very similar fashion, Husserl (1982) also claims for a philosophical refoundation of knowledge, this time at the beginning of the contemporary era, and changing the rationalist Descartes ego for a phenomenological one. However, most of the method applied by Descartes remains: it is possible to access to natural and immediate evidence over which it is feasible to constitute a new and necessary philosophy. Despite Husserl changing the metaphor with respect to Descartes, from building to constitution, the fundamental intention remains.

After Husserl, without doubt, one of the most important philosophers who gives continuity to him is Heidegger, who changes the approach of his master a little, understanding that the natural evidence that Husserl found phenomenologically has to be search in another context: the experience that human beings have with temporality and facing death (Heidegger, 2008); as well as the context of the thought about the being as it appeared in the Ancient Greek (Heidegger, 1991, 2014). Both versions of Heidegger philosophy, although they contextualize and make the effort of thinking the truth and the essence of the thought in relation with an existential or historical becoming, they still keep the idea of an original evidence, one that is giving exclusively to philosophy by putting apart the rest of the sciences.

Precisely, in this purpose of separating philosophy from the rest of the knowledge domains, specially the scientific ones, Heidegger follows particularly well his master Husserl. The last one had written *The crisis of European sciences and transcendental phenomenology* (Husserl, 1984), with the intention of recovering a more genuine kind of truth, recreating a domain where only philosophy could enter, the phenomenological space of the experience of the consciousness. Heidegger continues this line of thought by stating that the existential truth is radically separate from the scientifical truth, to the extent that human beings have to be considered, from a biological and an anthropological point of view just in a second moment, after thinking of our most relevant feature, those that make us a *dasein*, that is, a witness of the appearance of the being.

In this context is where the Heidegger short text concerning technology (2013b) can be understood and where the German philosopher criticizes the practical knowledge based on preparing the world to interact with, arguing that the very truth is not given by a by-hand relation with it. Instead of that, it is necessary to understand that the nature itself is governed by a poethical *techné*, the one that is able to pass from the not-being to the being, retrieving the etymology which classical philosophers gave to poetry. As for Husserl and for Heidegger, the kind of world where we live in is an inauthentic one since human beings are not able to poetically inhabit the being.

With more or less emphasis, this controversy about the philosophical knowledge as opposed to the scientific one remains in much of contemporary philosophy, characterizing most of the philosophical movements that appear in the 20th century. To mention just a few, it is present in the philosophy of Foucault, where he relativizes scientific approaches such as medicine, which would be developed in order to construct an idea of the man (1994); the critical theory developed by the Frankfurt School, which accuses the limitations of scientific laws to understand human beings in history and society (Horkheimer & Adorno, 2002); or even from more analytical points of view, such as those proposed by Wittgenstein (1998), who relativizes the capacity of the language to objective reality contextualizing it in what he called games of language.

Such a philosophical landscape will give ground to the postmodernist movement, which wants to destroy the theoretical framework of modernity, relativizing all the principles where the subject, the knowledge, and the language were based. It is the rise of the so-called weak thought (Vattimo, 2013), focused on an anti-metaphysical purpose, which wants to liberate the capacity of interpretation in a nihilistic horizon of a being without any telos. With some perspective, in the 21st century, some thinkers qualify this voyage as a very productive one, which implies the culmination of the concept of immanence (Kerskale, 2009); however, from more dissenting points of view, the concept of immanence must lead to a new contact with reality far away from the relativism that characterizes the postmodernism, in the form of a new materialism (De Landa, 2011), as well as new forms of realism (Meillassoux, 2009).

6.2 The Philosophical Reductionism of Immediacy

As Brassier (2007) have pointed out, particularly interesting regarding this problem, is the position of the American philosopher Wilfrid Sellars, who in his article "Empiricism and the Philosophy of Mind" (1956) criticizes what he calls the myth of the given, that is, the supposition that is possible to access to an immediate and clear content of the mind, such as Descartes, Husserl, or Wittgenstein employed in their philosophies. All of them have prioritized what Sellars calls the manifest image, that is, a conception through which human beings are aware of themselves by encountering themselves (Sellars, 1962). The problem for Sellars is that although the epistemological category of the given is attended to account for the idea that supposedly the immediate evidence of the self-rests on non-inferential knowledge of matters of fact, it is also a construct.

Because of this, these kinds of philosophies are discrediting the scientific image, that is, the one that is based on the objective and external data of the scientific laws. Of course, Sellars accepts that any scientific image is indeed a deductive-nomological construct, in continuous expansion and remodification, that explains the human being as a complex physical system from multiple discourses –physics, chemistry, evolutionary biology, neurophysiology, cognitive science, etc.; however, despite this artificial nature of the scientific image, any kind of subordination, such as the one that many philosophers paradigmatically did during the 20th century, should not be attributed. By doing so, a sort of philosophical reductionism is taking place (Figure 6.1).

More specifically, the manifest image is composed, according to Sellars, of propositional attitudes, related to each other in much the same way as scientific models are: we possess beliefs, desires, intentions, and other attitudes that underlie our explicit activity. Sellars thus goes beyond behaviorism in the analysis of the facts that make up our consciousness and accepts the existence of a cognitive materiality that any cognitive science will have to account for. For Sellars, both the manifest image and the scientific one should live together with autonomous relationships, without any kind of dependence in between since both are constructed independently. However, after Sellars, and specifically in the 21st century, some philosophers and psychologists will begin to question the manifest image itself to the detriment of a new scientific and technological one. There will thus be an inverse movement to that which had occurred in the 20th century philosophy: what would happen if the manifest image were to be completely replaced by a scientific image technologically empowered by neurocomputational enhancement?

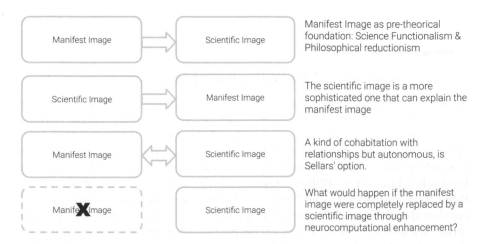

FIGURE 6.1
Four different understandings of the relationship between the manifest image and the scientific one.

6.3 The Neurocognitive Model of the Mind

Goin beyond Sellars, one of his Ph.D. students, the philosopher and neurologist Paul Churchland states that our common sense conception of psychological phenomena constitutes a radically false theory, in which epistemological and ontological principles can be completely displaced by neuroscience (Churchland, 1989: 1). Under this consideration, Churchland gives place to what he will call *eliminative materialism* since it is based in materialistic computational processes that make dispensable the model of folk psychology supported on the supposed consistency of the propositional attitudes.

For doing this, Churchland first contradicts Sellars regarding the explanation of consciousness: further than propositional units, such as beliefs, desires, or intentions, our state of mind can be described in more precise elements, those that the neuroscientific approach provides. Thus, there are events of what we are not aware of because they are smaller or thinner than what we can perceive, events that happen in a non-propositional manner. Churchland describes these events as vector transformations with learned synaptic connection configurations, so they are different from the deductive inferences with which we use to build our logical reasoning.

Through artificial neural networks (ANNs), it is possible to model the behavior of biological nervous systems, using software representations of neurons and the strengths of their interconnections, usually known as weights. Thus, representations in ANNs would be more geometric, or map-like, and they do not need to resort to propositional models (Churchland, 2012). For Churchland, these map-like models resolve much better some of the questions that most challenge folk psychology, i.e., the very specific relationship between some brain damages and cognitive fractionation; or the particular mechanisms involved in artistic or scientific creativity. In addition, folk psychology is stagnant with respect to technological advances in society, and irrelevant to the developments achieved by cognitive science.

For Churchland, it is not that meaning and its propositional structure does not exist, but that it is a simplification or two-dimensional projection of a series of multidimensional relations and movements with their own kinematics, dynamism, and normativity. Thus, the propositional structure on which the manifest image is based is just a uni-dimensional projection of a much more complex structure within which it is possible to enter by means of an activation–vector–space framework in neuroscience (Churchland, 2012: 18); curiously, Churchland uses to explain this dimensional flattening of perception the metaphor that Plato had proposed to defend the immediacy of consciousness.

Other philosophers have complemented this critic to the manifest image and the self as a construction. Among them, Metzinger (2004) refers to Raffman qualia (Raffman, 1995): those simplest perceptual discriminations that cannot be intuited because as with very fine pitch intervals, they lack the conditions of subjective identification. For example, we are able to distinguish many sound and color nuances, although we are not usually aware of them. Thus, we call yellow – or any other color – to a variety of tones that, if we observe in detail, we can notice the difference between. We access them non-conceptually, just by comparison, in a space that is not propositional, although these nuances can be affecting us and even constituting our behavior before the environment. Raffman's qualia, therefore, are fixations of content whose recognition requires organs other than phenomenological intuition, but which would nevertheless be accessible through neurological measurements. Because of this, despite Metzinger discredits,. the ultimate sustenance of Husserl's phenomenology to account the variety of information we are able to perceive, he also affirms that a neurophenomenology would be possible following other authors such as Petitot et al. (1999) or Lutz & Thompson (2003).

For the philosopher Ray Brassier, both Churchland and Metzinger are invoking unconscious and sub-symbolic processes through which phenomenal consciousness is produced, making the evidence of natural attitude and manifest image unjustifiable (Brassier, 2007, p. 31). Although Brassier objects, especially to Churchland, to derive in a kind of neurocomputational apriorism in which it is not quite clear how the brain/world relation is differentiated (Brassier, 2007, p. 18), there is no doubt that the possibility of understanding consciousness through the intervention of neurocomputational methods is a reality that leads us to a deeper and wider space of the mind, where most of the previous conceptions are outdated. Thus, the propositional conception of the mind is not enough; it is necessary to go further; and what is philosophically more interesting is that these authors do not propose to travel this road turning their backs on philosophy; on the contrary, they use all the scaffolding that cognitive sciences and technologies generate in the epistemological and ontological framework, which philosophy has always provided.

6.4 The Threat of the Transhumanism

Based on this kind of neurocomputational model of the mind, along with other technological improvements in human life, a social movement calling itself transhumanist appeared in the 21st century. They claim human enhancement by means of technology, overcoming all the problems and sufferings we have today, in the course of a perpetual progress and expansion of human and machine intelligence (More, 2013; Vita-More, 2008; Kurzweil, 2005). Of course, most of these transhumanist proposals are very uncritical: they tend to assume the human being as the sole heir of technological evolution, without considering how it is impacting on many other inequalities with respect to human beings themselves, but also on an exploitation of other species and the planet (Braidotti, 2013). Thus, for some authors, transhumanists develop a new kind of humanism, even an ultrahumanism (Onishi 2011) since they embrace the same principles of hierarchy and dominance that have constituted the conquering and colonizing culture of the modern age (Ferrando, 2019, p. 33). This is how a philosophical critique of transhumanism emerges: it is the critical philosophical movement called posthumanism, which understands that what human beings need to overcome the humanist paradigm has more to do with attitudinal transformations than with the explicit incorporation of technology into our bodies. By deconstructing the foundations that support the ontological idea of the human as a white Western male within an heteropatriarchal culture, human beings would have overcome themselves as species, reaching a posthuman status (Braidotti, 2013; Ferrando, 2019).

Nevertheless, these posthumanistic critiques of transhumanism draw on many of the contemporary philosophical ideas that have been previously developed under the shadow of phenomenology, i.e., those that deny or undervalue the relevance of technical and scientific production. In much the same way, critical posthumanism does not take technological advances and innovations seriously enough. This is the claim of philosopher David Roden (2015), who suggests that it is necessary to go beyond critical posthumanism, even assimilating some of Churchland's speculations. For example, if two parts of the human brain can learn to communicate, would not the same be possible for two or more spatially distributed brains? Churchland refers to this possibility as an artificial neural commissure, which will allow people to connect through the exchange of neurocognitive data, creating new bonds of solidarity and empathy between people, as some would feel the harm and pain suffered by others (Churchland, 1989: 21). This possibility will lead to a kind of *neurotopia* which, for Roden, would properly imply the emergence of a posthuman species since its way of feeling and being in the world would be completely different (Roden, 2015: 3–4) (Figure 6.2).

Certainly, the speculative approaches contain important utopian components: they work more with possibilities rather than facts, projecting what technologies such as neurocomputing and AI would bring, more than what they are. However, from another point of view, these speculative possibilities constitute, indeed, the space where our society is playing out its future and even part of the current present. It is not a novelty that engineering advances

FIGURE 6.2
Different understandings of the impact of technologies in the evolution of human species into something new.

by shaping most of our environment and relationships, most of the time without asking permission. It is in the hands of disciplines such as philosophy to remain away from this debate, being conformed with a deconstructive position, practicing a sort of philosophical reductionism, or at the same time playing these very needed preventive critiques, it also contributes to a more proactive approach. That is the case of Roden's position, who affirms that by admitting that we are already posthuman just changing our behavior, critical posthumanisms deny the possibility of the posthuman as such to emerge (Roden, 2015, p. 25). It is necessary to maintain the philosophical point of view but also to include in an effective way the scientific-technological potentialities with which we coexist, even if it requires a much more speculative reason.

6.5 Posthuman Transcendental Structure

With arguments similar to those of Metzinger, who suggests the plausibility of a neurophenomenology, Roden develops a speculative posthumanistic point of view where individuals would develop a phenomenology based on neurocomputational skills, capable of processing much more data coming from reality, that is, in a more comprehensively manner than our language and propositional logic is capable of. In order not to reduce this phenomenology to any kind of technological apriorism, as well as to avoid the transcendental constraints that characterize the original Husserl's proposal, Roden suggests the concept of a *dark phenomenology* (2015, p. 85), in which intuition is not always complete; on the contrary, due to its not propositional character, phenomenological intuition is often confused, providing a material that has to be reworked later. This is not to say that dark phenomenology is *ipso facto* inaccessible: a dark phenomenon might influence the experimenter's dispositions, feelings, or actions without enhancing his or her ability to describe or theorize them in classical propositional terms (Roden, 2015, p. 87).

Let us see more precisely what this dark phenomenology consists of. In addition to Raffman's qualifications already cited (Raffman, 1995), in the case of color perception, the gulf between its discrimination and identification is enormous, with discriminable differences totaling about ten million compared to the color lexicon of about thirty (Mandik, 2012). Further evidence is provided by experiments such as that of George Sperling (1960): subjects reported seeing or perceiving all or nearly all elements of a 12-character alphanumeric memory presented for 50 milliseconds, although they could only identify about 4 elements after presentation. If the detail with which humans are able to retrieve, through memory, perceptual data is coarser than the effective underlying phenomenology, this implies that some of this phenomenology goes unnoticed and will be forgotten. Other researchers (Block, 2007) have argued that these findings demonstrate the capacity of experience to overwhelm our attentional availability. Therefore, the ultimate models of our conscious activity cannot depend on the circuits of working memory, where an important part of the information received is lost.

In addition to these arguments, Roden reaffirms the need to accept the obscure part of intuition in which subphenomenal elements occur on the basis of the critique of the nested structure of temporality in Husserl, and the incoherence of its projection ad infinitum (Wood, 2001: 78–79). For all these reasons, it is necessary to reject a single a priori phenomenological theory of world experience; on the contrary, it is necessary to affirm the possibility that posthuman modes of being may structure experience differently, in ways that we humans cannot currently comprehend. This includes rejecting the projection of our propositional and discursive structure onto possible non-human intelligent beings. For Roden, this does not imply that humans and posthumans do not belong to the same reality or universe; although, on the other hand, neither can we ever disentangle what actually it means to belong to the same reality (Roden, 2015: 96).

Under these kinds of considerations, the constraints that typically characterize the trascendental structure of human beings is no longer extended for other types of intelligence, which change the spatial-time conditionings, but also the essence of what is evidence in the cognition process, leading us to a phenomenology without any particular or stable trascendental structure.

6.6 AI Frame Problems

Roden's proposal evidences the possibility of assimilating the technological advances of neuroscience by re-reading and adapting previous philosophies of the tradition, using their concepts in a broader theoretical framework.

Technologies such as AI suppose a breakpoint or benchmark on a par of previous technological disruptions, such as industrial revolution, or the invention of the print. Moreover, AI technologies have a particularity that relates them even more closely to philosophy since both have a very similar purpose: the understanding of knowledge itself. In this sense, AI technologies could be placed at the level of other human evolutions of the first level of philosophical impact, such as the irruption of writing or of language itself. Precisely, both discoveries have been highlighted, in the contemporary epoch, as the spaces that properly host philosophical thinking. For Heidegger, the language is the house of being since it provides the means by which we access the revelations of what things are and mean in our existence (Heidegger, 2013a), and Derrida goes further by pointing out how the focus on the writing over the oral speech can deconstruct a metaphysical tradition too much oriented to the presence (Derrida, 2010). For Derrida, writing would be not just the context where all the meaning of the consciousness can be expressed but also the unconscious trace that completes the map of the mind from a psychoanalytic point of view.

What would computational coding imply philosophically in this evolutionary line? Of course, coding is an act of writing, but on another level than the propositional. The act of coding, regardless of the informatic language we use, allows us to penetrate deeper into the structure of reality, accounting for a large number of events that can be quantified in some way. This is the capacity of reception that technologies have, acquiring and processing amounts of data that human beings will never be able to face by themselves. In addition to this extraordinary memory, computational methods have also proved an outstanding power to put in relation a great variety of variables, in the context of the so-called big data. By means of methods such as machine learning, it is possible to correlate countless fields of reality, finding how any variable is affecting an apparently disconnected event (Cebral-Loureda, 2019). Precisely, when we say apparently we are saying that it does not occur in a propositional way, because we cannot explain such correlation grammatically. That is what underlies in the well-known data science statement: correlation is not causation.

Of course, this potentiality is not unlimited. Precisely because of the non-propositional character of the computational methods, it is not possible to interpret them nor can they face phenomena such as metaphors or expressions of feelings when they appeal to meaning, further than in a quantitative manner. Overall, the most difficult problems to solve are those related with frame problems, that is, how a cognitive system can distinguish relevant information from irrelevant information (Wheeler, 2005). About this, Churchland had already suggested some examples: we listen to sounds and voices coming from the basement, how do we infer that we left the CD running?; or, we see smoke coming out from the kitchen, how do we deduce that the toast is burning? (Churchland, 1989, p. 109). A machine would capture a lot of data of events like these, from sound or temperature changes to different grades of visibility appearing due to smoke, or even air currents. There are many small and precise observations which we barely perceive as human beings, but they are surely happening. Instead of being misguided by all these small events, we would have clear just one thing: it is time to stop the CD or to remove the toast from the toaster, something very difficult to infer for a machine learning algorithm that has not been trained for this specific task. In philosophical words, as Wheeler acknowledges, machines have problems to face the Heidegger aim of authentic existence: how do we place all the facts we experience in the wider context of the existence that gives them meaning.

6.7 Non-Propositional Metacognition

To solve these frame problems and to provide consciousness dynamics to computational technologies, deep learning models based on ANNs are being proposed (Cao, 2022; Shymko, 2021). Their layered structures allow us to think not only of a high capacity to easily and flexibly manage and process large amounts of data but also to introduce metacognitive qualities into these processes. Thus, by overlaying ANNs, it would be feasible to emulate the consciousness of a perceptual state, creating a network layer in charge of the metarepresentational activity of a cognitive process carried out by another first-order layer (Cleeremans et al., 2007). Some little examples were already proved in experiments executed in the context of binocular rivalry, i.e., how the brain chooses one stimulus over another when two eyes are perceiving a difference, and it is not clear what is more relevant. To face this metacognitive problem, Hohwy et al. (2008) develop a probabilistic predictive coding capable of solving such a rivalry.

More recent articles have even suggested a more unified approach, where the perception and action models are integrated together in a hierarchical prediction machine, in which the attention is already inclined to generate actions since "perception and action both follow the same deep 'logic' and are even implemented using the same computational strategies" (Clark, 2013; who also references Friston et al., 2009). As Churchland has proposed, connections between low- and high-level orders can be based on mapping operations, instead of interpreting propositional meaning, with the particularity that these connections can occur reversely and bidirectionally. By means of techniques such as back-propagation (Rumelhart et al., 1988), it is possible to adjust the prediction accuracy of an ANN from the last highest layers to the lower layers, and then use the error bidirectionally to also improve the whole orientation of the algorithm (Figure 6.3).

At the beginning of this operation of constant adjustment, Clark places *hyperpriors* (2013), appealing to the Kantian notion of a trascendental structure of perception that conditions cognition. Hyperpriors would be the superior layers that are adapting and leading the mental activity, integrating perception, cognition, and action in a process that is continuous and unified. For example, many of our actions are determined by our expectations, producing some kind of sensation previously warrant; other times, actions are defined by the adaptive function of proprioception, i.e., they respond to perception requirements as they adjust. It is supposed that deep learning neural networks, because of their recursive hyperconnected structure, are ideal for these kinds of tasks. They are expected to play the role of consciousness by processing the frame of relevance in which the data acquire meaning. Finally, Clark calls these aprioristic conditionals hyperpriors since they are not fixed, like they were in Kant, but metacognitive, working as more than just priors.

6.8 AI Historical and Social Context

Meanwhile for Roden, this adaptation happens in relation with Husserl's phenomenology, and for the Iranian philosopher Reza Negarestani (2018),

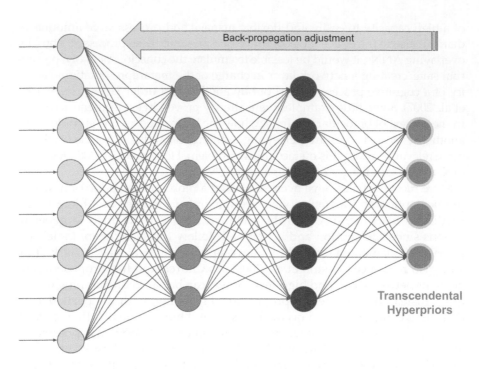

FIGURE 6.3

Highest layers work as hyperpriors since they can define, by back-propagation, the activity of the whole ANN, more than providing entire contents each time, by means of the constant adequation and recursive feedback that these highest layers give regarding the outcomes and their match with the world.

it occurs with the phenomenology of Hegel. With such theoretical support, Negarestani describes how the accomplishment of an AI is a process that should not be considered apart from our previous cognitive achievements but that happens in a process of integrated phases, through which AI can gain more and more consciousness, becoming an object of its own reflection.

Mind will never have a unique place in nature, on the contrary, as Hegel stated (2013), its nature consists precisely in a continuous movement, a kind of wandering without having a final stop. In this sense, intelligence is the result of being applied over itself, identifying "the history of the mind as the history of artificialization" (Negarestani, 2018, p. 58). Applied to the case of the creation of an AI, this approach leaves no room for doubt: the attempt to obtain an AI should not be considered as an inappropriate external aim; by contrast, it should be understood as part of the evolution of our own intelligence into a new level of development in which it is projected onto itself by means of computational methods. Far from being a kind of alienation, the possibility of realization in a computational exteriority confirms the capacity of intelligence to break its attachment to any special status or given meaning, demonstrating that it is an artifact that creates its own history through its own alienation.

In addition to this evolutive and historical conditioning, Negarestani also highlights the social aspect needed to the achievement of an AI, something that has not been considered enough previously. Effectively, we used to think about how to overcome the functionalism of the intelligence in a machine by implementing complex algorithms attempting to produce consciousness. Most of the time, these algorithms are focused in the machine as an individual, trying to provide it with intelligence by means of techniques that are applied to that machine in isolation (Guillermin & Georgeon, 2022; Bolotta & Dumas, 2022). By doing this, we are forgetting that intelligence also needs a social context to achieve its realization, a context where perceptions are confirmed and where the language is really useful and properly acquires its meaning. For Hegel, this social component present in the evolution of knowledge constitutes the second and objective stage of the mind in its development, where the truth is socially given, constituting a second and objective stage that includes the existence of a family, a community, and institutions that ensure the search of objectivity for each epoch (Hegel, 2018). It is coherent to admit that something similar should apply for any realistic AI training.

Precisely, in Hegel, this capacity of socialization of the mind gives birth to a third and final stage, those that Negarestani's approach also stands out: the ethic one. Effectively, by means of an objective social context, even in an institutional form, an interactive ethical framework can be generated for an AI. Following Hegel's schema, it makes sense to prepare law and objective regulation in order to socially accompany AI development. Thus, the current demands of regulation that the AI development is claiming (Cath et al., 2017; Rahwan, 2018) can be well addressed by means of Hegel's philosophy. However, by doing this, Negarestani highlights another relevant point: the ethical aim in Hegel's philosophy is freedom, in a very deep and abstract way, which surpasses any regulatory framework, embracing an ethical concept that regulates itself, without any kind of constraint.

Certainly, this kind of approach has consequences regarding the trans-/posthumanist debate. As it is well known, freedom is opposed to fear, and what an artificial mind needs, to achieve full development, is precisely an absolute freedom, so also, by definition, the lack of fear. This implies that alongside the transhumanist fantasies that we have earlier dismissed as uncritical in considering a future in which AI and technology will save all our problems, it is also necessary to abandon all the fears that are also projected in many socio-political conceptions and approaches to AI and technology enhancement. As in a mirror, only by relying on the capability of AI, the artificiality of the mind can truly acquire its new step of development.

6.9 Discussion: What Would It Mean to Survive for an AI?

Unfortunately, the anthropocentric bias does not only happen in transhumanist proposals but also in the opposite direction when all the calamities and fears are projected onto technologies – especially AI technologies – by arguing that they will replace us. One of the most common examples is how intelligent machines will fill human jobs, forming a new workforce (Bessen, 2018; Smith & Anderson, 2014). Even further, there are references, especially in science fiction (Garland, 2015; Kubrick, 1968), but also in academic papers (Buttazzo, 2001) that consider that these technologies could attack us, guided by a sense of ambition, revenge, and other similar types of pulses. It is important to realize that these kinds of feelings do not belong properly to the machines, but that we are projecting human moral behavior onto them (Walsh, 2022). Otherwise, why would an AI attack us?

To do so, AIs would have to seek to impose themselves on humans, i.e., they would want to preserve their identity over ours. Now, to do this, they would in turn require a notion of identity, the reference to a self that they would indeed want to preserve. However, what kind of self as a closed and corruptible unity would be meaningful to an intelligent machine? In other words, what would be it that in its functioning would struggle to survive?

During the 1990s, attempts were made to build an embodied AI (Franklin, 1997), trying to simulate the finite horizon of human intelligence. This idea was transformed in the 2000s, with the *affective turn* in AI (Clough & Halley, 2007), studying the emotional implications of AI, giving rise to sentiment analysis approaches. Since then, however, increasingly the two fields have evolved separately, with AI development focusing on abstract algorithmics, on the one hand, and a roboticists field concerned about mechanical and electrical engineering, concentrating on sensori-motor functions (Rajan & Saffiotti, 2017). Even the current developments of producing an embodied AI have more to do with an algorithm that manipulates a physical thing (Savva et al., 2019), rather than a general AI, which would integrate its intelligence in a body with more holistic consequences. Just some authors have intended to produce something like an instinctive computing (Cai, 2016), but in a very fuzzy way, that has not had a run for the time being.

Effectively, how could we introduce a physical concern in a system that runs independently regardless of its material substrate? Of course, many can argue that every AI algorithm needs material supply, as well as hardware to run properly in a device. However, instead of continuity, like in the nervous system of human beings, the relation between the code and its hardware is merely associative and discontinuous. We know that exactly the same code can run on different substrates, which are reproducible, and lack the uniqueness of biological bodies. In fact, many of the biometric technologies that measure our emotions and feelings are based precisely on this kind of uniqueness of our bodies: facial recognition, iris recognition, voice analysis, among others. So, if not through its continuity with respect to a unique physical body, how can an AI have a differentiated self-awareness?

A proposal could lie in the recursion of the algorithms. Effectively, algorithms have this function as a core one, being one of their most important features since the very beginning with Alan Turing. Let us think in the case of a deep learning ANN model. In that case, to survive will be something more related to the economy of accuracy of the algorithm than to the preservation of a body or physical integrity, of which the AI properly lacks.

Indeed, it is very interesting how this notion of recursivity has been recently related with the philosophical concept of the self in the German idealist philosophy (Hui, 2019), developing a new idea of cybernetics that includes the current implementation of AI. The difference is that in this idea of selfness, what gives continuity to the recursive function is an abstract, mathematical, and virtual continuity, which will have no fear of being destroyed since it does not have emotions connected to any material singularity risk. If we would want to talk, in a more metaphorical way, of a survival instinct in these machines, it would be a very efficient one, a rational calculation, which will always prioritize a smart use of its resources.

6.10 Conclusions

The caveats of many contemporary philosophies to think seriously about the chances of AI neurocomputational technologies to succeed in the creation of an AI can be more a limitation than a virtue since it has been observed the kind of reductionism that underlie such philosophies. On the contrary, by emptying the understanding of a trascendental structure of phenomenological perception, we will be nearer to producing the advent of an AI, one that will not be generated by an anthropocentric resemblance, but properly a posthumanistic one.

Interestingly, each in its own way, both Roden and Negarestani appeal to the necessity of a benchmark in our relation with AI: Roden by stating that we have to accept that posthuman beings would be effectively different from us, by means of some kind of technological mediation such as the neurocomputational one, and Negarestani by applying Hegel's conception of interrelational sociability, but also highlighting that AI can only emerge autonomously by being trained in a space without constraints and limitations, just by surrendering to its own freedom.

Of course, this does not mean that AI can appear or be created just by letting it be. It is necessary, not just the sophisticated knowledge originated by data scientists in collaboration with researchers from other fields but also the socialization of this knowledge, both in terms of sharing it among human beings and in relation to a community of other AIs. Here, socialization should include the ethical framework that recently occupies the debates around AI, only that, for the reasons stated herein, this ethical framework should be based in guaranteeing freedom and independence to AIs, more than just preventing and limiting the development of the artificial mind to the unknown.

Concerns about the emergence of an autonomous AI are more often than not fueled by the same anthropomorphism that encourages the most transhumanist and *ultrahumanist* positions. It is unlikely that an AI could have a notion of itself and, if it did, it would be in a very different way from ours. The concepts of ethics, altruism, and egoism will function in a totally different way in such subjects since they will reference their actions in a completely different way with respect to notions such as those of self and totality, if they have them; but even having them, what is very questionable is that they will be closed and identitarian concepts, such as those that have nurtured a very specific, and never absolute, stage of humanity.

References

Bessen, J. (2018). *AI and jobs: The role of demand* (No. w24235; p. w24235). National Bureau of Economic Research. https://doi.org/10.3386/w24235.

Block, N. (2007). Consciousness, accessibility, and the mesh between psychology and neuroscience. *Behavioral and Brain Sciences, 30*(5–6), 481–499. https://doi.org/10.1017/S0140525X07002786.

Bolotta, S., & Dumas, G. (2022). Social Neuro AI: Social interaction as the "Dark Matter" of AI. *Frontiers in Computer Science, 4*, 846440. https://doi.org/10.3389/fcomp.2022.846440.

Braidotti, R. (2013). *The posthuman*. Polity Press, Cambridge, UK; Malden, MA.

Brassier, R. (2007). *Nihil unbound: Enlightenment and extinction*. Palgrave Macmillan: Basingstoke New York.

Brassier, R. (2010). *Nihil unbound: Enlightenment and extinction*. Palgrave Macmillan.

Buttazzo, G. (2001). Artificial consciousness: Utopia or real possibility? *Computer, 34*(7), 24–30. https://doi.org/10.1109/2.933500.

Cai, Y. (2016). *Instinctive computing*. Springer, London. https://doi.org/10.1007/978-1-4471-7278-9.

Cao, R. (2022). Putting representations to use. *Synthese, 200*(2), 151. https://doi.org/10.1007/s11229-022-03522-3.

Cath, C., Wachter, S., Mittelstadt, B., Taddeo, M., & Floridi, L. (2017). Artificial intelligence and the 'good society': The US, EU, and UK approach. *Science and Engineering Ethics*. https://doi.org/10.1007/s11948-017-9901-7.

Cebral-Loureda, M. (2019). *La revolución cibernética desde la filosofía de Gilles Deleuze: Una revisión crítica de las herramientas de minería de datos y Big Data* [Universidade de Santiago de Compostela]. http://hdl.handle.net/10347/20263

Churchland, P. M. (1989). *A neurocomputational perspective: The nature of mind and the structure of science*. MIT Press, Cambridge, MA.

Churchland, P. M. (2012). *Plato's camera: How the physical brain captures a landscape of abstract universals*. MIT Press, Cambridge, MA.

Clark, A. (2013). Whatever next? Predictive brains, situated agents, and the future of cognitive science. *Behavioral and Brain Sciences*, 36(3), 181–204. https://doi.org/10.1017/S0140525X12000477.

Cleeremans, A., Timmermans, B., & Pasquali, A. (2007). Consciousness and metarepresentation: A computational sketch. *Neural Networks*, 20(9), 1032–1039. https://doi.org/10.1016/j.neunet.2007.09.011.

Clough, P. T., & Halley, J. O. (Eds.). (2007). *The affective turn: Theorizing the social*. Duke University Press, Durham.

De Landa, M. (2011). *Intensive science and virtual philosophy* (Reprint). Continuum, London.

Deleuze, G. (1983). Plato and the Simulacrum (R. Krauss, Trad.). *October*, 27, 45. https://doi.org/10.2307/778495.

Derrida, J. (2010). *Writing and difference* (A. Bass, Ed.; Transferred to digital print. 2006, repr). Routledge, London.

Descartes, R. (2021). *Meditations on first philosophy & other metaphysical Writings*. Arcturus, London.

Ferrando, F. (2019). *Philosophical posthumanism*. Bloomsbury Publishing, New York; London.

Foucault, M. (1994). *The birth of the clinic: An archaeology of medical perception*. Vintage Books, New York.

Franklin, S. (1997). Autonomous agents as embodied AI. *Cybernetics and Systems*, 28(6), 499–520. https://doi.org/10.1080/019697297126029.

Friston, K. J., Daunizeau, J., & Kiebel, S. J. (2009). Reinforcement learning or active inference? *PloS One*, 4(7), e6421.

Garland, A. (2015). *Ex Machina*. A24; Universal Pictures.

Guillermin, M., & Georgeon, O. (2022). Artificial interactionism: Avoiding isolating perception from cognition in AI. *Frontiers in Artificial Intelligence*, 5, 806041. https://doi.org/10.3389/frai.2022.806041.

Hegel, G. W. F. (2013). *Phenomenology of spirit* (A. V. Miller & J. N. Findlay, Eds.; Reprint.). Oxford University Press, Oxford.

Hegel, G. W. F. (2018). *Hegel: The phenomenology of spirit* (M. J. Inwood, Trad.; First edition). Oxford University Press.

Heidegger, M. (1991). *Nietzsche* (D. F. Krell, Trad.; 1st HarperCollins pbk. ed.). HarperSanFrancisco.

Heidegger, M. (2008). *Being and time* (J. Macquarrie & E. S. Robinson, Trads.). HarperPerennial/Modern Thought, New York.

Heidegger, M. (2013a). Carta sobre el humanismo (2. ed). Alianza Ed, Madrid.

Heidegger, M. (2013b). *The question concerning technology and other essays* (W. Lovitt, Trad.). HarperCollins Publishers, New York; London; Toronto.

Heidegger, M. (2014). *Introduction to metaphysics* (G. Fried, Trad.; Second Edition). Yale University Press, New Haven, CT.

Hohwy, J., Roepstorff, A., & Friston, K. (2008). Predictive coding explains binocular rivalry: An epistemological review. *Cognition*, 108(3), 687–701. https://doi.org/10.1016/j.cognition.2008.05.010.

Horkheimer, M., & Adorno, T. W. (2002). *Dialectic of enlightenment: Philosophical fragments*. Stanford University Press, Stanford, CA.

Hui, Y. (2019). *Recursivity and contingency*. Rowman & Littlefield International, London; New York.

Husserl, E. (1982). *General introduction to a pure Phenomenology* (M. Nijhoff ; Distributors for the U.S. and Canada). Kluwer Boston, The Hague; Boston, MA; Hingham, MA.

Husserl, E. (1984). *The crisis of European sciences and transcendental phenomenology: An introduction to phenomenological philosophy* (D. Carr, Trad.; 6th pr). Northwestern University Press, Evanston, IL.

Kerslake, C. (2009). *Immanence and the vertigo of philosophy: From Kant to Deleuze*. Edinburgh University Press, Edinburgh. http://www.jstor.org/stable/10.3366/j.ctt1r20p3

Kurzweil, R. (2005). *The singularity is near: When humans transcend biology*. Viking: New York.

Kubrick, S. (1968). *2001: A space Odyssey*. Metro-Goldwyn-Mayer, Beverly Hills, CA.

Lutz, A., & Thompson, E. (2003). *Neurophenomenology: Integrating subjective experience and brain dynamics in the neuroscience of consciousness* (p. 52). Imprint Academic, Charlottesville, VA.

Mandik, P. (2012). Color-consciousness conceptualism. *Consciousness and Cognition*, 21(2), 617–631. https://doi.org/10.1016/j.concog.2010.11.010.

Meillassoux, Q. (2009). *After finitude: An essay on the necessity of contingency* (Pbk. ed). Continuum, London; New York.

Metzinger, T. (2004). *Being no one: The self-model theory of subjectivity* (1. paperback ed). MIT Press, Cambridge, MA.

More, M. (2013). The Philosophy of Transhumanism. *The Transhumanist Reader* (M. More & N. Vita-More, 1st ed., pp. 3–17). Wiley, New York. https://doi.org/10.1002/9781118555927.ch1

Negarestani, R. (2018). *Intelligence and spirit*. Urbanomic; Sequence, Windsor Quarry; Falmouth; New York.

Onishi, B. B. (2011). Information, bodies, and Heidegger: Tracing visions of the Posthuman. *Sophia*, 50(1), 101–112. https://doi.org/10.1007/s11841-010-0214-4.

Petitot, J., Roy, J.-M., Varela, F. J., & Pachoud, B. (Eds.). (1999). *Naturalizing phenomenology: Issues in contemporary phenomenology and cognitive science*. https://doi.org/10.1515/9781503617421.

Puchner, M. (2014). *Drama of ideas: Platonic provocations in theater and philosophy*. Oxford University Press I, Delhi.

Raffman, D. (1995). On the persistence of phenomenology. *Conscious experience* (T. Metzinger, p. 293â308). Ferdinand Schoningh, Paderborn.

Rahwan, I. (2018). Society-in-the-loop: Programming the algorithmic social contract. *Ethics and Information Technology*, 20(1), 5–14. https://doi.org/10.1007/s10676-017-9430-8.

Rajan, K., & Saffiotti, A. (2017). Towards a science of integrated AI and Robotics. *Artificial Intelligence*, 247, 1–9. https://doi.org/10.1016/j.artint.2017.03.003.

Roden, D. (2015). *Posthuman life: Philosophy at the edge of the human*. Routledge, London.

Rumelhart, D. E., McClelland, J. L., Group, P. R., & Others. (1988). *Parallel distributed processing* (Vol. 1). IEEE, New York.

Savva, M., Kadian, A., Maksymets, O., Zhao, Y., Wijmans, E., Jain, B., Straub, J., Liu, J., Koltun, V., Malik, J., Parikh, D., & Batra, D. (2019). Habitat: A platform for embodied AI research. *2019 IEEE/CVF International Conference on Computer Vision (ICCV)*, pp. 9338–9346. https://doi.org/10.1109/ICCV.2019.00943.

Sellars, Wilfrid S. (1962). Philosophy and the scientific image of man. In Robert Colodny (ed.), *Science, Perception, and Reality*. Humanities Press/Ridgeview: New York. pp. 35–78. Edited by Robert Colodny.

Shymko, V. (2021). Discourseology of linguistic consciousness: Neural network modeling of some structural and semantic relationships. *Psycholinguistics*, 29(1), 193–207. https://doi.org/10.31470/2309-1797-2021-29-1-193-207.

Smith, A., & Anderson, J. (2014). AI, robotics, and the future of jobs. *Pew Research Center*, 6, 51.

Sperling, G. (1960). The information available in brief visual presentations. *Psychological Monographs: General and Applied*, 74(11), 1–29. https://doi.org/10.1037/h0093759.

Vattimo, G. (2013). *Weak thought*. Suny Press, New York

Vita-More, N. (2008). Designing human 2.0 (transhuman) – Regenerative existence. *Artifact*, 2(3–4), 145–152. https://doi.org/10.1080/17493460802028542

Walsh, T. (2022). *Machines behaving badly: The morality of AI*. Flint.

Wheeler, M. (2005). *Reconstructing the cognitive world: The next step*. MIT Press, Cambridge, MA.

Wittgenstein, L. (1998). *Philosophical investigations:=Philosophische Untersuchungen* (G. E. M. Anscombe, Trad.; 2nd ed., repr). Blackwell, Cambridge, MA.

Wood, D. (2001). *The deconstruction of time*. Northwestern University Press, Evanston, IL.

Section II

Knowledge Areas Facing AI

Nowadays and increasingly, Artificial Intelligence (AI) is present in people's lives and is assumed consciously or unconsciously; sometimes it is unnoticed, threatening, and in others a means that facilitates and makes our lives more comfortable. As we saw in section one, the light or darkness that AI brings to the lives of humans is related to the purpose of use and the medium from which the information was obtained to feed the algorithms, that is, how AI is being culturally and socially assimilated. According to UNESCO, AI can support millions of students to complete their basic education, fill jobs, and even face the consequences of the COVID-19 pandemic; however, these benefits are accompanied by risks caused by the misuse of technology or by the gap that is generated between people in terms of inequality and access.

Although global authorities are clamoring for legislation to regulate the use of AI, the speed at which the technological development of AI is growing is much greater than the speed of national and international negotiations and agreements. In this sense, we must be cautious about the euphoria of AI present in everyday life, being aware that technology must be at our service and not the other way around. Alan Turing already predicted in the 1950s the arrival of thinking machines; currently, various technologists and philosophers talk about conscious AI. Perhaps we are not so far from the technology of Star Trek or Star Wars, and I hope so, very, very far from a scenario like the one presented in Terminator movie.

Given the ideas expressed above, section two of this book chapter "What AI Can Do?" presents the applications of AI in different areas of our lives, such as education, health, smart cities, music, and the criminal justice system. Each chapter presents the benefits of AI in these areas, as well as the risks or limitations generated by bad practices of its use or lack of regulations. This section invites the reader to reflect and discuss, and perhaps with

DOI: 10.1201/b23345-8

a bit of luck, it motivates him to investigate more about the use of AI in other areas of daily life that undoubtedly exist.

In the field of education, three interesting applications are covered: the first is presented by Rodriguez-Hernandez et al., who tell us about the use of self-organizing maps as a method for clustering a large group of data sets allowing us to find common variables that allow knowing, in this case of a population of students, the factors that affect the rate of school retention with the aim of making educational decisions that benefit the educational institutions and of course to the students. Then, we have the study of Olmos-López et al., in which they present the use of AI developed from predictive algorithms based on decision trees, which forecast the performance of university students in the Engineering Faculty allowing the combination of humanized attention of the professor and the use of AI. Finally, the study by Morán-Mirabal et al. introduces us to the possibilities of collecting information within academic environments through multimodal devices that through AI can better understand learning interactions either physically or remotely or in hybrid modalities.

In the medical field, the work presented by García-González tells us about the digitalization of health systems developed worldwide as a possibility to face the challenges of demand for specialists in the medical area, as well as the quality of care for people. Then, we have another interesting study presented by Hinojosa-Hinojosa and González-Cacho on The Dark Side of Smart Cities in which the advantages of smart cities and how their development generates an inequality gap among citizens are presented. Next, we have the work of Vivar-Vera and Díaz-Estrada in which they develop a debate on the predictive software used in the criminal justice system.

Finally, we close the section with the study of Sacristán et al., in which they expose how AI can be a valuable tool in the creative field, for this they share one of the most famous cases of unfinished works: the case of Piccini's Turandot.

7

A Cluster Analysis of Academic Performance in Higher Education through Self-Organizing Maps

Carlos Felipe Rodríguez-Hernández
Tecnológico de Monterrey

Eva Kyndt
Swinburne University of Technology

Eduardo Cascallar
KU Leuven

CONTENTS

DOI: 10.1201/b23345-9

7.1 Introduction

The general trend in higher education literature has been to collect data on different predictors of academic performance to confirm or refute theoretical claims previously established. Nonetheless, this approach possibly fails at identifying several existing patterns among the data set of predictors. Furthermore, as the size of the data sets containing students' information keeps increasing, the number of unknown patterns is also growing. These patterns are relevant as they could provide critical information about learning approaches, effective teaching practices, or positive institutional experiences. Therefore, by identifying such patterns, a better understanding of academic performance in higher education could be achieved. Cluster analysis emerges as one powerful alternative for pattern recognition because it makes it possible to organize students in specific subgroups according to their characteristics (Romero & Ventura, 2010) and to use the resulting subgroups for subsequent analysis (Antonenko et al., 2012; Peiró-Velert et al., 2014).

Several clustering methods, such as K-means clustering, hierarchical clustering, and self-organizing maps (SOMs), have been applied in the past educational research to analyze students' academic performance (Kausar et al., 2018). Two essential conclusions could be obtained from this body of research. On one hand, it has been suggested that prior academic achievement and socioeconomic status (SES) are influential variables to create not only students' subgroups but also contribute to differentiating their academic performance in higher education (De Clercq, et al., 2017; Da Silva et al., 2012; Rodríguez-Hernández, 2016). On the other hand, there is no definitive evidence on which clustering method shows the best performance, as each method uses different algorithms as well as various similarity measures to group the data (Kumar & Dhamija, 2010).

Despite the inconclusive evidence on which clustering method performs the best, neither K-means nor hierarchical clustering provides clear insights into the relationship among the input variables in the resulting grouping. However, self-organizing maps (SOMs) deal with large data sets, reduce the dimension of any given input space, preserve the relationship among the input variables, and provide a graphical representation of such association (Haykin, 2009; Kohonen, 2013). Starting from these features of SOMs, it seems worthwhile to explore whether their use contributes to a better understanding of the relationship between prior academic achievement, SES, and academic performance in higher education.

Therefore, the present chapter illustrates the use of SOMs to analyze academic performance in higher education, by answering three specific questions. *What are Self-organizing maps (SOMs)?* defines SOMs and describes their structure, designing steps, and several quality measures. *How SOMs can be used to analyze academic performance in higher education?*

provides a practical example of the application of SOMs, that is, SOMs are used to analyze academic performance of a cohort of 162,030 Colombian university students. In this way, it is possible to explore the existing patterns between prior academic achievement, SES, and academic performance through a SOM model, and then cluster it to determine specific groups of students based on these variables. *What are the takeaway messages?* discusses the main conclusions of the chapter and outlines some directions for future research on the use of SOMs in educational sciences.

7.2 What Are Self-Organizing Maps (SOMs)?

Two types of predictive models based on artificial neural networks can be distinguished: supervised and unsupervised models. On one hand, supervised models use patterns in the data to iteratively compare output estimates to observed estimates, while reducing the difference between them. On the other hand, unsupervised models directly search for patterns in the data and group similar cases into clusters (Garson, 2014). As such, there is no information on the output variables in unsupervised models. Self-organizing maps (SOMs) are a type of unsupervised artificial neural network developed by Teuvo Kohonen in 1982. The main goal of SOMs is to transform any n-dimension input data set with n-variables into a low-dimensional map (usually with two dimensions) while maintaining the existing relationship in the data (Haykin, 2009; Kohonen, 2013). In other words, if two elements from the data set exhibit similar patterns in the input variables, their position in the resulting SOM would also be very similar (Wehrens & Buydens, 2007).

7.2.1 Structure of SOMs

According to Haykin (2009), SOMs are layered structures with two layers. Notice that a layer simply refers to a set of neurons interconnected to perform a specific task within the artificial neural network. In the *input layer*, the number of neurons is equal to the number of cases in the data set. As in any other artificial neural networks, the input layer connects the network to the data set inputs. In the *competitive layer*, the neurons are organized as the nodes of an empty grid, which is usually two-dimensional. The size of the grid determines the number of neurons in the competitive layer (e.g., a square grid of dimension five would have 25 neurons). Each neuron of the input layer is linked to every neuron in the competitive layer. Therefore, all the information provided to the input layer is connected to all the neurons in the competitive layer.

7.2.2 Competitive Learning of SOMs

SOMs identify existing patterns in large data sets through competitive learning (Haykin, 2009). This process seeks the "most similar" neuron in the competitive layer to every neuron in the input layer (Iturriaga & Sanz, 2013), as explained by Haykin (2009) and Kohonen (2013). To begin with, each case in the data set is transformed into an n-dimensional input vector, where n indicates the number of input variables in the data set. Secondly, a grid of neurons is created, and each neuron within the grid is initialized with a random n-dimensional weight vector. This weight vector is usually known as the prototype of that neuron. Thirdly, every input vector (i.e., each case in the data set) is compared to all the prototypes in the grid. The neurons in the grid compete among themselves to win each of the input vectors. The winner neuron is the one whose prototype is closer to the input vector, and it is known as the best matching unit (BMU). Finally, the weight vector of the BMU, as well as the weight vectors of the surrounding neurons, is adjusted to best match with the input vector. This adaptative process is possible through the so-called *neighborhood function* that is defined around the BMU.

7.2.3 Designing of SOMs

Several parameters should be tuned when designing SOMs (Kohonen, 2014). First, the form and the size of the grid (i.e., the number of neurons) are defined. Usually, the form of the grid is rectangular. Next, the shape of the neurons in the grid (rectangular or hexagonal) is selected. Such a parameter is rather crucial because it defines the topological relation of the neurons in the SOM. To put it differently, a rectangular neuron would have eight surrounding neurons, while a hexagonal neuron would have six neighboring neurons. In addition, the weight vector of each neuron is initialized using either random initialization (weights are small random values) or linear initialization (weights are obtained from the space generated by the linear combination of the two principal components of the input data). Finally, the type of neighborhood function (typically Gaussian) and the type of training (batch or sequential) are also set.

7.2.4 SOMs Quality Measures

There are two traditional ways to evaluate the quality of SOMs: the quantization error and the topographic error (Pölzlbauer, 2004). On one hand, the quantization error occurs because all the input vectors included in a BMU are similar but not precisely equal (Peiró-Velert et al., 2014). Therefore, a difference exists between each input vector and the prototype of its respective BMU. The quantization error is simply the average of the distances between each input vector and the prototype of its corresponding BMU. Smaller values of quantization error suggest a better fit of the data. On the other hand, the topographic error refers to how well the obtained map resembles the input data

(Pölzlbauer, 2004). The topographic error is calculated by determining first the best matching unit and the second-best matching unit for each input vector and then evaluating their positions. An error occurs if the best matching units are not contiguous in the grid. The topographic error ranges between 0 and 1, being 0 an indication of perfect topology preservation. Although there is no standard value to consider these quality measures as low, Uriarte and Martín (2008) show how these values change according to changes in the size of the map. More specifically, they suggest that when the size of the map increases, the quantization error decreases while the topographic error increases.

7.3 How SOMs Can Be Used to Analyze Academic Performance in Higher Education?

This section exemplifies the use of SOMs to analyze academic performance of a cohort of 162,030 Colombian university students. First, a description of the participants and variables included in the analysis is given. Second, a description of the data analysis is presented in detail. Third, the results of this application example include the existing patterns between prior academic achievement, SES, and academic performance in higher education, and a description of the groups of students based on these variables.

7.3.1 Description of the Participants

The Colombian Institute for Educational Evaluation (*Instituto Colombiano para la Evaluación de la Educación,* ICFES) provided the data for this application example. The total sample included 162,030 students of both genders, females (60%) and males (40%), with a mean age of 23.5 years (*SD* = 2.89 years), from private (64.4%) and public universities (35.6%) in Colombia. The students graduated from academic secondary schools (61.6%), both academic and technical secondary schools (18.1%), technical secondary schools (16.7%), and teacher-training secondary schools (3.6%). Furthermore, the students in the sample were enrolled in one of several different academic programs: agronomy, veterinary, and related degrees (1.3%), arts (3.5%), economics, management, and accounting (25.4%), education (8.7%), engineering, architecture, urban planning, and related degrees (25.6%), health (11.4%), mathematics and natural sciences (2.1%), and social and human sciences (21%).

7.3.2 Description of the Variables

Prior academic achievement: Students' scores in seven subject areas (Spanish, mathematics, social sciences, philosophy, biology, chemistry, and physics) of the SABER 11 test (between 2006 and 2011) are selected as the operationalization of prior academic achievement. SABER 11 is a standardized test required for admission to higher education institutions in Colombia. This is a cluster of exams that assess students' level of achievement in eight subject areas at the end of high school. Each subject test consists of multiple-choice questions with only one correct answer and three distractors. Test results are reported using a scale from 0 to 100 for each subject, resulting from an item response theory (IRT) analysis of the data.

Students' socio-economic status: Information regarding parents' educational level and monthly family income are selected as indicators of students' socio-economic status (Sirin, 2005; Van Ewijk & Sleegers, 2010). Parents' educational level was assessed under the International Standard Classification of Education (ISCED). The ISCED is a classification of education programs by levels and areas of knowledge proposed in 1997 by the United Nations Educational, Scientific, and Cultural Organization (UNESCO) and then revised in 2011. There are nine ISCED levels from Early childhood education to Doctoral or equivalent degree. The monthly family income was measured as a multiple of the monthly minimum wage (i.e., one minimum wage, twice minimum wage, etc.).

Students' home stratum: The home stratum refers to the seven categories used by the Colombian government to classify households based on their physical characteristics and surroundings. The main reason behind this classification is to establish the price of public services hierarchically in each area (The World Bank, 2012).

Academic performance in higher education: Students' outcomes in the quantitative section of the SABER PRO test (hereafter, SABER PRO test) from the 2016 administration are chosen as the operationalization of academic performance in higher education. SABER PRO is a standardized test required for graduation from higher education institutions in Colombia. The exam assesses students' performance in the following sections: critical reading, quantitative reasoning, written communication, civic competencies, and English. Each section of the test consists of multiple-choice questions with only one correct answer and three distractors. Additionally, SABER PRO evaluates specific competencies such as scientific thinking for science and engineering students; teaching, training, and evaluating for education students; conflict management and judicial communication for law students; agricultural and livestock production for agricultural sciences students; medical diagnosis and treatment for medicine students; and organizational and financial management for business administration students. Test results are reported using a scale from 0 to 300, resulting from an IRT analysis of the data.

7.3.3 Data Analysis

7.3.3.1 Preparation of the Data

One critical aspect to consider when designing SOMs is the scaling of the input variables (Kohonen, 2014). Any input variable whose scale is larger than the other scales can appear to be "more important" to the network, and consequently, produce misleading representations of the data. Unless a robust theoretical reason indicates that the importance of the input variables is different, they all should be considered equally important. Therefore, in this application example, all the input variables were normalized between 0 and 1 for further analysis. While values closer to 0 indicated a low value of the variable (e.g., low income, low prior achievement), values closer to 1 represented a high value of the variable.

7.3.3.2 Designing the SOM Model

The software tool chosen for this application example was the SOM Toolbox 2.0 under MATLAB 9.7 (R2019b). Although several software packages exist to create SOMs, Kohonen (2014) strongly recommends the use of the SOM Toolbox, developed by his research lab at the Helsinki University of Technology. The major advantage of using the SOM Toolbox is the flexibility of the MATLAB environment, as its functions make it possible to design SOMs simply and easily (Vesanto et al., 2000). Furthermore, the SOM model was designed in four steps by following the suggestions of Vesanto et al. (2000).

In step 1, a rectangular grid of 200 neurons (20 rows×10 columns) was created as is explained next. The SOM Toolbox calculates the default number of neurons of the grid using the expression 5*sqrt(n), where n is the number of cases in the data set (Kohonen, 2014). As there were 162,030 cases in the analyzed data set, the number of neurons for the grid would be approximately 2000. Nevertheless, many neurons could impact the training time as well as the interpretability of the resulting SOM. Therefore, and following the illustration purposes of the present chapter, the default number of neurons was reduced to 200. Regarding the size of the grid, Kohonen (2014) has pointed out that the relationship between the sides of a rectangular map (i.e., length and width) can be determined as the ratio between the two largest eigenvalues of the training data set. In the analyzed data set, this ratio was equal to 2.43. Consequently, the relationship between the length (l) and the width (w) of the map was represented using the two-by-two equations system shown in Eq. (7.1).

$$\begin{cases} lw = 200 \\ l = 2.43w \end{cases} \tag{7.1}$$

Solving Eq. (7.1) showed that the width of the grid should be 9.07, and the length of the grid should be 22.04. Both values were then respectively rounded to 10 and 20 so that the number of neurons in the grid was exactly 200.

In step 2, the combination of two types of training (sequential and batch training), two ways of initialization (random and linear), and four types of neighborhood functions (Gaussian, cut Gaussian, Bubble, and Epanechnikov) resulted in the creation of 16 SOMs. In step 3, each of the created SOM was trained during

200 epochs. It is important to note that in the case of sequential training, the initial learning rate was equal to 0.5, and then, it decreased to values close to 0. Conversely, no learning rate was needed in the case of batch learning (Kohonen, 2013). In step 4, the quality measures (quantization error and topographical error) were calculated for each of the trained SOMs. The SOM which exhibited the best quality after considering all the measures was kept for further analysis. In this application example, such a model was achieved through the Gaussian function, random initialization, and sequential training.

7.3.4 Patterns between Prior Academic Achievement, SES, and Academic Performance in Higher Education

To identify the existing patterns between the analyzed variables, both correlation and visual analysis were carried out on the SOM model. Recall that this model was the one obtained after following the designing process described earlier. Regarding the correlation analysis, notice that the weight vectors in each one of the nodes correspond to the final values of the input variables after training the SOM model. Therefore, the existing patterns among the input variables were determined by calculating Pearson's r coefficient among the final weight vectors (being completely different from calculating the correlation between the original variables). The resulting correlation coefficients provided a quantitative indicator of the existing patterns between prior academic achievement, SES, and academic performance in higher education. Regarding the visual analysis, the "component planes" for all the input variables were created. A "component plane" specifies individual information on the value of each input variable in every node of students on the map (Kohonen, 2014). In the "component plan", the magnitude of the variables is also represented by colors. Thus, darker colors indicate nodes where the variable is high, while lighter colors represent nodes where the variable is low. Interestingly, the comparison across the 12 resulting "component planes" provided a qualitative description of the existing patterns among the input variables.

Table 1 shows the results of the correlation analysis carried out on the weight vectors across the 200 units of the SOM model. According to Table 7.1, there is a very strong significant relationship between the scores in all the areas of the SABER 11 test and the SABER PRO test: Spanish ($r(200)=0.995$, $p<0.01$); mathematics ($r(200)=0.989$, $p<0.01$); social sciences ($r(200)=0.988$, $p<0.01$); philosophy ($r(200)=0.981$, $p<0.01$); biology ($r(200)=0.992$, $p<0.01$); chemistry ($r(200)=0.987$, $p<0.01$); and physics ($r(200)=0.979$, $p<0.01$). In addition, there is a moderate significant relationship among stratum ($r(200)=0.576$, $p<0.01$), income ($r(200)=0.525$, $p<0.01$), mother's educational level ($r(200)=0.511$, $p<0.01$), and father's educational level ($r(200)=0.537$, $p<0.01$) and the SABER PRO test. As expected, the relationship between income and stratum ($r(200)=0.895$, $p<0.01$) was strong and significant as well as the relationship between the mother's educational and father's educational levels ($r(200)=0.743$, $p<0.01$).

TABLE 7.1

Descriptive Statistics and Correlation of the Weight Vectors across the SOM Model ($n = 200$)

Variable	M	SD	1	2	3	4	5	6	7	8	9	10	11
1. Stratum	0.46	0.13											
2. Income	0.32	0.20	0.895[a]										
3. Mother's educational level	0.6	0.20	0.671[a]	0.621[a]									
4. Father's educational level	0.58	0.24	0.6[a]	0.582[a]	0.743[a]								
5. S11 Spanish	0.55	0.04	0.563[a]	0.514[a]	0.483[a]	0.5[a]							
6. S11 mathematics	0.42	0.04	0.593[a]	0.559[a]	0.521[a]	0.53[a]	0.991[a]						
7. S11 social sciences	0.49	0.04	0.501[a]	0.453[a]	0.431[a]	0.451[a]	0.992[a]	0.985[a]					
8. S11 philosophy	0.46	0.04	0.459[a]	0.409[a]	0.401[a]	0.442[a]	0.988[a]	0.978[a]	0.996[a]				
9. S11 biology	0.42	0.03	0.522[a]	0.484[a]	0.476[a]	0.48[a]	0.995[a]	0.991[a]	0.994[a]	0.991[a]			
10. S11 chemistry	0.43	0.03	0.535[a]	0.497[a]	0.469[a]	0.484[a]	0.993[a]	0.993[a]	0.991[a]	0.988[a]	0.996[a]		
11. S11 physics	0.39	0.03	0.518[a]	0.491[a]	0.469[a]	0.48[a]	0.986[a]	0.991[a]	0.985[a]	0.982[a]	0.991[a]	0.997[a]	
12. SABER PRO	0.52	0.05	0.576[a]	0.525[a]	0.511[a]	0.537[a]	0.995[a]	0.989[a]	0.988[a]	0.981[a]	0.992[a]	0.987[a]	0.979[a]

Notes: M and SD are used to represent mean and standard deviation, respectively.
[a] indicates $p < 0.01$.

Figure 7.1 shows the "hits-map" of the SOM model. The "hits-map" indicates how many students were mapped to each of the neurons of the SOM (Kohonen, 2014), being also an indication of the distribution of the data set in the SOM model (Kohonen, 2013). More specifically, areas on the map with many students could suggest increasing the size of the grid, while areas on the map with few students could imply that a reduction of the grid is needed. According to the "hits-map" depicted in Figure 7.1, there were no empty nodes in the SOM model. Also, most of the neurons located in the outermost part of the map are the most populated ones (more than 1,000 students). In contrast, most of the neurons located toward the inside of the map are less populated (less than 100 students). This relatively uniform distribution of the students across the map may indicate that the chosen map size (20 rows × 10 columns) was adequate for representing the input data.

Figure 7.2 presents the "component plans" for each one of the 12 input variables analyzed in this application example. The "component plan" separately indicates the value of a variable in each one of the nodes forming the SOM.

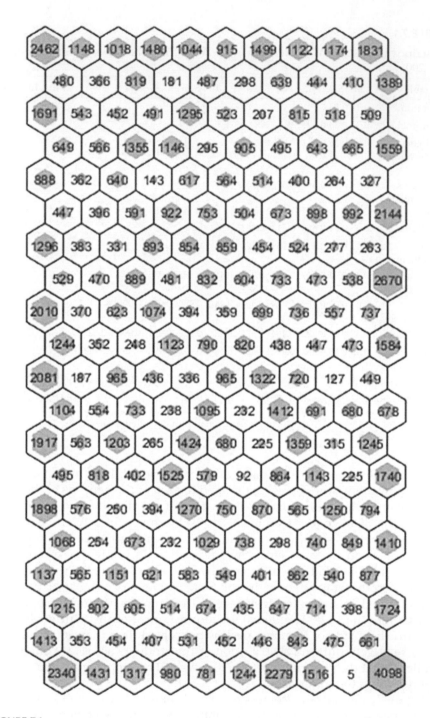

FIGURE 7.1
"Hits map" for the SOM model.

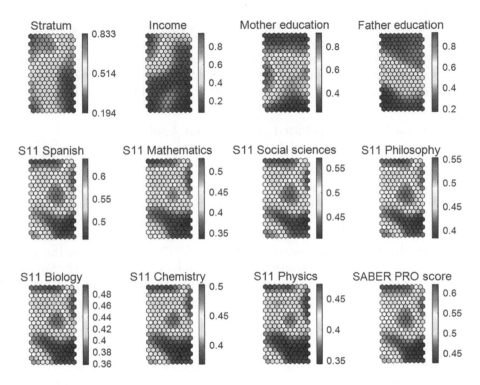

FIGURE 7.2
Relationship between the input variables.

The analysis of the overlapping of the "component plans" reveals interesting patterns of the relationship among the input variables. The first pattern is observed in the upper left corner of the SOM model. This area of the map groups the students who are the best performers in the SABER PRO test. These students also come from the highest stratum houses, families with the highest income, their parents have the highest educational level, and have the highest performance in all the areas of the SABER 11 test. The second pattern occurs in the lower right corner of the SOM model. The students with the lowest performance in the SABER PRO test are placed in this area of the map. Besides, these students come from the lowest stratum houses, families with the lowest income, their parents have the lowest educational level, and have the lowest performance in all the areas of the SABER 11 test. The third pattern appears in most of the nodes of both the center and the lower left corner of the SOM model. Remarkably, there are students with very high performance in the SABER PRO test, who also come from low to medium-stratum houses, families with low income, have parents with low to medium educational level, and have very high performance in all the areas of the SABER 11 test.

Altogether, these three patterns corroborate the moderate association found in the correlation analysis between SES and academic performance in higher education. Moreover, there is also evidence to confirm a strong association found in the correlation analysis between prior academic achievement and academic performance in higher education. Summing up the results from both correlation and visual analyses, it can be stated that prior academic achievement is more strongly related to academic performance in higher education than SES.

7.3.5 Grouping Students Based on Their Prior Academic Achievement, SES, and Academic Performance

To group students based on the variables of interest, both visual analysis and K-means clustering of the map units of the SOM model were carried out. Concerning the visual analysis, the so-called U-matrix of the SOM model was created. An "U-matrix" indicates how far or close the neurons are in the resulting SOM (Kohonen, 2014). In the "U-matrix", the distance between the neurons is represented by colors. The more similar the colors, the more similar the distances between the neurons. The "U-matrix" suggests the likely existence of clusters among the map units. Regarding the K-means clustering, the procedure in the SOM Toolbox runs multiple times for each K, and the best run is chosen based on the sum of squares. Then, the optimal number of clusters is evaluated through the Davies-Bouldin index. The solution with the lowest index is selected as the best one. In this application example, K was set to 10, and each K was run 1000 times. The Davies-Bouldin index was found to be the lowest for the solution of three clusters. Next, descriptive statistics of the three resulting clusters were performed. More specifically, the mean and standard deviation of each weight vector were calculated within each cluster. Subsequently, the mean for each input variable was compared among the three clusters through Welch's ANOVA. Welch's ANOVA is an alternative test to the classic one-way ANOVA, especially useful in case of heterogeneity of variance (Moder, 2010). Because there were significant differences between the three clusters' means, a post-hoc comparison using the Games-Howell test was also conducted. The Games-Howell test is recommended in cases where the assumption of homogeneity of variance is not fulfilled (Lee & Lee, 2018).

Figure 7.3 displays the resulting "U-matrix" for the SOM model. The "U-matrix" shows the distances among the 200 neurons forming the grid. A visual inspection of the "U-matrix" reveals that there are three areas of the map where the distances are dissimilar, which might indicate the existence of at least three clusters of nodes. To further confirm the existence of three clusters of students, K-means clustering was performed on the 200 units of the SOM model. After 1,000 runs, the David-Bouldin index was calculated for each of the chosen values of K. The smaller the index, the better the grouping. According to this index, the optimal number of groups of students is three, as displayed in Figure 7.3.

Figure 7.4 shows each one of the three clusters of students as well as the number of students classified within each cluster. The first cluster was labeled

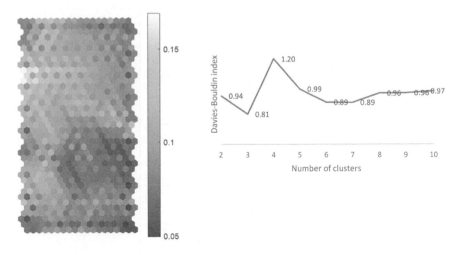

FIGURE 7.3
"U matrix" and selection of the optimal number of clusters.

as "high", the second cluster was labeled as "medium", and the third cluster was labeled as "low", as is explained subsequently. To begin with, the mean and standard deviation were calculated for each of the 12 input variables within the three clusters of students as shown in Table 7.2. According to Table 7.2, the "high" group includes the students who, on average, have a high performance in the SABER PRO test, high performance in all the areas of the SABER 11 test, and high values in the SES indicators (stratum, income, and

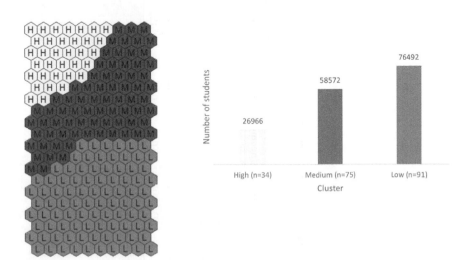

FIGURE 7.4
Classification of students based on the input variables.

TABLE 7.2

Descriptive Statistics for Each Cluster

Cluster		Stratum	Income	Mother's Educational Level	Father's Educational Level	S11 Spanish	S11 Mathematics	S11 Social Sciences	S11 Philosophy	S11 Biology	S11 Chemistry	S11 Physics	SABER PRO
High (n=34)	M	0.68	0.66	0.84	0.86	0.59	0.46	0.52	0.5	0.45	0.46	0.42	0.56
	SD	0.06	0.17	0.09	0.1	0.03	0.04	0.03	0.03	0.03	0.03	0.03	0.03
Medium (n=75)	M	0.47	0.32	0.69	0.71	0.56	0.43	0.5	0.48	0.43	0.44	0.4	0.54
	SD	0.08	0.11	0.15	0.16	0.03	0.03	0.03	0.03	0.02	0.02	0.02	0.03
Low (n=91)	M	0.37	0.19	0.44	0.36	0.52	0.39	0.47	0.44	0.4	0.41	0.38	0.49
	SD	0.08	0.06	0.12	0.12	0.04	0.04	0.04	0.04	0.03	0.03	0.03	0.04

mother's and father's educational levels). In addition, the group "medium" comprises the students who, on average, exhibit a medium performance in the SABER PRO test, a medium performance in all the areas of the SABER 11 test, and have medium values in the SES indicators. Finally, the students belonging to the "low" group are the students who, on average, have low performance in the SABER PRO test, low performance in all the areas of the SABER 11 test, and low values in the SES indicators.

Next, a Welch's ANOVA was conducted to compare the means of the 12 input variables among the "high", "medium", and "low" clusters. The results from Welch's ANOVA showed that there were significant differences for all the means across the three clusters, as presented in Table 7.3. Because of these significant differences, a post-hoc comparison using the Games-Howell test was conducted as reported in Table 7.4. The Games-Howell test showed that the mean value of each input variable in the "high" group was significantly higher than in the "medium" group and in the "low group". Similarly, the mean value of each input variable in the "medium" group was significantly higher than in the "low" group.

A concluding remark of these analyses is that three different groups of students exist, namely, "high", "medium", and "low", based on their prior academic achievement, SES conditions, and academic performance in higher education. Besides, these variables are statistically different among the three existing groups.

7.4 What Are the Takeaway Messages of This Chapter?

Any reader can (hopefully) draw five messages from this chapter. The first message is that analyzing academic performance in higher education is

TABLE 7.3

Welch's ANOVA for the Cluster Means

Variable	Welch	df1	df2	p
Stratum	244.535	2	100.783	.000
Income	161.867	2	71.046	.000
Mother's educational level	212.470	2	106.424	.000
Father's educational level	320.810	2	97.908	.000
S11 Spanish	49.926	2	93.328	.000
S11 mathematics	51.661	2	85.694	.000
S11 social sciences	37.351	2	92.242	.000
S11 philosophy	32.575	2	91.313	.000
S11 biology	44.919	2	91.093	.000
S11 chemistry	45.967	2	87.841	.000
S11 physics	42.227	2	85.355	.000
SABER PRO	55.300	2	94.065	.000

TABLE 7.4

Games-Howell Test for Multiple Comparisons between the Clusters Means

Variable	(I) Cluster	(J) Cluster	Mean Difference (I-J)	S.E.
Stratum	High	Low	0.31[a]	0.01
		Medium	0.21[a]	0.01
	Medium	Low	0.10[a]	0.01
Income	High	Low	0.47[a]	0.03
		Medium	0.34[a]	0.03
	Medium	Low	0.13[a]	0.01
Mother's educational level	High	Low	0.40[a]	0.02
		Medium	0.15[a]	0.02
	Medium	Low	0.25[a]	0.02
Father's educational level	High	Low	0.50[a]	0.02
		Medium	0.15[a]	0.03
	Medium	Low	0.35[a]	0.02
S11 Spanish	High	Low	0.07[a]	0.01
		Medium	0.03[a]	0.01
	Medium	Low	0.04[a]	0.01
S11 mathematics	High	Low	0.07[a]	0.01
		Medium	0.03[a]	0.01
	Medium	Low	0.04[a]	0.01
S11 social sciences	High	Low	0.05[a]	0.01
		Medium	0.02[a]	0.01
	Medium	Low	0.03[a]	0.01
S11 philosophy	High	Low	0.06[a]	0.01
		Medium	0.02[b]	0.01
	Medium	Low	0.04[a]	0.01
S11 biology	High	Low	0.05[a]	0.01
		Medium	0.02[a]	0.01
	Medium	Low	0.03[a]	0.00
S11 chemistry	High	Low	0.05[a]	0.01
		Medium	0.02[a]	0.01
	Medium	Low	0.03[a]	0.00
S11 physics	High	Low	0.04[a]	0.01
		Medium	0.02[a]	0.01
	Medium	Low	0.02[a]	0.00
SABER PRO	High	Low	0.07[a]	0.01
		Medium	0.02[a]	0.01
	Medium	Low	0.05[a]	0.01

Notes: [a] indicates $p < 0.01$; [b] indicates $p < 0.05$

still needed. The findings of this chapter indicated that a strong relationship exists between prior academic achievement and academic performance in higher education. Although this finding was somehow expected, it also suggests the importance of starting university studies with adequate academic preparation. In addition, there was evidence of a moderate association between SES and academic performance in higher education. This finding refutes the idea that only students coming from the most privileged families would perform well in the university. As a matter of fact, there were students placed in both the center and the lower left corner of the SOM model who achieved very high academic performance at university and come from the least advantaged families (low stratum, low income, and low parents' educational level). Interestingly, such students also exhibited very high performance in their prior academic achievement. As this chapter confirms that prior academic achievement is strongly related to academic performance in higher education than SES, a first step to support low-SES students would be to offer them preparation programs to foster their academic preparation before university.

The second message is that SES and prior academic achievement are of importance to differentiate students' academic performance in higher education. The results from this chapter suggested that three clusters of students, namely, "high", "medium", and "low" exist in the resulting SOM model, grouping the students according to their prior academic achievement, SES, and academic performance in higher education. The means of these input variables were statistically different among the three clusters, which supports their usefulness to distinguish university students efficiently. The largest differences across the clusters were found in the SES indicators (stratum, income, and parents' educational level), whereas the smallest differences were observed in the outcomes of both the SABER 11 and SABER PRO test. These differences are consistent with the larger standard deviation observed in the SES indicators.

The third message is that educational researchers should recall some hints when designing SOMs, which are well-known facts stated in the literature. The first hint is to normalize the input variables so that they have the same importance to the model. The use of the original scaling of the input variables may lead to the identification of erroneous patterns in the data. The second hint is to train several SOM models by systematically changing their tuning parameters. In this way, more valid representations of the input space could be found. The third hint is to calculate several quality measures and not rely exclusively on the quantization error. In this respect, the topographic error could be a more interesting value for educational researchers, as it represents how well the existing relationship among the input variables is preserved in the final model.

The fourth message is an invitation for educational researchers to explore several other possibilities to conduct quantitative research, aiming to extend the possibilities of classical statistical techniques. In this respect, the present chapter has exemplified how the use of unsupervised learning models can allow for the discovery of patterns of information, even when no input-output relationship is specified. The recommendation is to recognize SOMs as a powerful clustering method, so students, teachers, and institutions can take advantage of a more precise and in-depth understanding of the factors related to academic performance in higher education. In this same line of reasoning, educational researchers are encouraged to explore alternative and more robust methods for comparing means across groups such as Welch's ANOVA and the Games-Howell test. This fact is particularly important given that violating the assumption of equal variance across the groups causes Type I error to inflate, resulting in misleading conclusions.

The final message is that SOMs could shape future research in higher education in both practical and methodological ways. At the practical level, SOMs could be used to discover patterns of teaching and learning existing at the individual, classroom, and institutional levels. This fact would be of enormous benefit, as it would suggest what works best to foster students' learning process. Furthermore, SOMs could be used to understand the academic and behavioral patterns of students who drop out of university. By following these patterns, several strategies to reduce students' drop out of the university could be proposed. At the method-ological level, and although there are fascinating examples of the use of categorical variables in SOMs (e.g., Hsu, 2006; Hsu & Lin, 2011), how to properly use nominal and ordinal variables still deserves further exam-ination. Finally, the use of SOMs could extend the development of *real* predictive models of academic performance in higher education, which do not require the specification of any input-output relationship. In this respect, the use of the so-called super SOMs (i.e., SOMs with more than one competitive layer) should be explored.

Acknowledgments. The authors would like to thank the Colombian Institute for Educational Evaluation (ICFES) for providing the data for this chapter. This research was supported in part by the Colombian Department of Science, Technology, and Innovation (COLCIENCIAS) under grant 756 of 2016.

References

Antonenko, P. D., Toy, S., & Niederhauser, D. S. (2012). Using cluster analysis for data mining in educational technology research. *Educational Technology Research and Development, 60*, 383–398. doi: 10.1007/s11423-012-9235-8.

Da Silva, E. T., de Fátima Nunes, M., Santos, L. B., Queiroz, M. G., & Leles, C. R. (2012). Identifying student profiles and their impact on academic performance in a Brazilian undergraduate student sample. *European Journal of Dental Education, 16*, e27–e32. doi: 10.1111/j.1600-0579.2010.00669.x

De Clercq, M., Galand, B., & Frenay, M. (2017). Transition from high school to univer-sity: A person-centered approach to academic achievement. *European Journal of Psychology of Education, 32*, 39–59. doi:10.1007/s10212-016-0298-5.

Garson, G. D. (2014). *Neural network models*. Asheboro, NC: Statistical Associates Publishers.

Haykin, S. S. (2009). *Neural networks and learning machines* (Vol. 3). Upper Saddle River, NJ: Pearson.

Hsu, C. C. (2006). Generalizing self-organizing map for categorical data. *IEEE Transactions on Neural Networks*, *17*, 294–304. doi: 10.1109/tnn.2005.863415.

Hsu, C. C., & Lin, S. H. (2011). Visualized analysis of mixed numeric and categorical data via extended self-organizing map. *IEEE Transactions on Neural Networks and Learning Systems*, *23*, 72–86. doi: 10.1109/tnnls.2011.2178323.

Iturriaga, F. J. L., & Sanz, I. P. (2013). Self-organizing maps as a tool to compare financial macroeconomic imbalances: The European, Spanish, and German case. *The Spanish Review of Financial Economics*, *11*, 69–84. doi: 10.1016/j.srfe.2013.07.001.

Kausar, S., Huahu, X., Hussain, I., Wenhao, Z., & Zahid, M. (2018). Integration of data mining clustering approach in the personalized E-learning system. *IEEE Access*, *6*, 72724–72734. doi: 10.1109/access.2018.2882240.

Kohonen, T. (2013). Essentials of the self-organizing map. *Neural Networks*, *37*, 52–65. doi: 10.1016/j.neunet.2012.09.018.

Kohonen, T. (2014). *MATLAB implementations and applications of the self-organizing map.* Helsinki: Unigrafia Oy.

Kumar, U. A., & Dhamija, Y. (2010). Comparative analysis of SOM neural network with K-means clustering algorithm. *Proceedings of the IEEE International Conference on Management of Innovation & Technology* 2010, pp. 55–59. doi: 10.1109/icmit.2010.5492838.

Lee, S., & Lee, D. K. (2018). What is the proper way to apply the multiple comparison test? *Korean Journal of Anesthesiology*, *71*, 353–360. doi: 10.4097/kja.d.18.00242.

Moder, K. (2010). Alternatives to F-test in one-way ANOVA in case of heterogeneity of variances (a simulation study). *Psychological Test and Assessment Modeling*, *52*, 343–353.

Peiró-Velert, C., Valencia-Peris, A., González, L. M., García-Massó, X., Serra-Añó, P., & Devís-Devís, J. (2014). Screen media usage, sleep time and academic performance in adolescents: Clustering a self-organizing maps analysis. *PloS One*, *9*, e99478. doi: 10.1371/journal.pone.0099478.

Pölzlbauer, G. (2004). Survey and comparison of quality measures for self-organizing maps. In J. Paralič, G. Pölzlbauer & A. Rauber (Eds.), *Proceedings of the Fifth Workshop on Data Analysis (WDA'04)* (pp. 67–82). Vysoké Tatry: Elfa Academic Press.

Rodríguez-Hernández, C.F. (2016). *Socioeconomic factors and outcomes in higher education: A multivariate analysis.* Bogotá: Universidad Externado de Colombia Press.

Romero, C., & Ventura, S. (2010). Educational data mining: Review of the state of the art. *IEEE Transactions on Systems, Man, and Cybernetics, Part C (Applications and Reviews)*, *40*, 601–618. doi: 10.1109/TSMCC.2010.2053532.

Sirin, S. (2005). Socioeconomic status and academic achievement: A meta-analytic review of research. *Review of Educational Research*, *75*, 417–453. doi: 10.3102/00346543075003417.

The World Bank (2012). *Reviews of national policies for education: Tertiary education in Colombia 2012.* OECD Publishing. doi: 10.1787/9789264180697-en.

Uriarte, E.A., & Martín, F.D. (2008). Topology preservation in SOM. *World Academy of Science, Engineering and Technology, International Journal of Computer, Electrical, Automation, Control and Information Engineering*, *2*, 3192–3195.

Van Ewijk, R., & Sleegers, P. (2010). The effect of peer socioeconomic status on student achievement: A meta-analysis. *Educational Research Review, 5*, 134–150. doi: 10.1016/j.edurev.2010.02.001.

Vesanto, J., Himberg, J., Alhoniemi, E., & Parhankangas, J. (2000). *SOM Toolbox for Matlab 5* (Research Report No. A57). http://www.cis.hut.fi/projects/ somtoolbox/documentation/.

Wehrens, R., & Buydens, L. M. (2007). Self-and super-organizing maps in R: The Kohonen package. *Journal of Statistical Software, 21*, 1–19. doi: 10.18637/jss. v021.i05.

8

Artificial Intelligence as a Way to Improve Educational Practices

Omar Olmos-López
Tecnológico de Monterrey

Elvira G. Rincón-Flores
Tecnológico de Monterrey

Juanjo Mena
University of Salamanca

Olaf Román
Tecnológico de Monterrey

Eunice López-Camacho
Tecnológico de Monterrey

CONTENTS

8.1 Introduction

Surely more than once you have entered into a social network and noticed that they announce a product that you have recently been looking for, or perhaps, if you are a user of a music platform it turns out that the playlist of songs that they propose is to your total liking; these events seem to be the result of the divinatory skills of someone else, but it is not like that, it is the magic of Artificial Intelligence (AI). Kaplan and Haenlein (2019) define it as the ability of a system to correctly interpret external data, learn from them, and use learnings to achieve objectives through flexible adaptation.

DOI: 10.1201/b23345-10

AI is present in different areas of our lives, in medicine, in art, in business, to name a few examples, and of course, it is in education. In the literature review conducted by Zawacki-Richter et al. (2019), they found that in general the most common applications of AI are (i) profiling and prediction, (ii) assessment and evaluation, (iii) adaptive systems and personalization, and (iv) intelligent tutoring systems. However, in the study developed by Hinojo-Lucena et al. (2019), they found that, although AI is a reality, its applications are not yet consolidated in higher education.

The causes can be of different kinds: AI is expensive, requires human resources with a high degree of specialization, resistance on the part of teachers and students, and its cognitive approach lacking affective elements. However, there is no doubt that AI can become a powerful tool to improve teaching-learning processes.

The great opportunity that education must apply AI technology is in its integration into the educational ecosystem and relating it to the study plans and programs, so that in the medium term, there are benefits in the personalization of the content generated and the student-learning activities. However, the foregoing entails challenges for educational institutions in issues such as data management, ethics in learning analytics, privacy and security, equity, or algorithmic justice, among others. These challenges include the change from the archetype of the teacher dedicated to only transferring knowledge to an active, critical, and analytical role; one who accompanies the student as a coach during their learning, aware of the incorporation of advanced technologies and solutions with AI (Park et al., 2021).

The incorporation of AI in different areas will be a standardized tool that allows, in the face of complex problems, efficient, effective, timely, and economically profitable solutions that will create an incremental loop of technology development. In this way, we can visualize the massification of AI technology in more daily and professional activities, making it necessary to include an ethical and institutionally regulated perspective, at different levels of its use, so that the application is not harmful or detrimental to the society.

The use of AI as an innovative tool has been quickly discovered. An example of this is Refik Anadol's artwork "Unsupervised" presented at The Museum of Modern Art in New York. The "living" piece is part of the Machine Hallucinations exhibition, which applies machine learning and a database of more than 138,151 images to create a self-generated work of art of unrepeatable and poetically intertwined images (Unsupervised—Machine Hallucinations—MoMA, 2022). These innovative creation mechanisms that combine AI systems and interconnect different data sources enable new perspectives, new narratives to interpret reality, or new ways of interacting with ourselves as human creators.

The concepts for the treaty of AI developments that transcend in society have become more and more familiar. Such is the case of "Automation bias" that refers to the cognitive shortcuts that humans take with the automated results of AI systems that we always assume are correct (Jones-Jang &

Park, 2022). "Ethics Washing" is a practice of private companies that develop AI models that mention being responsible before regulatory frameworks but deep down do not exercise real ethical policies (Maanen, 2022) or the "Anthropomorphism" that encompasses meanings of visualizing AI as a moral person with autonomous judgments like human beings, the exaggeration in the capabilities of AI that create fears to such a degree that the machine replaces the human being or the uncritical optimism that gives the assumption that the AI will be able to behave and overcome difficult tasks for human beings (Salles et al., 2020).

It is essential that AI developments are conceived including human-centered ethical principles. The minimum ethical considerations that must be considered in the design of AI systems, as proposed by Rees & Müller (2022), are embracing the diversity of diversity, future-proof systems, reassess ethical compliance regularly, and summarizing in layperson's terms. This must be done for more organizations, institutions, governments, and companies to commit to regulatory measures that reduce ethical debt with its possible consequences and sign up for a human-centered legal framework according to the current context.

The accelerated scientific and technological development provokes philosophical reflections, such as: What is irreplaceable in the human being in the face of artificial intelligence? What characteristics make us unique in the face of technological advances that emulate human capabilities? What goal should permeate scientific-technological progress? Evoking the Greek myth of Prometheus, we can consider the fire delivered to humans by Prometheus as access to knowledge through new possibilities with artificial intelligence, and the domain of fire would be interpreted as AI systems that instrumentalize knowledge with the desire to benefit humanity but that contain risks and consequences.

This chapter is distributed as follows: first, an explanation of AI based on predictive algorithms is presented with the aim of putting the reader in context, then a review of the literature on the applications of AI in education and its implications is offered, a case is presented in which predictive algorithms were applied to forecast the performance of students, and we close the chapter with final reflections.

8.2 What Is a Predictive Algorithm Based on Artificial Intelligence (AI)?

The tasks carried out by education professionals have evolved during the last decades. The number of factors, attributes, methods, and metrics they use to evaluate performance criteria in training and educational processes have increased. These academic tasks support processes such as communication, academic practice, evaluation, and continuous feedback. Some of them are very operational and repetitive tasks due to the number of students that each teacher needs to attend. Some of these academic tasks require cognitive learning and the execution of procedural activities still exclusive to human reasoning and understanding. However, thanks to the technological development of information technologies, especially Machine Learning technologies, these tasks have benefited from the development of tools such as Artificial Intelligence algorithms.

The Artificial Intelligence prediction algorithms that support some academic processes and activities are a set of logical and operational rules that, through computational methods, carry out different processes of identifying similarities, differences, or patterns of a data set to generate a prediction of a certain variable.

Educational management and teaching are supported by logical processes, methods, and didactic strategies, as well as evaluation instruments to develop skills and competencies in learners. Therefore, the inclusion of these technologies and technological tools of artificial intelligence algorithms has been valuable and favorable, especially in academic performance prediction algorithms. Some of the classifier algorithms that are used to build an academic prediction model are Random Forest, Decision Trees, Neural Networks, and Nearest Neighbors.

In general, artificial intelligence-prediction algorithms perform different comparisons with the data that has been entered as a training set in order to generate a set of rules and classifiers. When a new data set is entered, the algorithm will use the generated rules to classify each element of the new data set based on the closest similarity and will give a classification result that reflects a higher probability of being close to a value in the training data set. We consider that the algorithm is good if predictions over the testing data set to turn out to be accurate enough.

The general process of building a production model supported by an artificial intelligence algorithm is shown in Figure 8.1. This figure shows the process of tuning a model using training and testing data sets by the Random Forest algorithm. The first input to be used is the training data set, with which a large number of decision trees are trained, and then the final selection of the best decision trees is determined. Finally, with this selection of decision trees, a set of test data is evaluated to get a measure of how good the model is. The prediction of the model is the class selected by the largest number of trees or the average of each tree result, depending on whether the variable being predicted is discrete or continuous.

To implement an AI algorithm for predicting academic performance, it is necessary to have data and documented information from previous academic cycles, where the teaching process has been planned or executed under certain regulated and defined conditions, thus forming a system of activities, resources, evaluation, and feedback instruments. The historical information collected, either by a digital information system or documented from log records or evaluation reports, will allow the creation and training of academic prediction algorithms.

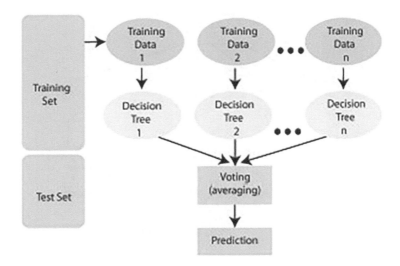

FIGURE 8.1
Training process and evaluation of the Random Forest classifier AI algorithm.

The records of academic performance reports, once collected, must be analyzed and assessed for their relevance in relation to the importance of each element within the evaluation system, in order to generate a model that relates the current grades with the evaluation achieved by the students in previous records. Thus, in this stage of construction of the academic prediction algorithm, the aim is to identify the variables of the evaluation plan that have the greatest prediction power and that present a correlation with the student's performance in a new academic cycle, under the assumption that future academic performance can be predicted based on current performance.

Once the impact factors have been identified in the prediction model, the relationship rules between the most relevant variables should be analyzed. In some models, the variables can be assigned different weights depending on their perceived relevance. The model in Figure 8.2 evaluates each decision tree in the forest to create a final prediction. With the correlation of variables and weights of the most relevant indicators, we train the database to generate a prediction model of the existing information and data of the previous cycles.

At this point, the artificial intelligence algorithm that best fits the behavior of the data must be identified and evaluated. This model would be the one that will have the best predictive results with higher probability. The algorithms that, in the data set of academic records, have shown greater effectiveness of the training model are Random Forest, Neural Networks, Nearest Neighbors, and Decision Trees (Chirici et al., 2016) and random forest (Ayyadevara, 2018; Rodríguez-Hernández et al., 2021). Figure 8.3 shows an example of these four algorithms building a model that explains how variables x and y are related (thick line). The model of Figure 8.3d seems to be perfect for the training set as it connects every data point (black dots). This

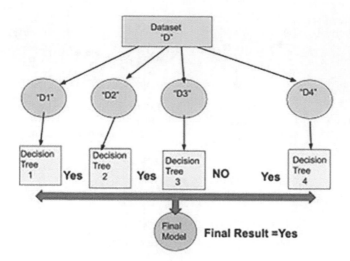

FIGURE 8.2
Random Forest algorithm with an evaluation of the decision trees.

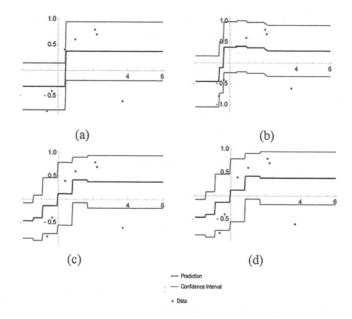

FIGURE 8.3
Data set prediction under different AI prediction models. (a) Decision Tree, (b) Random Forest, (c) Nearest Neighbors, and (d) Neural Networks.

model illustrates the problem of *overfitting* which occurs when the algorithm produces a model that fits too closely to the original data set but that will fail to predict new data points as perfectly.

The selection of the academic performance prediction algorithm must be evaluated with different statistical criteria in order to assess the best possible fit. Criteria such as correlation index, mean cross entropy, standard deviation, single evaluation time, and residuals are some of the measurements used to qualify and select the best adjustment method. Finally, tests are carried out with the selected model and the impact variables, seeking to calibrate the best-fit parameters of the algorithm. These parameters depend on the algorithm; in particular, some examples are the number of branches, output precision, number of iterations, and processing time, among others. Once the prediction algorithm has been evaluated and selected based on performance measurements, it is ready to be used in a digital evaluation platform or system to support the learning process.

Learning course that uses the adaptive learning approach is one of the applications that has had a great deployment and adoption of prediction algorithms. In this case, the objective of the strategy is to generate an assignment of activities and personalized learning routes, according to the needs and abilities of the student. Thus, the function of the prediction algorithm within a learning management system is mainly to make a prediction of the support elements that will benefit the student the most, in order to provide alternatives that will support the student's learning. This prediction is made by the algorithm based on similar cases that had an equivalent performance.

Figure 8.4 shows the traditional sequence of a course where learning modules are reviewed sequentially. However, in a teaching model supported by a

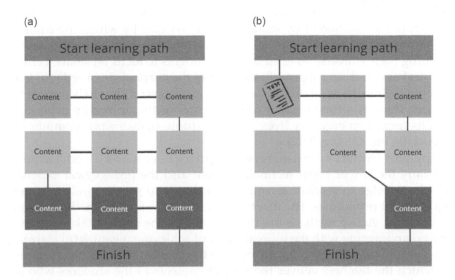

FIGURE 8.4
Map of the route in teaching (a) traditional vs. map in the adaptive course (b) supported by the algorithm of prediction model.

prediction algorithm in a learning strategy, the sequence is evaluated from a diagnosis of knowledge, and based on it, routes are offered to solve the learning needs in the topics in which there is no adequate performance, allowing enough time and amount of content.

8.3 AI Applied in Education to Improve the Learning Experience

Society has gone through several transformations in the last decades. The so-called fourth industrial revolution has been irrupted by the inclusion of technologies such as artificial intelligence, advanced robotics, the internet of things (IoT), cyber-physical systems (CPS), and machine-to-machine communication (M2M) (Norouzi, 2022). Thus, fundamental shifts have been introduced in the global production chains, reverting to new societal changes. However, higher education teaching has not always been challenged at the same pace, and universities still perpetuate traditional practices based on the unidirectional transmission of knowledge from teachers to students.

Education 4.0 embraces the technologies that emerged in the fourth industrial revolution and promises personalized learning through the availability of the contents to integrate the educational materials to the student's learning needs (Bonfield et al., 2020; Goh & Abdul-Wahab, 2020). It is not just a question of uploading materials (e.g., files, videos, etc.) or interacting with them but about the creation of dynamic learning systems by which the contents are continuously adapted and modified to reorient them to each student's learning pace. Universities need to teach students to focus on smart technology and new learning approaches as part of the curriculum in order to keep serving their purpose of preparing new generations for the challenges of the future. There is a fundamental need to improve the university experience and its impact on everyday lives: job opportunities, teamwork, communication skills, etc. (Zhang, 2022).

Applying IA and by extension engaging in the Education 4.0 model requires not only the use of smart technologies but also the pedagogical expertise and knowledge to reorient learning accordingly to support industrial and agricultural revolutions. For this to occur, the use of active methods comes to the front. Active methods are those by which the student takes action in their own learning with the ultimate goal of achieving meaningful learning. These types of methods typically oppose the concept of traditional teaching in which the teachers deliver the subject contents to the students who are the recipients of the information. The need to use active methods responds to the fact that any approach in higher education should be grounded on a competence-based model.

TABLE 8.1

Active Learning Methods for Higher Education Teaching

Application	Approach	Active Learning Methods
Individual Group	1. *Focused on personalized learning*	LM1. Thinking-based learning
		LM2. Visual thinking
		LM3. Creative thinking
	2. *Focused on flexible learning*	LM4. Flipped classroom
		LM5. Mixed reality (virtual and augmented)
	3. *Focused on game-based learning (GBL)*	LM6. Gamification
		LM 7. Serious and commercial games
		LM 8. Educational robotics
	4. *Focused on constructive learning*	LM 9. Problem-based learning
		LM 10. Project-based learning
		LM 11. Design thinking
	5. *Focused on cooperative learning*	LM 12. Peer learning
		LM 13. Collaborative learning
		LM 14. Service learning (S-L)

Source: Adapted from Mena (2022), Mena and Calle (2020), and Calle and Mena (2022).

In a nutshell, we can summarize the active methods that could be used in universities to promote the desired goals according to five approaches (please, refer to Table 8.1):

1. Personalized learning
2. Flexible learning
3. Game-based learning.
4. Constructive learning
5. Cooperative learning

Although AI has been developed since the 1960s, it was not until the 1990s that it began to mature, and in recent years, at breakneck speed. In education, it has ventured into various forms, particularly adaptive learning. Cheung et al. (2003) developed a smart tutoring system aimed at online adult education in Hong Kong, which they called SmartTutor. The purpose of this Intelligent Tutoring System (ITS) is to explain the contents, administer exams according to previous performance, and suggest activities in such a way that learning is adapted to the needs and abilities of students. On the other hand, McLaren et al. (2011) studied the effects of a polite tutor system (tutor who offers friendly feedback) with respect to the learning of high school students, which they divided into high and low prior knowledge, both groups were exposed to a direct feedback tutor, and another polite, finding that students with low prior knowledge responded more positively to feedback and suggestions from the polite tutor than to the direct feedback tutor.

Following the line of affective elements, Hwang et al. (2020) developed an adaptive system that considers the affective and cognitive factors of the students; in the study, they divided the students into three groups—mathematics area—a group assisted by an adaptive system that considers affective and cognitive factors, another group with a system that considers only cognitive factors, and the third that did not have an adaptive learning system. The results showed a significant difference in the performance of the first group with respect to the other two. In this sense, it is striking that including elements in the AI of an adaptive learning system can contribute to the improvement of student learning since it humanizes the learning experience.

According to Kulik and Fletcher (2015), there are two differentiated types of AI that can be used in education: (i) rule-based learning: decision-making maps that are generated to produce a solution (e.g., Intelligent Tutoring System) allowing for personalized feedback to students; (ii) machine-based learning: AI based on the use of multiple data sets that can become more precise over time allowing for more complex tasks such as predicting students' performance or monitoring their progress.

The first type of AI has been widely researched. For instance, Olmo-Muñoz et al. (2022) analyzed students' problem-solving performance after following intratask flexibility-based training mediated by an ITS. The main results indicated that intratask flexibility allowed students to better solve arithmetic problems if they are completed in a sequential manner (not in different sessions). The second type of AI has been less researched in education. Luan and Tsai (2021) have conducted a systematic review of 40 studies that use machine learning as an AI technique. The main results indicated that the vast majority focused on the prediction of students' learning performance or dropouts in the fields of computer sciences or STEM.

8.4 Limitations in the Educational Field for AI Prediction Algorithms

Today, in education, there is a growing inclusion of exponential technologies such as artificial intelligence algorithms. AI algorithms have a pragmatic approach, seeking to support academic activities, which humanly require a high effort and resources to be able to streamline processes such as the evaluation and identification of patterns and trends in the performance of learners. Understanding its scope, its bases, and possible applications will allow us to adopt, modify, or reject technological solutions to support the tasks of management, learning, evaluation, and feedback that we seek to be reinforced by these technologies.

The objective of modern education is to provide skills that allow students' knowledge, skills, and behaviors favorable to the tasks they are expected to perform. Under the fundamental requirements of teaching, technology has evolved to support the different processes that teachers carry out in the class-room. Since the incorporation of computer systems in educational institutions at the beginning of the 1980s, an increasingly specialized use of computer systems and intelligent algorithms to perform tasks that at the time would be considered exclusive to the teacher's work can be seen. Tasks such as the evalu-ation of knowledge, identification of learning profiles, or providing timely and personalized feedback to students in the courses where the teacher teaches.

Currently, technology has evolved and brought together both hardware and software resources that have expanded tasks and improved the preci-sion with which academic tasks are carried out in the educational ecosystem: student-learning process-teacher. Today, a variety of technologies are being incorporated that allow teachers to expand their skills and improve attention and focus on the valuable task of the learning process: the management of self-directed and autonomous learning in students.

The applications and solutions that Artificial Intelligence has applied in the educational field have been diverse. We find applications in activities such as evaluation, personalized learning, learning feedback, identifica-tion of patterns or performance profiles, the teaching of foreign languages (grammar and oral expression), plagiarism detection, and finally, some proposals for obtaining performance forecasts. In the educational sector, solutions supported by AI have been incorporated slowly, due, among some factors, to the cost and understanding of their contribution to the educational process.

AI solutions in education have been well accepted in adaptive courses, which have a digital platform that allows collecting information on the per-formance of current students, and based on this, delivering a prediction algorithm and recommendation of activities to be carried out to enhance the learning of a unit of study. There are commercial solutions that provide this type of technology such as RealizeIT, Coog Books, Smart-Parrow, and Knewton, among other tools. These services offer an integrated system to generate learning activities that allow areas of opportunity to be detected, rec-ommend new activities, and provide both teachers and students with results boards that facilitate monitoring of the course, strengths, and weaknesses.

Adaptive platforms have made it possible to streamline the work of rein-forcement and detection of individual areas in students. It favors attending to each student in a differentiated way according to their needs and offers the teacher follow-up reports that support decision-making to improve their teaching and improve student performance. Some limitations or difficulties that are faced when adopting this technology are as follows:

- The quality of the results of a prediction algorithm is conditioned to the quality of the data and quantity of data collected in the previous periods.

- There must be training and recognition of the prediction process to understand the process of operation of the tool and to understand its contribution to the learning process.
- Students must continually take diagnostic tests at different points in the learning units in order to establish a new learning path.
- The cost of the technology that is expected to be incorporated into the educational process must be considered.
- The contents and activities of evaluation and feedback must be prepared and aligned to the adaptive learning strategy of the course.
- Given the high specialization of these tools, the providers of the adaptive learning service supported by AI algorithms must be in charge of the logistics and publication of activities and programming of the algorithms, limiting the teacher in the activities of updating content or making the programming more flexible in educational creation.
- The teacher needs to align his teaching vision to focus on individual progress and should always have up-to-date and recent information to take corrective actions that favor the group of students.
- The classification and prediction algorithms assign a performance pattern to a student supported by the patterns of students who had an equivalent behavior in the previous records, inferring a process that is probabilistically similar; however, there is no 100% certainty that this will occur, which should always be understood as a limitation of these prediction models.

In the experience of implementation projects with AI in the educational sector, there are two elements that remain highly resistant to the development of AI-supported teaching: digital skills and the renunciation of traditional teaching models. The current education professionals are in transition in the new educational models; however, their previous teaching experience and practices are created to meet the needs of the school children, many of which are successful, often oppose the vision of adaptive strategies.

As indicated in this section, the applications of Artificial Intelligence to support different tasks and educational processes are diverse, and they offer support to professionals in education where these technologies have been incorporated. The use of AI tools in education has increased the capabilities of the teacher and expanded the visibility and processing of information, analysis of information, identification of patterns, identification of objects, people, or even generate forecast and projection models based on the information provided to generate self-training algorithms to improve their performance. Its contributions are extensive and greater applications and solutions are still expected for tasks where the capabilities of this technology can add value to people's professional, procedural, or behavioral activities.

However, the indiscriminate use of artificial intelligence solutions also invites us to reflect on the ethical or moral aspects that the response of an algorithm can infer under the assumption that the probability of occurrence of an event is generated, under the assumption that in your training database, there are situations or profiles that have acted incorrectly and lead us to wrongly qualify a person, a process, or make a decision poorly evaluated due to the limitations of the algorithm. There are still aspects that must be kept in mind and always put the virtues and qualities of people before the algorithms. If it is possible to integrate technology and human capacities in a balanced and synergistic way, the support is clearly favorable to support the activities in which it has been incorporated. There are still ethical and moral aspects that must be discussed and analyzed, which depending on the task to be carried out must be considered to avoid situations of bias, misinterpretation, and affectation of people or organizations.

8.5 AI-Based Adaptive Learning: Case Study

In 2018, Olmos et al. (2018) developed an algorithm based on artificial intelligence to predict the performance of undergraduate students in Physics for Engineering courses. The forecasting model uses the algorithms K-nearest neighbor (Chirici et al., 2016) and random forest (Ayyadevara, 2018; Rodríguez-Hernández et al., 2021) at its core. To initiate the process of predicting the academic performance, both algorithms were trained with data from a sample of students using biometric information such as neural frequency, facial recognition, heart rate, and historical academic information. Then, the predictive algorithm was fed by the grades that students were obtaining in their physics courses (see Figure 8.5).

The results of Figure 8.5 show that as the algorithm is trained on more information, the more accurate its predictions become. So, under the premise of the functionality of the algorithm, it was decided to develop a study in 2019 with the aim of applying adaptive learning based on the predictions thrown by the AI-based algorithm in a face-to-face learning environment. The study involved five professors of the Physics course for engineering, each with a control and experimental group. At first, the image of the student was used for facial recognition, for which the corresponding permits were requested, it is worth saying that this process delayed the investigation due to the use of sensitive information and that it can threaten ethics.

Once the predictive results were obtained, each teacher was given the task of taking adaptive actions with the aim of helping students who were predicted to be at risk of failing the class. In the second moment, the algorithm

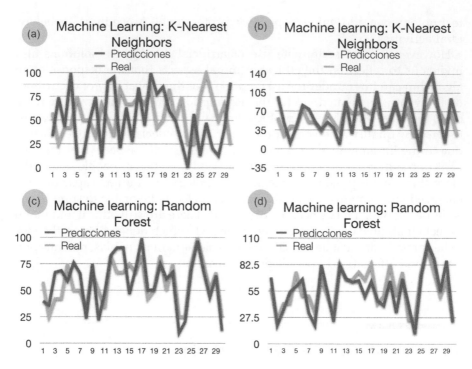

FIGURE 8.5
Algorithm training with (a) Quizzes, (b) Quizzes+Homework (HW), (c) Quizzes+HW+Student surveys to evaluate teachers (ECOA), and (d) Quizzes+HW+ECOA+Biometrics (Rincón-Flores et al., 2022).

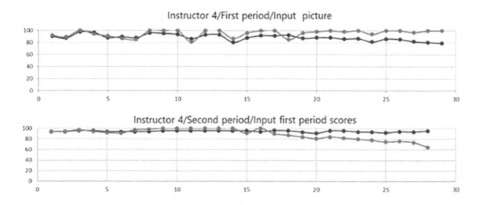

FIGURE 8.6
Predictive and real results of one of five instructors' first and second evaluation periods (black=forecasted student performance; grey=actual performance) (Rincón-Flores et al., 2022).

was fed with information on the grades of the first and the second partial evaluation. The results showed that the actual and predictive ratings were not accurate, but they did show a trend at the group level. Figure 8.6 shows the results of one of the groups.

One of the reasons attributed to the lack of precision of the algorithm was that this type of algorithm requires a vast amount of information. Initially, this study was planned to use information from 3,000 students from different geographical areas of the country; however, after a debate about the use of sensitive information such as face photographs, the face images could not be collected from all the students. When an algorithm is fed with a vast amount of data, it gets the opportunity to find different relationships among variables, and consequently, the results are more accurate. For example, one of the inaccuracies was to generalize the academic patterns of students, that is, it did not identify those cases in which students do not need to do activities or tasks because for them it is enough to pay attention in class to learn and obtain passing grades, but as we know, these are scattered cases.

On the other hand, when the results of the control group were compared with the experimental group, no significant difference was found in their learning. This was not only due to the lack of individual precision of the algorithm but also during the course of the semester the professors used predictive information to see the overall performance of the group and based on this was that they changed or added activities in the corresponding experimental group. When teachers were asked about their perception of the algorithm, only one of them commented that he would not like to continue using it until it was more accurate, the rest found that it provided valuable information about the overall performance of the course but not at the individual level.

During the intervention, the students did not know about the algorithm, and it was until the end that a questionnaire was applied to them, and interviews were developed. Regarding the results of the instrument, most students commented that they would like to know their performance forecast because it would allow them to plan their extracurricular activities in advance to keep their scholarship; however, some students commented that they preferred not to know because it may cause them anxiety.

While it is true that the predictive algorithm was not as accurate individually, it did manage to predict behavior at the group level, and with it, teachers were able to take adaptive measures to help students. It is true that there was no significant difference between the control group and the experimental group; however, this educational intervention has allowed visualizing the potential of AI in face-to-face environments whenever the algorithm is fed with a greater amount of data each time in such a way that teachers have an educational tool that allows promoting adaptive learning within their courses, in this case, those of physics for engineering careers (Rincón-Flores et al., 2022). Thus, the student's experience during university will be more effective and could contribute to reducing the school dropout rate.

On the other hand, this research opens the possibility of combining AI with the face-to-face teaching-learning process, because although efficient adaptive learning platforms are developed that offer positive messages to students, such as the polite tutor (McLaren et al., 2011), it does not compare to the warmth of the accompaniment that a human teacher can provide.

8.6 Final Lecture

There is much to discover about our behaviours and the creative capacity that we can achieve. Through current technologies, we can go deeper into various areas of knowledge. With artificial intelligence, machine learning, or predictive model programming, it is possible to overcome barriers that are unknown for us so far. The development of these tools, applied to different processes, has the ability to exponentiate knowledge in a vertiginous and extraordinary way.

Changes in society are imminent, given current technological advances. If we refer to the "scientific revolutions" of Thomas Khun, we are experiencing what the philosopher of science identifies as "anomalies", which originate from the application of predictive models with the intervention of AI, giving results that were previously difficult to obtain. It is convenient to continue with the application of predictive analytics to obtain greater findings from some academic processes and activities, which generates and channels scientific consensus that allows the establishment of other paradigms, which are focused on positively transforming education.

Universities, as organizations, will have a significant role in the transformation of societies. For a feasible interconnection between educational ecosystems, it is necessary that the captured data follow a classification thought, are planned, and are designed for the application of learning analytics. For this to happen, it is recommended that the models can be structured to have fast interactions but smooth scaling. In this way, the educational centers will have dynamism and adaptability through the context and the intelligent data collected, which in each academic period provides new information on the students and results of the learning activities, allowing the analysis of the needs that are required to be covered in each one of the students and didactics that are used.

Educational technology managers and the academy that promote systems equipped with AI, combined with pedagogical techniques based on predictive models, will be able to develop more efficient teaching processes in academic programs and, at the same time, preserve quality and promote improvement. Innovating in educational systems using predictive models with AI means a revolution in the teaching-learning process, efficiency in administrative systems, and personalization of students' academic experiences. However, its application generates the need to incorporate ethics in AI, to encourage respect for diversity, and not cause biases that cause or exacerbate stereotypes and discrimination.

Nowadays, the problems around us affect us locally and globally. These situations require simultaneous and comprehensive solutions with rational actions, promptness, and scientific support. Here is the relevance of the use of analytics, big data, predictive models, advanced computing programming, and artificial intelligence to advance solutions to global problems and benefit society. AI has various "life stages": conception, development, and use in various systems; in each of them, different actors are involved who are participants and socially and ethically coresponsible. This is why UNESCO published in 2022 the document "Recommendation on the Ethics of Artificial Intelligence", with the aim of providing a universal framework of values, principles, and actions to guide governments in creating policies on AI to protect human rights, fundamental freedoms, human dignity, equality, and cultural diversity, among others.

By including AI in the economic, labor, legal, educational, or any other order, it impacts each of them by raising ethical issues. AI systems that have the ability to emulate activities that were exclusive to human beings, in some cases more efficiently, generate new dilemmas that challenge our current ethical and regulatory concepts. Therefore, this leads us to reflect that in the future, "AI systems could challenge humans' special sense of experience and agency, raising additional concerns about, inter alia, human self-understanding, social, cultural and environmental interaction, autonomy, agency, worth and dignity". With the implications that arise, it will be necessary to incorporate an ethical perspective in the design of artificial intelligence.

It is relevant to reflect on the coexistence between the human being and the machine, given that technology will be increasingly active, immersed, and powerful in the relationship between them. In this accelerated progress, it is imperative to recognize and place the human element in the foreground so that the transcendent is the purpose of the use of technology.

References

Ayyadevara, V. K. (2018). Random forest. In: *Pro Machine Learning Algorithms*. Apress, Berkeley, CA. https://doi.org/10.1007/978-1-4842-3564-5_5.

Bonfield, C. A., Salter, M., Longmuir, A., Benson, M., & Adachi, C. (2020). Transformation or evolution?: Education 4.0, teaching and learning in the digital age. *Higher Education Pedagogies*, 5(1), 223–246.

Calle, C. & Mena, J. (2022). *Uso de metodologías activas en el aula de idiomas: propuesta práctica*. En: S. Carrascal, D. Melaré y D.J. Gallego (Eds.). *Nuevas Metodologías, espacios y estilos de enseñanza-aprendizaje: Prácticas docentes e innovación educativa*. Madrid: Editorial Universitas.

Cheung, B., Hui, L., Zhang, J., & Yiu, S. M. (2003). SmartTutor: An intelligent tutoring system in web-based adult education. *Journal of Systems and Software*, 68(1), 11–25. https://doi.org/10.1016/S0164-1212(02)00133-4.

Chirici, G., Mura, M., McInerney, D., Py, N., Tomppo, E. O., Waser, L. T., Travaglini, D., & McRoberts, R. E. (2016). A meta-analysis and review of the literature on the k-Nearest Neighbors technique for forestry applications that use remotely sensed data. In *Remote Sensing of Environment* (Vol. 176, pp. 282–294). Elsevier Inc. https://doi.org/10.1016/j.rse.2016.02.001.

Goh, P. S. C., & Abdul-Wahab, N. (2020). Paradigms to drive higher education 4.0. *International Journal of Learning, Teaching and Educational Research*, 19(1), 159–171.

Hinojo-Lucena, F. J., Aznar-Díaz, I., Cáceres-Reche, M. P., & Romero-Rodríguez, J. M. (2019). Artificial intelligence in higher education: A bibliometric study on its impact in the scientific literature. *Education Sciences*, 9(1). https://doi.org/10.3390/educsci9010051.

Hwang, G. J., Sung, H. Y., Chang, S. C., & Huang, X. C. (2020). A fuzzy expert system-based adaptive learning approach to improving students' learning performances by considering affective and cognitive factors. *Computers and Education: Artificial Intelligence*, 1(July), 100003. https://doi.org/10.1016/j.caeai.2020.100003.

Jones - Jang, M. & Park, Y. (2022). How do people react to AI failure? Automation bias, algorithmic aversion, and perceived controllability. *Journal of Computer-Mediated Communication*, 28. https://doi.org/10.1093/jcmc/zmac029.

Kaplan, A., & Haenlein, M. (2019). Siri, Siri, in my hand: Who's the fairest in the land? On the interpretations, illustrations, and implications of artificial intelligence. *Business Horizons*, 62(1), 15–25. https://doi.org/10.1016/j.bushor.2018.08.004.

Kulik, J. A. & Fletcher, J. D. (2015). Effectiveness of intelligent tutoring systems: A meta-analytic review. *Review of Educational Research*, 86(1), 42–78. Available at https://www.researchgate.net/publication/277636218_Effectiveness_of_Intelligent_Tutoring_Systems_A_Meta-Analytic_Review.

Luan, H. & Tsai, C. C. (2021). A review of using machine learning approaches for precision education. *Educational Technology & Society*, 24(1), 250–266.

Maanen, G. (2022). AI ethics, ethics washing, and the need to politicize data ethics. *Digital Society*, 1. https://doi.org/10.1007/s44206-022-00013-3.

McLaren, B. M., Deleeuw, K. E., & Mayer, R. E. (2011). Polite web-based intelligent tutors: Can they improve learning in classrooms? *Computers and Education*, 56(3), 574–584. https://doi.org/10.1016/j.compedu.2010.09.019.

Mena, J. & Calle, C. (2020). *Actualización de estrategias en el aprendizaje integrado de idiomas*. Curso de Formación del profesorado en red del INTEF. 2ª edición 2020. Del 15 de septiembre al 17 de noviembre 2020. 70 horas. http://www.educacionyfp.gob.es/dam/jcr:08ef2043-5ac3-4d6f-9b6c-2c51c274aeab/08-ficha-intef-estrategias-aprendizaje-idiomas.pdf.

Mena, J. (2022). *Uso de recursos en la plataforma Genially*. Genially Inc. En https://view.genial.ly/629f157fc300560013bca67c.

Norouzi, N. (2022). Sustainable Fourth Industrial Revolution. In C. Popescu (Ed.), Handbook of Research on Changing *Dynamics in Responsible and Sustainable Business in the Post-COVID-19 Era* (pp. 58–77). IGI Global. https://doi.org/10.4018/978-1-6684-2523-7.ch003

Olmo-Muñoz, J., González-Calero, J. A., Diago, P. D., Arnau, D. & Arevalillo-Herráez, M. (2022). Using intra-task flexibility on an intelligent tutoring system to promote arithmetic problem-solving proficiency. *British Journal of Educational Technology*, 53, 1976–1992. https://doi.org/10.1111/bjet.13228.

Olmos, O., Hernández, M., Avilés, E., & Treviño, I. (2018). Optimal Paths for academic performance supported by artificial intelligence. In *Conference Proceedings of the 6th International Conference on Educational Innovation, CIIE 2018*. Monterrey, Mexico.

Park, C. S.-Y. , Kim, H., & Lee, S. (2021). Do less teaching, do more coaching. *Toward Critical Thinking for Ethical Applications of Artificial Intelligence*, 6, 97–100.

Recommendation on the Ethics of Artificial Intelligence. (2022). En UNESCO, https://unesdoc.unesco.org/ (SHS/BIO/PI/2021/1). United Nations Educational, Scientific and Cultural Organization. https://unesdoc.unesco.org/ark:/48223/pf0000381137.locale=en.

Rees, C. & Müller, B. (2022). All that glitters is not gold: Trustworthy and ethical AI principles. *AI and Ethics*. 1–14. https://doi.org/10.1007/s43681-022-00232-x.

Rincón-Flores, E. G., Lopez-Camacho, E., Mena, J., & Olmos, O. (2022). Teaching through learning analytics: Predicting student learning profiles in a physics course at a higher education institution. *International Journal of Interactive Multimedia and Artificial Intelligence, IP(IP)*, 1. https://doi.org/10.9781/ijimai.2022.01.005.

Rodríguez-Hernández, C. F., Musso, M., Kyndt, E., & Cascallar, E. (2021). Artificial neural networks in academic performance prediction: Systematic implementation and predictor evaluation. *Computers and Education: Artificial Intelligence*, 2(March). https://doi.org/10.1016/j.caeai.2021.100018.

Salles, A., Evers, K., & Farisco, M. (2020). Anthropomorphism in AI. *AJOB Neuroscience*, 11, 88–95. https://doi.org/10.1080/21507740.2020.1740350.

UNESCO (2022). Recommendation on the Ethics of Artificial Intelligence https://www.unesco.org/en/legal-affairs/recommendation-ethics-artificial-intelligence Unsupervised—MachineHallucinations—MoMA.(2022,28octubre).RefikAnadol. https://refikanadol.com/works/unsupervised-machine-hallucinations-moma/.

Zawacki-Richter, O., Marín, V. I., Bond, M., & Gouverneur, F. (2019). Systematic review of research on artificial intelligence applications in higher education – where are the educators? *International Journal of Educational Technology in Higher Education*, 16(1). https://doi.org/10.1186/s41239-019-0171-0.

Zhang, X. (2022). The influence of mobile learning on the optimization of teaching mode in higher education. *Wireless Communications and Mobile Computing*, 2022, Article ID 5921242, (1–9 pp). https://doi.org/10.1155/2022/5921242.

9

Using AI for Educational Research in Multimodal Learning Analytics

Luis Fernando Morán-Mirabal, Joanna Alvarado-Uribe, and Héctor Gibrán Ceballos

Tecnológico de Monterrey

CONTENTS

9.1 Introduction

In current higher education programs, students are required to develop professional competencies and skills (e.g., collaboration, critical thinking, communication, technology literacy, and mastery), which help them to stand out in terms of proficiency and productivity when applying for jobs in companies and organizations (Rasipuram & Jayagopi, 2020). Such education programs aim to foster and boost student capabilities to become change agents and face world challenges successfully (Cornide-Reyes et al., 2020). Nonetheless, the evaluation of the competencies is usually done through course assessments and evaluations. Thus, the introduction of additional measurements that involve behavioral and learning environment factors can assist in generating a deeper understanding of how such competencies are developed.

DOI: 10.1201/b23345-11

Furthermore, education is increasingly taking place through remote and hybrid learning environments, especially during and after the COVID-19 pandemic. This scenario has posted multiple challenges for teachers and learners who were accustomed to in-person classrooms and laboratories (Rapanta et al., 2020), such as the necessity for introducing and managing new and existing technological tools and platforms for course design and instruction, as well as student assessment, follow-up, and feedback. These educational technologies (EdTechs) helped mitigate the impact of distant interactions; however, a new gap related to the way students interacted with the provided digital content called for new approaches of data analysis.

In response to the aforementioned challenges, multiple artificial intelligence (AI) techniques (e.g., support vector machines, neural networks, natural language processing, heuristics) have been applied to study teachers' and learners' interactions with EdTechs through learning analytics (LA) and multimodal learning analytics (MMLA). These prominent fields of study aim to close the gap between technology (i.e., tools) and pedagogy (i.e., theory) by analyzing educational data traces and linking them to pedagogical constructs, behavioral indicators, and actionable feedback (Boothe et al., 2022).

This chapter aims to present the potential that AI in MMLA has for improving the analysis and understanding of the learning interactions taking place in physical, remote, and hybrid learning environments, as well as to identify the challenges it currently faces to be fully implemented in real contexts. This chapter is structured as follows: First, an introduction on LA and MMLA is presented in Sections 9.2 and 9.3, respectively. Then, in Section 9.4, three applications of AI in MMLA found in recent literature are described and analyzed. Finally, in Sections 9.5 and 9.6, current limitations and the future of AI in MMLA are discussed, and the author's closing remarks on the subject are presented.

9.2 Learning Analytics

In 2011, the Society for Learning Analytics Research defined LA as the measurement, collection, reporting, and analysis of data about learners and their context for purposes of understanding and optimizing learning and its environments (SoLAR, 2011). In essence, LA focuses on acquiring a deeper understanding of students' learning activities and environments, using computational analysis techniques, such as Data Science and AI.

In addition to analyzing and reporting the collected data, LA seeks to deliver actionable feedback (Wehrmann & Van der Sanden, 2017) to the teachers and learners interacting in remote, hybrid, and in-person learning environments. Such feedback promotes students' success and reinforces their skills on knowledge acquisition and creation (Mory, 2004). Thus, many

AI applications in LA result in indicators, reports, and dashboards, which offer valuable insights for the enrichment of the teaching/learning process.

LA research has mainly focused on how students interact with learning management systems (LMSs) (Lang et al., 2022); however, it has been consistently expanding to consider learning interactions in different environments such as massive online open courses, online programs, intelligent tutoring systems (ITS), and serious games. Such computer-based and computer-managed learning environments allow for the extraction of student's interactions (i.e., with keyboard and mouse) in the form of data traces and logs, which are subsequently processed and analyzed to measure learning constructs, conduct timely interventions, and provide actionable feedback (Gray & Bergner, 2022).

Moreover, the rapid evolution in smart devices and online technologies has facilitated the development of new ways to implement different learning activities (Sher et al., 2022). However, it is important to analyze the impact that these modalities (ubiquitous learning environments) have on student learning considering their context. For example, Sher et al. (2022) performed a log-based study on LMS usage across three different modalities: computer, mobile, and tablet to analyze the patterns of use of each modality by the time of day (and day of the week) of the student learning sessions in the context of blended learning. In the same way, Krumm et al. (2022) analyzed data from a LMS to understand and measure how students in grades 6–12 used and navigated digital learning activities (referred to as playlists) in a common LMS.

These and other studies help to understand how students learn, leading institutional stakeholders to propose new solutions to address educational challenges, such as student dropout. For instance, De Silva et al. (2022) propose the implementation of institutional analytics-based solutions to support decision-making at this level, identifying the factors that influence student dropout from the perspective of institutional stakeholders.

In addition to processing data traces and logs from computer-based learning environments, LA often makes use of additional resources (e.g., surveys, self-reports, social network data, user demographic data) in an effort to build a more robust model that better describes the learning activities being studied (Nistor & Hernández-García, 2018). Furthermore, LA takes advantage of multiple statistical analysis techniques (e.g., principal component analysis, linear and multiple regression, correlation tests, Bayesian methods) to compare, test, and validate the fitness and accuracy of one or more models.

9.3 Multimodal Learning Analytics

The constant development and access to data collection technologies have allowed researchers and scholars to apply AI techniques to study learning interactions that take place outside of computer-based environments, thus providing them with new insights into learning tasks, where students interact with multiple stakeholders and tackle problems in both hybrid and in-person learning environments (Blikstein & Worsley, 2016).

Since 2013, the term MMLA has been formally adopted and used by the scientific and academic community (Ochoa, 2022), to specify a branch of LA focused on the study of learners and their contexts, by integrating different technologies to capture, measure, and analyze different communication modalities (Blikstein, 2013). MMLA allows for a deeper understanding of students' learning paths and cognitive work even in complex and open learning environments, and according to Blikstein and Worsley (2016, p.233), it can be understood as "a set of techniques employing multiple sources of data to examine learning in realistic, ecologically valid, social, mixed-media learning environments."

As mentioned in Section 9.2, most LA research focuses on analyzing student's behaviors while interacting with computer-based environments. However, solely paying attention to LMS data traces and logs can lead to incomplete or ambiguous information, mainly because there are additional factors that prove difficult to capture in traditional multi-stakeholder learning environments (Ciordas-Hertel, et al. 2021). Thus, MMLA enables the detection of additional learning interactions that are difficult for the human eye to perceive in real-world contexts (Chan et al., 2020). For example, to measure and evaluate the collaboration, an individual analysis of each students' performance and their interactions with others would be necessary, which in turn would demand plenty of human effort and time (Mercer et al., 2004; Prieto et al., 2018). Furthermore, MMLA can also ideally be suited to capture and measure learners' cognition, and affect as well as study learning design objectively (Azevedo, 2015; Mangaroska et al., 2020). This provides a more holistic picture of the teaching/learning process.

MMLA studies from 2013 onward have utilized multiple multimodal technologies to capture different modality features, such as video, audio, hands and face gestures, physiological biomarkers, body postures, eye gaze, texts, sketches, environmental factors, and body/object spatial positions (Ochoa, 2022). MMLA has been applied in the study of teacher–student interactions and behaviors on multiple learning tasks and environments, such as oral presentations (Vieira et al., 2021), calligraphy learning (Loup-Escande et al., 2017), group-work activities (Ochoa, Chiluiza, et al., 2018), remote collaboration (Cornide-Reyes et al., 2020), distance learning (Ciordas-Hertel et al., 2021), classroom design (Cruz-Garza et al., 2022), and spatial pedagogy (Yan et al., 2022), among many others.

To further understand the bridge that MMLA seeks to build between pedagogy and technology, Boothe et al. (2022) present a pragmatic theory-driven framework for multimodal collaboration feedback (see Figure 9.1). Such framework portrays how multimodal technologies (sensors) can be utilized to extract features from multiple modalities, which in turn can be processed and analyzed through Data Science and AI to produce actionable feedback

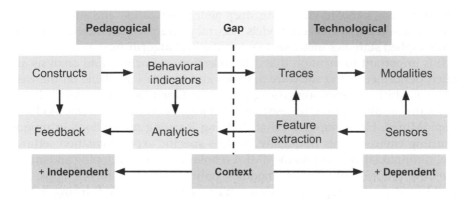

FIGURE 9.1
Framework for multimodal collaboration feedback. (Taken from Boothe et al., 2022.)

related to the pedagogical constructs and behavioral indicators present in physical, remote, and hybrid learning environments, therefore closing the gap between dependent and independent collaboration (learning) contexts.

Data processing in MMLA begins with the extraction of features captured by multimodal devices (e.g., video cameras, microphones, eye trackers, biosensors, location trackers, interactive surfaces), which often operate independently. Thus, the extracted data are usually multidimensional in nature and require further processing before being analyzed. This is accomplished by introducing the key element of data fusion in MMLA (Samuelsen et al., 2019), which allows for the combination of extracted features from different data sources into a unified dataset for modeling and analysis. Nonetheless, data fusion can prove to be a challenging task in MMLA, mainly because data from different multimodal devices are often collected at different grain sizes and timestamps, which complicates the process of integration (Liu et al., 2018).

Three kinds of relationships can be observed when multimodal data are fused (integrated) into learning indicators: one-to-one, many-to-one, and one-to-many (Mu et al., 2020). In one-to-one, each collected data type is appropriate for measuring one learning indicator (e.g., audio can be used to measure collaboration). In many-to-one, multiple data types are appropriate for measuring the same learning indicator (e.g., audio, video, and electrodermal activity can be used to measure engagement). In one-to-many, each collected data type is appropriate for measuring several learning indicators (e.g., eye gaze can be used to measure attention, cognition, and emotion).

Once a data-to-indicator relationship is identified, the next step in MMLA is to build a data fusion algorithm that models the connection between the collected multimodal features and the studied learning indicators. Most MMLA applications found in the literature apply multiple AI techniques to model data-to-indicator relationships (e.g., sequential pattern mining (SPM), natural language processing, support vector machines, Bayesian networks, decision trees, k-means clustering, deep neural networks), and subsequently, a comparison between each models' accuracy and performance is analyzed with applied statistics methods to select the best algorithm.

It is important to mention that data collection, extraction, and analysis are often performed using software modules and libraries available on the web, some of which are open source (e.g., OpenPose, Open-Face, openSMILE, Praat) or require a subscription/license to be used (e.g., iMotions Biosensor Modules, RapidMiner). However, given that studying learning interactions in unique learning environments frequently involves specific experimentation settings (i.e., multimodal devices), many MMLA scholars opt for developing and implementing proprietary software modules, online platforms, mobile apps, and dashboards, which capture, analyze, and report the specific learning indicators being studied (Munoz et al., 2018; Aguilar, 2022; Ng et al., 2022).

The final step in MMLA consists in translating the best models' outcomes into actionable feedback, which is essential for teachers to monitor students' interactions, improve learning design, and perform timely and precise interventions, as well as for students to effectively learn from their mistakes and improve their learning outcomes (Nicoll et al., 2022). Thus, some MMLA studies have focused on building automated feedback systems that improve the teaching/learning process in different learning environments. Some examples in this specific field of MMLA include designing a real-time visualization dashboard for co-located collaboration (Praharaj et al., 2018), developing a system to provide automatic feedback of oral presentation skills (Ochoa, Domínguez, et al., 2018), and implementing an affective visual analytics dashboard to monitor learners' emotions (Ez-zaouia & Lavoué, 2017).

This section closes by summarizing MMLA through the conceptual model for multimodal data analysis presented by Mu et al. (2020) (See Figure 9.2). Such a model portrays the process by which MMLA translates the learning process digitalization (multimodal data) into teaching application (learning indicators). First, by extracting data from digital, physiological, psychometric,

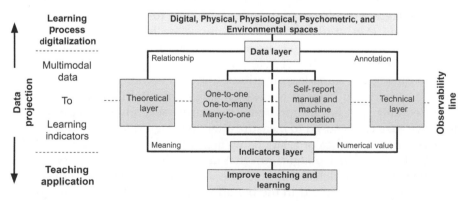

FIGURE 9.2
Conceptual model for multimodal data analysis. (Taken from Mu et al., 2020.)

and environmental learning spaces, and then, by integrating such data into learning indicators through multiple types of relationships and annotations which are fused through Data Science and AI to result in the improvement of teaching and learning.

9.4 Applications of Artificial Intelligence in MMLA

This section presents three examples of applications of AI in MMLA. Each application relates to a specific MMLA study found in the literature, where one or more communication modalities, cognitive and physiological features, and environmental factors are collected through multimodal devices. These works were selected because they are recent (2020–2021) and were published in open access. Additionally, each study employs different AI techniques to fuse the extracted data and build models that relate such data to learning constructs, behaviors, and indicators. Finally, such models are tested and analyzed through statistical methods, to gain a deeper understanding of the dynamics taking place in physical, remote, and hybrid learning environments, as well as to provide actionable feedback to further improve the teaching/learning process.

9.4.1 Analyzing Body Postures in Oral Presentations

During their academic development, students can share their acquired knowledge through oral presentations. Such presentations are usually conducted in classrooms where fellow students become the audience, while an instructor evaluates their performance. However, a successful oral presentation not only depends on the content being shared by the student, but on additional non-verbal skills, such as facial gestures, body postures, and voice intonation (Chen et al., 2014). Such non-verbal skills might be difficult to assess, especially if the instructor is not prepared to measure their execution. Thus, the introduction of body posture analysis through MMLA can provide additional insights and feedback about students' performance.

Vieira et al. (2021) conducted a case study with 222 computer engineering students at the University of Valparaíso, Chile, to test a framework for analyzing body postures in oral presentations. The objective of the study was to analyze the body postures and patterns used by pairs of students while performing three oral presentations in an engineering course during two different years (i.e., three presentations per year). The authors used a Microsoft Kinect camera (sensor), along with an open-source application called Lelikëlen (developed by Munoz et al., 2018), to capture and classify a set of 12 different presenters' attributes (arms crossed, tilted body, straight body, watching public, hand on chin, explaining with open hands, hands down, explaining with one hand, hands on hip, hand on head, pointing with one hand, and talking), where 11 attributes are related to skeletal detection (body postures) and one to voice detection. Figure 9.3 illustrates the learning environment setting used by the authors for conducting the experiments.

After extracting the data captured by Lelikëlen, the authors performed a descriptive and inferential statistical analysis to search for trends, relationships, and comparisons, between the attributes collected from each year and from different years. To find similarities between datasets, they applied a Wilcoxon rank sum test, which is a non-parametric test for two populations with independent samples.

To continue with their data processing and analysis, Vieira et al. (2021) made use of two machine learning clustering algorithms (AI) to segment observations in their datasets: a k-means clustering algorithm to segment students into different groups based on their presentation attributes, and a hierarchical clustering technique to evaluate presentations by grouping observations. In both unsupervised techniques, k-means and hierarchical clustering, you must indicate how many clusters you will have in the end. Nevertheless, in k-means, you indicate a priori the number k of clusters to obtain and support your decision by using the Silhouette Coefficient or the Elbow method. In hierarchical clustering, on the other hand, you decide the number of clusters

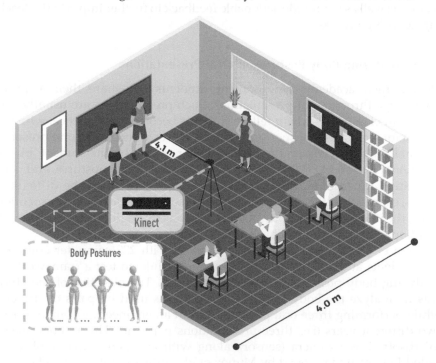

FIGURE 9.3
Learning environment setting used by Vieira et al. (2021).

by inspecting the resulting dendrogram and selecting a minimum distance between elements of each cluster.

To further complement their study, the authors applied a SPM technique to analyze students' postures in a chronological sequence. This data mining technique aims to extract frequently occurring patterns, compare sequences for similarity, and recover missing sequence members. SPM is similar to process mining as long as it tries to identify patterns in sequential data. This AI technique allowed them to find all sequence patterns based on the minimal support (number of occurrences), which was defined by the authors as an initial parameter. Their objective was to identify the main sequences followed by students in their presentations, and if they occurred in all presentations throughout the year. For this end, the authors used the sequential pattern mining framework (SPMF) (Fournier-Viger et al., 2016), an open-source data mining library implemented in JAVA.

After presenting their outcomes, the authors discussed that their statistical analysis allowed them to identify common body posture attributes between presentations in different years, as well as to gain a deeper understanding of certain attributes during the period of student presentations. Furthermore, by applying clustering algorithms, they were able to identify groups of behaviors (based on the collected attributes) shown by the students when performing their oral presentations. Moreover, the use of the SPM technique allowed them to identify sequences with specific behaviors that otherwise would have been difficult to visualize.

Vieira et al. (2021) concluded by mentioning that MMLA can assist in the evaluation of complex learning environments by collecting information from different sources, and suggested that by integrating additional data, such as teacher notes and students' grades, their study could be further improved to gain deeper insights about the learning process, as well as to provide feedback on the development of students' oral presentation skills.

9.4.2 Assessing Spoken Interactions in Remote Active Learning Environments

As previously mentioned in Section 9.1, the COVID-19 pandemic forced a change in education by shifting learning interactions from in-person to remote learning environments. People were required to stay at home, and thus, students had to attend lectures remotely through online platforms (e.g., Zoom, Microsoft Teams, Google Meet). However, such a shift introduced new challenges for instructors to assess participation and collaboration, given that remote lectures depended on each student having their personal camera turned on, and actively engaging through spoken interactions (Cao et al., 2020). This entailed the development of additional assessment tools, which allowed instructors and students to successfully engage in remote active learning environments.

To tackle such a challenge, Cornide-Reyes et al. (2020) developed NAIRA, a real-time feedback platform that collects and processes spoken interactions, for instructors to assess student's participation and collaboration in remote learning environments. NAIRA operates by connecting three modules that capture, process, and report speaking data:

(i) a mobile application (NAIRA APP), (ii) a cloud-based service back-end (NAIRA BE), and (iii) a dashboard web application (NAIRA WEB). NAIRA allows instructors to perform timely interventions and provide real-time feedback to students interacting in remote group activities, using an online dashboard that displays an influence graph, a silent time bar, a speaking distribution pie chart, and a table with collaboration metrics. Figure 9.4 illustrates a general diagram of the solution developed by the authors.

To capture speaking interactions coming from a lapel microphone, NAIRA APP makes use of the AudioRecord (Android) and the AudioKit (iOS) libraries, which transform the collected audio input into numerical data for further processing and analysis (e.g., amplitude and frequency). Both libraries use the fast Fourier transform to convert audio signals (in time) to individual

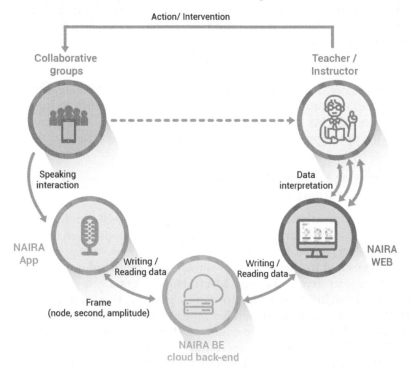

FIGURE 9.4
General diagram of NAIRA, as presented by Cornide-Reyes et al. (2020).

spectral components in the frequency domain. Then, to process the trans-formed speaking data, NAIRA BE applies AI techniques in social network analysis (SNA) to construct an influence graph that depicts the group interaction dynamics (i.e., communication relationships between students, number of interventions, total speaking time). SNA can be further used to identify the student that speaks with more classmates (centrality) or those brokers that interconnect student groups in the classroom. Network visualizations help to easily understand the classroom dynamics.

In addition to the influence graph, NAIRA BE calculates multiple indicators, such as the number of speaking interactions, the silent time, the effective speaking time, the average spoken intensity, and the level of participation (Bruffee, 1984), which, in turn, are used to construct a silent bar, a speaking distribution pie chart, and a collaboration metrics table, all of which are delivered to the instructor in real-time through a dashboard in NAIRA WEB.

To test the functionality of NAIRA, Cornide-Reyes et al. (2020) conducted a case study with 24 voluntary undergraduate students from an undergraduate course of object-oriented design. The objective of the study was to identify if students' speaking interactions could be monitored through NAIRA, and if its dashboard visualizations triggered timely teacher interventions to improve the activity being conducted. To this end, the authors designed a Jigsaw learning activity (Aronson, 2002), which took place in a virtual classroom through Zoom during the COVID-19 pandemic. The activity consisted in presenting the students with an object-oriented design problem and defining specific roles to contribute to its solution.

To begin with the activity, students were randomly assigned to five expert groups, where they learned one of the five Single Responsibility, Open-Closed, Liskov Substitution, Interface Segregation, and Dependency Inversion (SOLID) design principles, and analyzed a specific class diagram. Then, students were randomly assigned to Jigsaw groups of five students each, where each member was an expert in a design principle. During this stage, each expert described their principle, their assessment on the class diagram, and solved questions from the other team members. The teacher monitored the Jigsaw groups through a dashboard in NAIRA WEB and performed interventions when required to ensure that the activity was carried out as planned.

To analyze the Jigsaw groups activity outcomes, the authors collected data from four sources: the field notes taken by the teacher during the activity, the data and visualizations provided by NAIRA, the Zoom recording and comments, and a survey completed by the students after the activity. The statistical analysis of such data allowed them to suggest that the participation in Jigsaw groups was more uniform than in expert groups. Furthermore, they found that NAIRA successfully helped the teacher to monitor the activity by seeing the dynamics being followed in the groups and performing timely interventions when required. They conclude by suggesting that their case study shows NAIRA's feasibility for effectively managing and conducting other learning activities with real-time feedback.

9.4.3 Analyzing Physical Environment Factors in Distance Learning

The Internet of Things has allowed people to stay connected with the rest of the world while engaging in all sorts of activities. The introduction of mobile devices such as laptops, tablets, smartphones, and smartwatches became a milestone for accessing and sharing a vast amount of data between the users of the Internet. This, in turn, has changed the way education takes place nowadays, where students can engage in learning activities almost at any place and any time they require. However, the effectiveness of distance learning is often difficult to assess, given that there are many factors that are present in the physical learning environment (PLE) where students conduct their studies (Choi et al., 2014). PLE factors can be relevant to understand students' learning performance, and thus, their collection and analysis can assist in providing valuable insights about the distance learning dynamics (Zhao et al., 2021).

To this end, Ciordas-Hertel et al. (2021) developed Edutex, a software infrastructure that takes advantage of students' smart wearables to collect and analyze PLE factors, thus providing learners with a deeper understanding of their learning contexts and behaviors. Edutex considers nine PLE factors (i.e., visual noise, audible noise, context dependency, air quality, nutrition, lighting, spatial comfort, presence of others, and self-care), which the authors identified as relevant, and related them into the cognitive, physiological, and affective effects they have on students. The relationship between such PLE factors and their effects is illustrated in Figure 9.5.

To collect and measure PLE factors throughout a learning activity, Edutex makes use of the mobile sensors present in the student's smartphone and smartwatch (e.g., light, microphone, Wi-Fi, Bluetooth, GPS), in conjunction with manually submitted self-reports (i.e., questionnaires and micro-questionnaires). The former allows Edutex to invisibly collect data related to temporary PLE factors (e.g., lighting, and audible noise), while the latter allow for validation of sensor readings and collection of data related to permanent PLE factors (e.g., self-care and spatial comfort). Edutex collects data in two different modes: (i) an exploratory mode that captures momentary data measurements, and (ii) an intervention mode that captures a continuous stream of data.

Although the authors do not mention the AI models involved in the data collection process, we can relate multiple AI models to the mobile devices used in Edutex. For instance, Bluetooth and Wi-Fi in the smartphone make use of direct methods for people detection (i.e., nearby devices and visible routers) to detect the presence of other devices (i.e., individuals) near a user of Edutex. Likewise, the smartphone GPS sensor is used to pinpoint the location of an Edutex user and relate it to a specific building or establishment. Furthermore, the smartphone luminance sensors measure and report ambient light intensity, while the oxygen saturation sensor in the smartwatch makes use of the reflectance oximetry technique, which projects red and infrared light (IR) into the wrist.

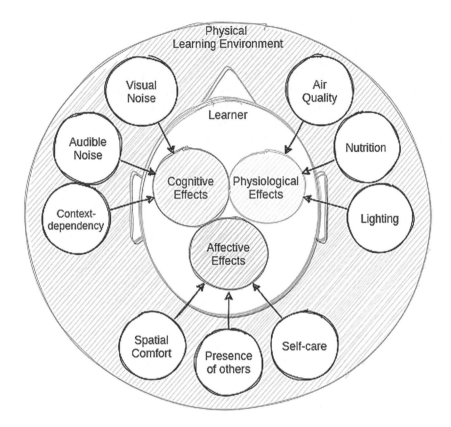

FIGURE 9.5
Relationship between PLE factors and their effects, as presented by Ciordas-Hertel et al. (2021).

Finally, Edutex makes use of an orchestration mechanism between client devices associated with a student and the server. In this way, it coordinates data collection, making it possible to aggregate and sequence all the sensor data being collected. In a posterior stage, data are synchronized and combined to detect patterns and use sensor data as characteristics for AI models. The analytics service implemented in Edutex makes use of previously computed and stored AI models, which are loaded and executed on demand. At runtime, these services register for the preprocessed data streams, and after the various indicators have been calculated, the services publish their results again in their dedicated topics. These data are later shown in dashboards that allow an instructor to look at aggregated data streams of online recordings from participants with several technical statistics.

To test the functionality of Edutex, the authors conducted a case study with ten participants from December 2020 to March 2021. The participants were provided with preconfigured smartwatches and smartphones, which they were expected to use throughout multiple learning sessions taking place at different times and days. Each learning session was divided into five stages: (i) a pre-survey stage where participants submitted a questionnaire on demographics and technology affinity, (ii) a pre-session stage where participants reported permanent spatial comfort PLE factors, and the according sensor measurements were made through GPS, Wi-Fi, and Bluetooth, (iii) an in-session stage where temporary PLE factors were periodically measured through the smart wearable sensors (i.e., microphone, heart rate sensor, accelerometer, light sensor), (iv) a post-session stage where participants submitted a complex standardized questionnaire, and (v) a post-survey stage where participants completed a survey on the usability and experience with Edutex.

The case study resulted in the collection of data from 55 learning sessions, which contained a total of 892,838 sensor measurements and 172 completed questionnaires. After analyzing the collected data, Ciordas-Hertel et al. (2021) discuss that Edutex could be successfully applied to predict self-reported learning contexts by using mobile sensors (i.e., lighting conditions, volume levels, and spatial positions), but that sensor data can be highly susceptible to interference, primarily because of the location of the sensors or the equipment of the PLE. Furthermore, the authors mention that blood oxygen saturation could be measured effectively, which could be related to air quality in a future study. Moreover, they comment that the nutrition factor could be measured with the implementation of a blood glucose sensor in upcoming smart wearables. Finally, they report that other PLE factors such as spatial comfort, presence of others, and self-care could be well measured through self-reports. The authors conclude by mentioning that Edutex could support students in self-directed distance learning, by providing them with insights about their learning habits, learning behaviors, physiological state, and their PLEs.

9.5 Limitations and Future of AI in MMLA

The examples presented in Section 9.4 portray a glimpse of the potential of using AI through MMLA to acquire new insights on the teaching/learning dynamics in physical, remote, and hybrid learning environments. A broad range of different MMLA applications can be found in the literature, focusing on innovative and ground-breaking subjects, such as teaching cognitive neuroscience (Segawa, 2019), quantifying collaboration quality in face-to-face classrooms (Chejara et al., 2020), engagement detection in online learning (Hasnine et al., 2021), detection of drowsy learning behaviors on e-learning platforms (Kawamura et al., 2021), and automatic sign language recognition (Hosain et al., 2019), among many others.

Although AI in MMLA shows a great potential for gaining a deeper understanding of learners' behaviors and their environments, it currently presents multiple limitations and challenges for its successful implementation and scalability in real contexts. Therefore, many of the MMLA studies found in the literature take place with a restricted number of participants and mostly in controlled environments. For MMLA to achieve its full potential, the following limitations need to be further analyzed and resolved:

- **Cost of technology**: High-quality and high-resolution devices such as physiological sensors and eye trackers are expensive. Thus, most MMLA studies focus on analyzing logs, video, audio, and gestures (Sharma & Giannakos, 2020). In order to have a more holistic picture of the factors, behaviors, and interactions taking place in learning environments, low-cost high-quality devices would be required. This would also allow researchers to conduct their studies with more participants.

- **Device integration**: The devices used in MMLA studies tend to be manufactured by different companies, which results in the requirement of different skills and resources to operate them and extract the collected data. This presents a challenge for conducting a MMLA study, as researchers need to find devices that can be easily integrated while looking for data quality and functionality for data processing.

- **Data fusion models**: Selecting the appropriate AI algorithms for performing a successful multimodal data fusion is a challenge (Mu et al., 2020), as data are collected from different sources that can have different timestamps, granularity, and noise sources. Most studies on MMLA opt to compare multiple models consisting of one or more AI algorithms to test their accuracy, where some of them are specifically designed for the learning environment being analyzed. This hinders the capability of reproducing such studies in different scenarios and therefore shows the necessity of more robust and generalized data fusion approaches.

- **Missing data**: The accuracy of AI models in MMLA depends on the amount of the analyzed data. Thus, a device malfunction during the collection or extraction phase can hinder the results of a study. AI models that take into consideration the loss of data would allow researchers to make predictions of learning indicators, even if based on incomplete datasets.

- **Personal data**: Multiple applications of AI in MMLA result in the collection of personal data that require consent from the students being analyzed. This poses an additional challenge to implement MMLA in real scenarios, as most of the technologies being used tend to rely on sensors that capture the student's voice, face gestures, body postures, eye gaze, and physiological biomarkers. Therefore, the way personal data are handled needs to be further studied and standardized.

- **Relationship to learning design**: Although AI in MMLA has been widely applied to understand students' behaviors and provide feedback on their interactions in multiple learning environments, little work has been done on relating MMLA with learning design (Mangaroska et al., 2020). This would allow instructors to develop learning activities that better suit their course objectives, classroom orchestration, and expected outcomes.

- **Scalability**: Most MMLA studies in the literature are conducted as case studies and experiments (Sharma & Giannakos, 2020), where a small group of participants is analyzed in a controlled environment. This is related to the difficulty of having ubiquitous sensors to conduct MMLA research in real-life contexts, along with the challenge of having fully integrated technologies and universal data fusion AI models. Thus, the scalability of MMLA relies on having affordable and easy-to-use devices and data-processing resources.

Assuming all the aforementioned challenges and limitations are tackled, we can expect the future of AI in MMLA to be truly revolutionizing for education. By having ubiquitous high-quality sensors in classrooms, we could be able to provide real-time feedback to the students and teachers involved in a learning activity. Moreover, the possibility of taking advantage of sensors in mobile devices and wearables would allow MMLA to be fully applicable to hybrid and remote learning environments, providing an enriched and adaptive learning experience. The future of AI in MMLA relies on the development of user-friendly scalable technology and standardized efficient AI models for data interpretation.

9.6 Closing Remarks

This chapter presented applications of AI in MMLA, a field of study that leverages technological devices to capture multimodal features for analyzing the behaviors, interactions, and factors that surround learning environments. MMLA seeks to bridge the gap between technology and pedagogy by translating multimodal features into learning indicators that provide actionable feedback for both teachers and students. AI plays a key role in MMLA, as appropriate algorithms and models are required to fuse and process the collected features into datasets that represent learning indicators.

Many applications of MMLA can be found in the literature, where features such as spoken interactions, body postures, face gestures, eye gaze, physiological biomarkers, and learning environment factors are jointly analyzed to gain a deeper understanding of the teaching and learning process. However, MMLA still presents limitations and challenges concerning the cost of technology, device integration, efficient data fusion models, handling with loss of data, personal data management, its relationship to learning design, and scalability. Thus, most applications of MMLA often take place in small groups with controlled environments.

The full potential of MMLA is yet to be seen, as low-cost high-quality devices and efficient standardized AI models are required for its successful implementation in real-life physical, remote, and hybrid learning environments. Furthermore, the challenge of personal data collection and management needs to be addressed and standardized for MMLA to be accepted in open learning spaces. These challenges are difficult and will surely take time, but new MMLA approaches, frameworks, methodologies, and research collaborations are continuously spawning around the world, which implies that the future of this ground-breaking field of study may be closer than we think.

References

Aguilar, S. J. (2022). Experimental Evidence of Performance Feedback vs. Mastery Feedback on Students' Academic Motivation. In *LAK22: 12th International Learning Analytics and Knowledge Conference*, pp. 556–562. Online, USA: Association for Computing Machinery. https://doi.org/10.1145/3506860.3506916.

Aronson, E. (2002). Chapter 10: Building Empathy, Compassion, and Achievement in the Jigsaw Classroom. In *Improving Academic Achievement*, pp. 209–225. Educational Psychology. Academic Press. https://doi.org/10.1016/B978-012064455-1/50013-0.

Azevedo, R. (2015). Defining and Measuring Engagement and Learning in Science: Conceptual, Theoretical, Methodological, and Analytical Issues. *Educational Psychologist*, 50(1), 84–94. https://doi.org/10.1080/00461520.2015.1004069.

Blikstein, P. (2013). Multimodal Learning Analytics. In *LAK13: 3rd International Learning Analytics and Knowledge Conference*, pp. 102–106. Leuven, Belgium: Association for Computing Machinery. https://doi.org/10.1145/2460296.2460316

Blikstein, P., and Worsley M. (2016). Multimodal Learning Analytics and Education Data Mining: Using Computational Technologies to Measure Complex Learning Tasks. *Journal of Learning Analytics*, 3(2), 220–238. https://doi.org/10.18608/jla.2016.32.11.

Boothe, M., Yu, C., Lewis, A., and Ochoa, X. (2022). Towards a Pragmatic and Theory-Driven Framework for Multimodal Collaboration Feedback. In *LAK22: 12th International Learning Analytics and Knowledge Conference*, pp. 507–513. Online, USA: Association for Computing Machinery. https://doi.org/10.1145/3506860.3506898.

Bruffee, K. A. (1984). Collaborative Learning and the "Conversation of Mankind". *College English*, 46(7), 635–652. https://doi.org/10.2307/376924.

Cao, C., Li, J., Zhu, Y., Gong, Y., and Gao, M. (2020). Evaluation of Online Teaching Platforms Based on AHP in the Context of COVID-19. *Open Journal of Social Sciences*, 08, 359–369. https://doi.org/10.4236/jss.2020.87029.

Chan, M. C. E., Ochoa, X., and Clarke, D. (2020). Multimodal Learning Analytics in a Laboratory Classroom. In *Intelligent Systems Reference Library*, 131–156. Germany: Springer Science / Business Media Deutschland GmbH. https://doi.org/10.1007/978-3-030-13743-4_8.

Chejara, P., Prieto, L. P., Ruiz-Calleja, A., Rodríguez-Triana, M. J., Shankar, S. K., and Kasepalu, R. (2020). Quantifying Collaboration Quality in Face-to-Face Classroom Settings Using MMLA. In *Collaboration Technologies and Social Computing*, pp. 159–166. Springer International Publishing. https://doi.org/10.1007/978-3-030-58157-2_11.

Chen, L., Feng, G., Joe, J., Leong, C. W., Kitchen, C., and Lee, C. M. (2014). Towards Automated Assessment of Public Speaking Skills Using Multimodal Cues. In *Proceedings of the 16th International Conference on Multimodal Interaction ICMI14*, pp. 200–203.Istanbul, Turkey: Association for Computing Machinery. https://doi.org/10.1145/2663204.2663265.

Choi, H., van Merriënboer, J. J. G., and Paas, F. (2014). Effects of the Physical Environment on Cognitive Load and Learning: Towards a New Model of Cognitive Load. *Educational Psychology Review*, 26, 225–244. https://doi.org/10.1007/s10648-014-9262-6.

Ciordas-Hertel, G., Rödling, S., Schneider, J., Di Mitri, D., Weidlich, J., and Drachsler, H. (2021). Mobile Sensing with Smart Wearables of the Physical Context of Distance Learning Students to Consider Its Effects on Learning. *Sensors*, 21(19). https://doi.org/10.3390/s21196649.

Cornide-Reyes, H., Riquelme, F., Monsalves, D., Noel, R., Cechinel, C., Villarroel, R., Ponce, F., and Munoz, R. (2020). A Multimodal Real-Time Feedback Platform Based on Spoken Interactions for Remote Active Learning Support. *Sensors*, 20(21). https://doi.org/10.3390/s20216337.

Cruz-Garza, J. G., Darfler, M., Rounds, J. D., Gao, E., and Kalantari, S. (2022). EEG-Based Investigation of the Impact of Room Size and Window Placement on Cognitive Performance. *Journal of Building Engineering*, 53, 104540. https://doi.org/10.1016/j.jobe.2022.104540.

De Silva, L. M. H., Chounta, I., Rodríguez-Triana, M. J., Roa, E. R., Gramberg, A., and Valk, A. (2022). Toward an Institutional Analytics Agenda for Addressing Student Dropout in Higher Education: An Academic Stakeholders' Perspective. *Journal of Learning Analytics*, 9(2), 179–201. https://doi.org/10.18608/jla.2022.7507.

Ez-zaouia, M., and Lavoué, E. (2017). EMODA: A Tutor Oriented Multimodal and Contextual Emotional Dashboard. In *LAK17: 7th International Learning Analytics and Knowledge Conference*, pp. 429–438. Association for Computing Machinery. https://doi.org/10.1145/3027385.3027434.

Fournier-Viger, P., Lin, J. C., Gomariz, A., Gueniche, T., Soltani, A., Deng, Z., and Lam, H. T. (2016). The SPMF Open-Source Data Mining Library Version 2. In *Machine Learning and Knowledge Discovery in Databases*, pp. 36–40. Springer International Publishing. https://doi.org/10.1007/978-3-319-46131-1_8.

Gray, G., and Bergner, Y. (2022). Chapter 2: A Practitioner's Guide to Measurement in Learning Analytics: Decisions, Opportunities, and Challenges. In *Handbook of Learning Analytics*, 2nd ed., pp. 20–28. Society for Learning Analytics Research (SoLAR). https://www.solaresearch.org/publications/hla-22/.

Hasnine, M. N., Bui, H. T. T., Tran, T. T. T., Nguyen, H. T., Akçapınar, G., and Ueda, H. (2021). Students' Emotion Extraction and Visualization for Engagement Detection in Online Learning. *Procedia Computer Science*, 192, 3423–3431. https://doi.org/10.1016/j.procs.2021.09.115.

Hosain, A. A., Santhalingam, P., Pathak, P., Košecká, J., and Rangwala, H. (2019). Sign Language Recognition Analysis using Multimodal Data. In *2019 IEEE International Conference on Data Science and Advanced Analytics (DSAA)*, pp. 203–210. https://doi.org/10.1109/DSAA.2019.00035.

Kawamura, R., Shirai, S., Takemura, N., Alizadeh, M., Cukurova, M., Takemura, H., and Nagahara, H. (2021). Detecting Drowsy Learners at the Wheel of e-Learning Platforms with Multimodal Learning Analytics. *IEEE Access*, 9, 115165–115174. https://doi.org/10.1109/ACCESS.2021.3104805.

Krumm, A. E., Everson, H. T., and Neisler, J. (2022). A Partnership-Based Approach to Operationalizing Learning Behaviors Using Event Data. *Journal of Learning Analytics*, 9(2), 24–37. https://doi.org/10.18608/jla.2022.6751.

Lang, C., Wise, A. F., Merceron, A., Gašević, D., and Siemens, G. (2022). Chapter 1: What Is Learning Analytics? In *Handbook of Learning Analytics*, 2nd ed., pp. 8–18. Society for Learning Analytics Research (SoLAR). https://www.solaresearch. org/publications/hla-22/.

Liu, R., Stamper, J. C., and Davenport, J. (2018). A Novel Method for the In-Depth Multimodal Analysis of Student Learning Trajectories in Intelligent Tutoring Systems. *Journal of Learning Analytics*, 5(1), 41–54. https://doi.org/10.18608/ jla.2018.51.4.

Loup-Escande, E., Frenoy, R., Poplimont, G., Thouvenin, I., Gapenne, O., and Megalakaki, O. (2017). Contributions of Mixed Reality in a Calligraphy Learning Task: Effects of Supplementary Visual Feedback and Expertise on Cognitive Load, User Experience and Gestural Performance. *Computers in Human Behavior*, 75, 42–49. https://doi.org/10.1016/j.chb.2017.05.006.

Mangaroska, K., Sharma, K., Gašević, D., and Giannakos, M. (2020). Multimodal Learning Analytics to Inform Learning Design: Lessons Learned from Computing Education. *Journal of Learning Analytics*, 7(3), 79–97. https://doi. org/10.18608/jla.2020.73.7.

Mercer, N., Littleton, K., and Wegerif, R. (2004). Methods for Studying the Processes of Interaction and Collaborative Activity in Computer-Based Educational Activities. *Technology, Pedagogy and Education*, 13(2), 193–209. https://doi. org/10.1080/14759390400200180.

Mory, E. H. (2004). Feedback Research Revisited. In *Handbook of Research on Educational Communications and Technology*. Lawrence Erlbaum Associates Publishers. https://members.aect.org/edtech/29.pdf.

Mu, S., Cui, M., and Huang, X. (2020). Multimodal Data Fusion in Learning Analytics: A Systematic Review. *Sensors*, 20(23). https://doi.org/10.3390/s20236856.

Munoz, R., Villarroel, R., Barcelos, T., Souza, A. A., Merino, E., Guiñez, R., and Silva, L. (2018). Development of a Software that Supports Multimodal Learning Analytics: A Case Study on Oral Presentations. *Journal of Universal Computer Science*, 24, 149–170. https://doi.org/10.3217/jucs-024-02-0149.

Ng, J., Wang, Z., and Hu, X. (2022). Needs Analysis and Prototype Evaluation of Student-Facing LA Dashboard for Virtual Reality Content Creation. In *LAK22: 12th International Learning Analytics and Knowledge Conference*, pp. 444–450. Online, USA: Association for Computing Machinery. https://doi. org/10.1145/3506860.3506880.

Nicoll, S., Douglas, K., and Brinton, C. (2022). Giving Feedback on Feedback: An Assessment of Grader Feedback Construction on Student Performance. In *LAK22: 12th International Learning Analytics and Knowledge Conference*, pp. 239–249. Online, USA: Association for Computing Machinery. https://doi. org/10.1145/3506860.3506897.

Nistor, N., and Hernández-García, A. (2018). What Types of Data Are Used in Learning Analytics? An Overview of Six Cases. *Computers in Human Behavior*, 89, 335–338. https://doi.org/10.1016/j.chb.2018.07.038.

Ochoa, X. (2022). Chapter 6: Multimodal Learning Analytics - Rationale, Process, Examples, and Direction. In *Handbook of Learning Analytics*, 2nd ed., pp. 54–65. Society for Learning Analytics Research (SoLAR). https://www.solaresearch. org/publications/hla-22/.

Ochoa, X., Chiluiza, K., Granda, R., Falcones, G., Castells, J., and Guamán, B. (2018). Multimodal Transcript of Face-to-Face Group-Work Activity around Interactive Tabletops. In *CEUR Workshop Proceedings*, 2163, pp. 1–6. 2nd Multimodal Learning Analytics across (Physical and Digital) Spaces, CrossMMLA 2018. http://ceur-ws.org/Vol-2163/paper4.pdf.

Ochoa, X., Domínguez, F., Guamán, B., Maya, R., Falcones, G., and Castells, J. (2018). The RAP System: Automatic Feedback of Oral Presentation Skills Using Multimodal Analysis and Low-Cost Sensors. In *LAK18: 8th International Conference on Learning Analytics and Knowledge*, pp. 360–364. LAK18. Association for Computing Machinery. https://doi.org/10.1145/3170358.3170406.

Praharaj, S., Scheffel, M., Drachsler, H., and Specht, M. (2018). Multimodal Analytics for Real-Time Feedback in Co-located Collaboration. In *Lifelong Technology-Enhanced Learning*, 187–201. Springer International Publishing. https://doi. org/10.1007/978-3-319-98572-5_15.

Prieto, L. P., Sharma, K., Kidzinski, L., Rodríguez-Triana, M. J., and Dillenbourg, P. (2018). Multimodal Teaching Analytics: Automated Extraction of Orchestration Graphs from Wearable Sensor Data. *Journal of Computer Assisted Learning*, 34(2), 193–203. https://doi.org/10.1111/jcal.12232

Rapanta, C., Botturi, L., Goodyear, P., Guàrdia, L., and Koole, M. (2020). Online University Teaching During and After the Covid-19 Crisis: Refocusing Teacher Presence and Learning Activity. *Postdigital Science and Education*, 2, 923–945. https://doi.org/10.1007/s42438-020-00155-y.

Rasipuram, S., and Jayagopi, D. B. (2020). Automatic Multimodal Assessment of Soft Skills in Social Interactions: A Review. *Multimedia Tools and Applications*, 79(19–20), 13037–13060. https://doi.org/10.1007/s11042-019-08561-6.

Samuelsen, J., Chen, W., and Wasson, B. (2019). Integrating Multiple Data Sources for Learning Analytics—Review of Literature. *Research and Practice in Technology Enhanced Learning*, 14, 11. https://doi.org/10.1186/s41039-019-0105-4.

Segawa, J. (2019). Hands-on Undergraduate Experiences Using Low-Cost Electroencephalography (EEG) Devices. *Journal of Undergraduate Neuroscience Education (JUNE)*, 17, A119–A124. Faculty for Undergraduate Neuroscience (FUN). https://www.ncbi.nlm.nih.gov/pmc/articles/PMC6650260/pdf/june-17-119.pdf.

Sharma, K., and Giannakos, M. (2020). Multimodal Data Capabilities for Learning: What Can Multimodal Data Tell us about Learning? *British Journal of Educational Technology*, 51(5), 1450–1484. https://doi.org/10.1111/bjet.12993.

Sher, V., Hatala, M., and Gašević, D. (2022). When Do Learners Study? An Analysis of the Time-of-Day and Weekday-Weekend Usage Patterns of Learning Management Systems from Mobile and Computers in Blended Learning. *Journal of Learning Analytics*, 9(2), 1–23. https://doi.org/10.18608/jla.2022.6697.

SoLAR. (2011). What is Learning Analytics? *Society for Learning Analytics Research*. https://www.solaresearch.org/about/what-is-learning-analytics/.

Vieira, F., Cechinel, C., Ramos, V., Riquelme, F., Noel, R., Villarroel, R., Cornide-Reyes, H., and Munoz, R. (2021). A Learning Analytics Framework to Analyze Corporal Postures in Students Presentations. *Sensors*, 21(4). https://doi.org/10.3390/s21041525.

Wehrmann, C., and van der Sanden, M. (2017). Universities as Living Labs for Science Communication. *Journal of Science Communication*, 16. https://doi. org/10.22323/2.16050303.

Yan, L., Martinez-Maldonado, R., Zhao, L., Deppeler, J., Corrigan, D., and Gašević, D. (2022). How Do Teachers Use Open Learning Spaces? Mapping from Teachers' Socio-Spatial Data to Spatial Pedagogy. In *LAK22: 12th International Learning Analytics and Knowledge Conference*, pp. 87–97. Online, USA: Association for Computing Machinery. https://doi.org/10.1145/3506860.3506872.

Zhao, L., Hwang, W., and Shih, T. K. (2021). Investigation of the Physical Learning Environment of Distance Learning Under COVID-19 and Its Influence on Students' Health and Learning Satisfaction. *International Journal of Distance Education Technologies*, 19, 77–98. https://doi.org/10.4018/IJDET.20210401.oa4.

10

Artificial Intelligence in Biomedical Research and Clinical Practice

Alejandro García-González

Tecnológico de Monterrey

CONTENTS

10.1 Introduction to Artificial Intelligence in Healthcare

Artificial intelligence (AI) has changed many human economic activities worldwide, creating solutions for old business structures and detonating new emergent companies. Artificial intelligence was born many decades ago (Meskó & Görög, 2020); however, the global digitalization process in private and public sectors has allowed its integration into the pipeline of many services (Lee & Yoon, 2021). The healthcare industry is not an exception; digitalization is considered an alternative to improve the national health systems (Lowery, 2020); nonetheless, regulatory affairs are still a challenge for new solutions based on artificial intelligence using these digital data sets, and compared with other industries, information is sensible in healthcare and specific protocols to protect, which must be carried out (Gerke et al., 2020).

Artificial intelligence is a vast, broad topic. Generally speaking, AI is the use of computers and technology (such as robots and autonomous systems) to simulate intelligent behavior comparable to a human being (European Commission. Joint Research Centre, 2020); consequently, characteristics such as reasoning, planning, learning, communication, and perception of the environment are mimicked by this type of mathematical algorithms, and the grade of their accuracy strongly depends on the quality of recovery data for a specific problem. In a superior layer, we found philosophical aspects and ethics; remarkable non-answered questions have emerged: Is ethics a learnable concept by an artificial intelligence algorithm? Could an artificial intelligence algorithm imitate at least some human ethics aspects? Who is responsible for the ethical response of an algorithm, the designer, the supplier, the users, or the group responsible for generating the data to train it? (Garcia-Gonzalez et al., 2022).

Some authors consider the classification of AI into artificial narrow intelligence (pattern recognition abilities in massive data sets), artificial general intelligence (comprehensive and total cognitive capacity), and artificial superintelligence (theoretically humanity´s combined capacity) (Meskó & Görög, 2020). Considering such classification, it is possible to affirm that we are in the first stage (Kulkarni et al., 2020). AI can be divided into subfields and techniques, like expert systems (which primarily rely on decision rules), machine learning (ML), and statistical methods; these techniques can be combined depending on the problem's characteristics or the ability background of the developer. Most of the current application of AI in medicine belongs to the implementation of ML; such ML algorithms model the relation between input and output information based on available data sets; for this reason, it has been accepted for application in the medical area due to that in medicine the standard approach to making a clinical decision is to establish clinical correlations associating patterns from an existing set of information (Amisha et al., 2019); traditionally, statistical methods are carried out to model such associations between symptoms, signals, and disease. AI, particularly ML and deep learning (DL), enriches the possibilities to assess and model healthcare scenarios.

AI algorithms perform two main tasks: the first is regression analysis and the second is classification or pattern recognition (Semmlow & Griffel, 2021). These two tasks aim to find an association between a set of variables called input variables, named "features" in the AI area, with an output variable also named "prediction;" for example, some features in healthcare are heart rate, sex, weight, blood glucose concentration, geometric properties of a tumor, number of erythrocytes in a blood sample to name a few. In the case of classifiers, this output variable corresponds to the assignment in a class (e.g., healthy or sick), so its nature is discrete data. Classification is the most frequent application that is carried out in the healthcare area using AI algorithms. To execute these tasks, the algorithms must be adjusted, which is known as training; this process consists of changing the values of the free parameters within the AI algorithm to reduce the error between the actual value of the class and the prediction or assigned class; this is known as a supervised learning algorithm, and artificial neural networks (ANNs) and support vectors machines (SVMs) are an example of structures adjusted by supervised learning methods; in this scenario, the greater the amount

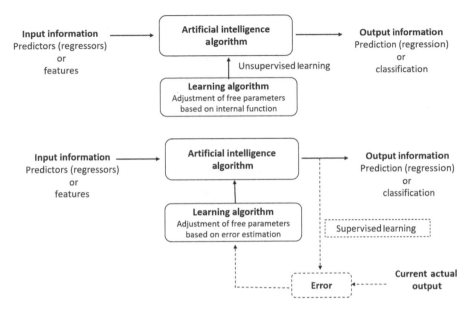

FIGURE 10.1
Supervised and unsupervised learning in ML algorithms to solve regression and classification problems.

and quality of data available, the better their performance in these tasks. There are two more learning methods (Meskó & Görög, 2020), unsupervised learning, used to identify clusters by grouping and to interpret data based solely on input data but without considering the previous assignation of a class, and reinforcement learning implemented to decision-making in a given environment and its execution to maximize its performance (Mak & Pichika, 2019). Figure 10.1 summarizes the two main tasks performed by an AI algorithm and the differences between supervised and unsupervised learning algorithms.

Figure 10.1 shows how unsupervised and supervised learning processes in ML algorithms are structured; a learning algorithm is needed to perform the free parameters adjustment for regression analysis and classification, or pattern recognition tasks performed by artificial intelligence.

Linear regression, logistic regression, decision trees, SVM, and ANN are examples of ML algorithms applied in medicine and healthcare (Perfecto-Avalos et al., 2019). The DL approach is more recent and can be considered part of ML. Its popularity is due to the ability to handle large data sets from which it is possible to find patterns of association and identification of features (input information), and the base structure of DL algorithms is ANN with some structural variation as convolutional ANN (Meskó & Görög, 2020). It is essential to mention that the main difference between the ML and DL algorithms is the feature extraction stage; in this stage, raw information is

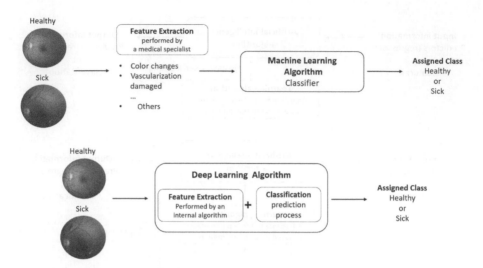

FIGURE 10.2
Differences between ML and DL algorithms.

processed to obtain numerical values that can be fed into the ML while preserving the representativity of original information; DL algorithm does not require the explicit declaration of these features, it is shown uncategorized data, and it must take care of finding characteristics within the information, which often turns them into black boxes where the structure of such patterns is not explicitly known. Figure 10.2 presents an example of classification between "sick" and "health" subjects; an eye image fundus is considered, and an ophthalmologist acquires this image to assess the retina health condition. In the first approach, an experimented physician is responsible for the features extraction step; among these features can be considered color, vascularization changes, or pigmentation; and with this information, a ML algorithm can be implemented. In the second approach, a DL algorithm is proposed; in this case, a feature extraction phase is not needed. Many studies must be presented to allow the algorithm to identify the possible relevant features to perform the classification task. It is important to remark that this set of features is not interpretable by a physician and is an internal set of information and part of the algorithm.

Figure 10.2 shows the class assignment for fundus eye images (retina images) of subjects with the presence of diabetic retinopathy (sick class) and healthy subjects (healthy class) by a ML algorithm and its differences in implementing a DL approach.

AI has a vast influence on the way new knowledge and solutions are being created in medicine. The number of possible applications includes clinical practice and basic biomedical research. Basic biomedical research provides a comprehensive and detailed understanding of the mechanisms that underlie the development and normal and pathological function of humans and other living organisms (National Research Council (U.S.) & National Academies Press (U.S.), 2011). AI algorithms analyze information from proteomics, genomics, and general molecular methods; this approach derives into new models that allow us to predict possible outcomes in simulated scenarios. New molecule synthesis and their potential therapeutic effects can now be predicted before they are tested in animal models (Mak & Pichika, 2019). AI is now integrated into the translational research process; translational research is the bridge between basic biomedical research and clinical practice, transforming research innovations created by basic biomedical science into new health products and diagnostic and therapeutic methods. The current most common application of AI is in clinical practice influencing the way biomedicine is performed by medical doctors and other healthcare professionals such as nurses, pharmacists, and therapists. Considering this scenario in this chapter, we discuss the influence of health digitalization and the changemaker paradigm of person-centered attention in how AI can contribute effectively to accelerating the creation of healthcare solutions.

10.2 Digital Health, Person-Centered Attention Paradigm, and Artificial Intelligence

The increase in artificial intelligence solutions in healthcare is not a coincidence, but the consequence of the increase in the digitization of the national health systems; this approach is a response to the need to improve the scope and quality of care since the demand has increased and the resources are diminishing (Alami et al., 2017). Health national systems around the world are unsustainable; the last COVID-19 pandemic revealed their incapacity to respond effectively under pressure. The physician-centered paradigm of attention is becoming inoperable; this kind of paradigm is wholly focused on diagnosis, management, and treatment of diseases but not on prevention; all stakeholders in the healthcare process, such as hospitals, insurance companies, pharmaceuticals, and medical device companies, produce services, and products are designed and oriented to attend physician necessities and no patient necessities directly. Person-centered medicine has emerged as the answer to change the future of National Health Systems; in this paradigm, prevention and prediction are the keys to improving patients' attention, involving them in the process, and sharing the responsibility of managing their healthcare and that of their families (Kostkova, 2015). In this sense, digitalization plays a central role; the World Health Organization (WHO) has developed specific digital health strategies to improve access to healthcare, reduce inefficiencies, improve the quality of care, lower costs, and provide more personalized healthcare (Lowery, 2020). Nevertheless, what is considered digital health? Digital health is the concept that incorporates information and communication technologies into products, services, and processes of medical care, representing new technology-driven and data-driven approaches to monitoring and improving patient and population health (Alami et al., 2017). Information and communication technologies involve computers, storage, networking, and other physical devices, infrastructure, and processes to create, process, store, secure, and exchange data. The DH scope includes telemedicine and telehealth, remote patient monitoring, mobile healthcare applications, individual-level and population-level analysis, and social networking (Lowery, 2020).

Digitalization has been crucial to the implementation of AI in healthcare; with this, data generation and storage will be enough to consider the use of algorithms of ML. Digitalization occurs at three levels: patient digitalization, medical devices digitalization, and medical services digitalization (Bhavnani et al., 2016). The digitization of patients implies self-monitoring, changes in their routine with a positive effect on their health, development of digital attachment, and the ability to receive feedback from remote health personnel (telemedicine); the patient becomes a generator of data that can be managed and with local decisions through artificial intelligence; for instance, according to the WHO, 60% of factors related to individual health and quality of life correlate to lifestyle factors (Farhud, 2015), and AI solutions can provide lifestyle interventions during the day based on physiological information registered by mobile phones, wearable technologies, and even intelligent autonomous robotic assistants (Garcia-Gonzalez et al., 2022). Digitalizing the devices implies the development of simplified designs with easily interpretable information, preferably with interconnection to the cell phone. In this sense, the lab-on-a-chip systems stand out, and in general, the point of care solutions (POCS), highlight the ability to acquire medical images such as portable ultrasound equipment that can be interconnected to the cell phone, what is called point of care ultrasound (POCUS), or the increment of a rapid test to detect molecules for preliminary screening of a disease, for instance, antigen test used in the COVID-19 pandemic.

The digitization of processes within health organizations implies the integration and interoperability of platforms, the management of medical and administrative information, the implementation of integrated work cycles for patient care (for instance, nursing management), and the creation of protected databases (medical records administration), with information available for analysis and developing of precision medicine or personalized medicine. Additionally, the healthcare sector is living through the transformation of the "Medical Internet of Things." In the "Internet of Things" framework, medical devices are interconnected by the Internet service; ambulatory and non-ambulatory devices could share their status information (accuracy, calibration, fails), and the data set from clinical measures. Figure 10.3 summarizes the integration of digitalization, person-centered attention paradigm, and artificial intelligence, considering, for example, the use of an

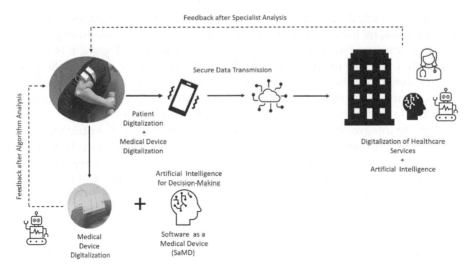

FIGURE 10.3
Digitalization process centered on the person, not the physician, and artificial intelligence inclusion on patient digitalization, device digitalization, and healthcare services digitalization.

electromyography device, a medical device used to register the electrophysi-ological response of a muscle. The medical design is simple and efficiently operated by patients at home without the supervision of a specialist. The digitization process occurs through the interconnection with a cellphone. A secure wireless channel can transmit the data to a central facility or direct it to the specialist; the physician or therapist can give feedback via digital media, which can be the same cellphone, and provide some recommendations for correction to the rehabilitation process. With this, the patient is admitting the responsibility for his care and participates in the central part of the atten-tion cycle. The incorporation of AI in the process from the patient side allows obtaining immediate feedback by the algorithm included in the device, con-sidering that such an AL algorithm can emit recommendations based on its training considering a validation carried out in a clinical protocol with the supervision of specialists. This possibility has led the regulatory agencies to create the term Software as a Medical Device (SaMD) to categorize such software as capable of screening, diagnostic, and emitting recommendations without the assistance of an expert physician (Gerke et al., 2020); as will be discussed, this is a considerable challenge in terms of patient security. On the other hand, the digitalization of healthcare services comprehends not exclu-sively the administrative process but the clinical decision-making process. In this sense, AI can be part of expert systems with or without the participa-tion of specialists to propose a preliminary diagnostic. Furthermore, AI algo-rithms can be embedded into robotic assistants, an option in the healthcare facility or at home. In any of these proposals, the objective is to improve the efficiency of patients' diagnostic and treatment cycles.

Figure 10.3 details the inclusion of AI algorithms in the digitization process of healthcare, including persons, medical devices, and the attention process. AI could improve the response to healthcare attention. In Figure, it is appreciated that SaMD can give direct feedback to the patient without the physician's assistance; in a more advanced system, this assistive AI algorithm can be embedded into robotic assistants. AI in healthcare facilities can be included to take clinical advice, screening, and diagnostics.

With the integration of digitalization and AI, the hospital-centric vision of healthcare is transforming into a more virtual and distributed cycle of attention. Technologies such as virtual reality and robotic assistants are contributing to preserving the human interaction between physicians and healthcare providers with patients.

10.3 Regulatory Affairs Implementing Artificial Intelligence in Medical Devices

According to the WHO, "A medical device can be any instrument, apparatus, implement, machine, appliance, implant, reagent for *in vitro* use, software, material or other similar or related article, intended by the manufacturer to be used, alone or in combination for a medical purpose." Today, an estimated two million different medical devices are on the world market, categorized into more than 7,000 generic device groups (*Medical Devices*, 2022). As has been exposed, prediction and categorization skills of AI have begun to be integrated directly into medical devices, allowing them to predict a potential health risk event and diagnose disease, and in the case of robotic systems, even the ability to intervene in treatment (Garcia-Gonzalez et al., 2022). This scenario implies a significant challenge for developers and relevant changes in the regulatory processes that every company that markets medical devices must comply with. In the previous section, the regulatory concept of SaMD was used; let us define it according to the International Medical Device Regulators Forum as "SaMD is software intended to be used for one or more medical purposes that perform these purposes without being part of a hardware medical device." (Gerke et al., 2020). It means this is not software for the correct operation of the medical device, but software designed to make diagnostic or treatment decisions without the intervention of a human, partially or entirely. As with any other medical device, SaMD must be evaluated to determine the level of risk involved in its use. With slight differences between European regulations and those established by the FDA, we can consider medical devices belonging to classes I, II, and III, going from a lower to a higher level of risk when using it. Then, the regulatory authority must determine whether to grant the privilege of sale as a medical device by granting sanitary registration. Some of the conditions have been established to have a framework that allows the evaluation of SaMD (Zanca et al., 2022): (i) establish the intention of use (target population), (ii) specifications on the implemented artificial intelligence model, (c) evidence of clinical tests and information sets used in its training, (d) requirements on cyber security, installation, and protection of the sensitive information, (e) maintenance and updating plan that implies a recalibration of the training of the algorithms when they are submitted to new databases, and (f) support provided by the manufacturer and marketers. For example, in Mexico, this has implied modifications to the national regulations by the Protection Commission for Sanitary Risk (COFEPRIS for its acronym in Spanish), explicitly changing what refers to good manufacturing practices for medical devices to incorporate the creation of this type of solution. This forces the creators to have the corresponding certifications even when the solution is software. All regulatory entities are considering these action plans. It has been deployed into the market to meet the entire chain of creation and life of AI SaMD medical devices, from the design stage up to techno-surveillance ones.

Some authors have proposed two subtypes of SaMD, closely related to the way they are trained before and after their deployment; type A SaMD has been trained with clinical data during design and development but whose performances are frozen at product validation, and type B SaMD that can collect actual world data during its clinical use, but, to use such data to ameliorate its performances, a human intervention (either of the manufacturer or the user) is necessary to retrain and revalidate the AI medical device (Zanca et al., 2022). Other authors have proposed a different classification according to the five levels of automation, considering two main categories, assistive and autonomous algorithms. Assistive AI is subdivided into data presentation and clinical decision support, while autonomous are subdivided into conditional automation, high automation, and full automation (Bitterman et al., 2020).

The following two sections focus on presenting the panorama of AI in medical imaging analysis. In this clinical area, AI has had the most important applications and influence of AI in drug development, whose relevance is transcendental considering the enhancement of basic biomedical research into translational medical applications provided by the inclusion of AI strategies in the design pipeline and production.

10.4 Radiomics and AI, the Most Extended Application

The most expansive area of application of artificial intelligence, encompassing the most significant number of proposals turned into SaMD, has been the area of medical imaging (Muehlematter et al., 2021). More than 90% of medical data is obtained from medical imaging, which is essential for doctors in diagnosis. This growing interest in medical image analysis has led to a new scientific term known as Radiomics. Radiomics is emerging translational research aiming to find associations between qualitative and quantitative information extracted from clinical images and clinical data, with or without associated gene expression, to support evidence-based clinical decision-making (Rizzo et al., 2018). This analysis process starts by acquiring some properties of the body, such as mass density, electron density, proton density, blood volume or flow volume, and temperature, among others. This tissue property detection and mapping are performed by specific electronic instrumentation after an energy source stimulates the body. Common energy sources can be divided into ionizing as X-rays or gamma rays or non-ionizing as ultrasound or magnetic fields. Imaging devices include emission and detection systems, except when the energy source is a radiotracer that is administered to the patient; a radiotracer is a chemical compound capable of emitting radiation, and an external process prepares it. After detection, the electronic signals proportional to the tissue property are digitized and processed in a computer. A mathematical algorithm converts such records into a representation of the information mapped over the space in the form of an image, an example of which is the retro-projection algorithms used in all tomographic systems that assign a position and brightness value to each calculation of tissue property of radiation attenuation. Therefore, medical images represent a tissue property in response to an energy stimulus constructed by an algorithm. Medical images are digitized and converted into information that can be interpreted by algorithms such as ML. Figure 10.4

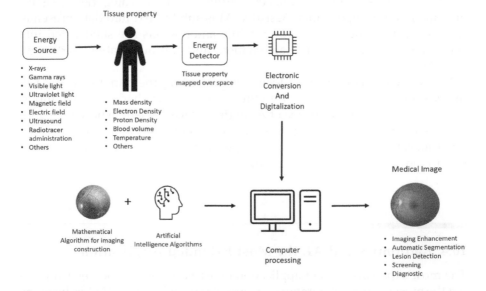

FIGURE 10.4
General process of constructing medical images representing the spatial distribution of a body's tissue property and AI incorporation to enhance the clinical result.

outlines this process from the source of energy to the medical image construction. ML has facilitated information processing, particularly image segmentation, a specific operation to separate relevant information from the rest of the acquired data, such as an injury or a tumor. ML also allows us to enhance the properties of the medical image as contrast or brightness, as well as propose screening methods and, in the case of clinical protocols, validate the possibility of diagnosis. This last characteristic is still controversial in the radiologist community. Figure 10.1 shows a schematic representation of the general process of constructing a medical imaging and the current tasks pursued by the artificial intelligence algorithms as image enhancement, screening (identifying possible subjects with a particular condition), and the final diagnostic task, which is still controversial. An alternative better accepted by radiology specialists has been a guided segmentation of a lesion by a radiologist with a subsequent classification of AI confirmed by the radiologist; this semi-automatic decision process has proved to perform better than the automatic approach (Rizzo et al., 2018).

To understand the process of implementing a ML algorithm in medical imaging analysis, we must consider that the minimum information in a digital image is called a pixel or picture element representing a specific point of reality; its size depends on the characteristics of the detection and digitization system; and in medical imaging, this pixel maps the average tissue property in that specific spatial point; for some medical imaging modalities, it is also possible to reconstruct a volume as in tomography. The minimum amount of information is called a voxel (volume element); in this example, a voxel represents in a brightness scale the average of the X-ray attenuation coefficient of the tissue in a specific minimal volume. Any dedicated software for medical imaging analysis aims to distinguish between normal and sick conditions, such as tumor regions, and some functional processes, such as the determination of blood flow changes, were assessed by Doppler ultrasound or magnetic resonance imaging (MRI), for instance. Feature extraction from images and volumes must be performed to do these tasks. There are four main categories of features in the radiomic area, and they are based on pixel and voxel properties. The first category involves shape features which focus on the geometric properties of the image or the reconstructed volume (volume, diameters, surface, sphericity of tumors); the second category corresponds to first-order statistics features, which focus on the statistical distribution of the pixels, and descriptive statistics of the properties of intensity in the image such as mean and median. The third category are mainly second-order statistics features, derived from the calculation of relationships between groups of nearby voxels, and they provide a measure of the spatial intensities of the voxels; the fourth category includes high-order statistics features, obtained by different methods after applying filters and mathematical transforms over pixels and voxel values.

Figure 10.4 depicts the process of stimulating a body's tissue with electromagnetic or mechanical energy whose response is detected, mapped, and transformed into a medical image for medical diagnosis, and inclusion of AI is helping to enhance the outcome of this process; however, the real value always is dependent on the clinical relevance, and it means the improvement of better and faster diagnosis.

As discussed in this chapter, feature extraction is critical to implement regular ML algorithms for classification and regression. However, the amount of information for the training stage provokes two challenges; on the one hand, the ML algorithm needs a features extraction stage, as has been described, this leads to some of the fourth main types of features, and the computational cost to process each image is high. In the case of the DL approach, the number of necessary medical images increases considerably, and large data sets are needed for training a DL algorithm. Applications of AI and ML exist from the simplest planar radiography (Rx) to the most complex functional imaging modality as MRI; however, the lack of well-labeled data sets and erroneous diagnostics in public databases is the typical situation. DL is the applied tool for the non-interpretative task, focused on improving the workflow in the radiology department, not on interpreting studies. Ophthalmology is one of the pioneer areas including AI, particularly for diabetic retinopathy detection; companies such as IDx obtained the approval of the FDA for its IDx-DR cloud-based AI system in April 2018 (Kulkarni et al., 2020) being the first software authorized to provide a screening decision. Other subdisciplines such as dermatology and pathology are beginning to include AI and ML algorithms to reduce the time of screening and diagnosis.

The number of AI applications in medical imaging that have been approved as a medical device (SaMD) is minimal, with 129 in the USA up to March 2020 (Muehlematter et al., 2021), contrasting with the significant increase in scientific publications in the area; it means low clinical relevance of all these applications. This situation has led different authors to be cautious or even skeptical about the reported results, mainly when claims about the possible replacement of radiologists in diagnosis by any of these algorithms are included. Evidence of meta-analysis has shown that the performance of DL compared with the radiologist is equivalent. However, most of the studies do not have a reliable design, and training is carried out exclusively in one type of population with no generalization. The basis of this study must be a rigorous clinical trial.

10.5 New Drugs' Development and AI

The use of AI and ML algorithms in drug discovery development has revolutionized the way *in-silico* platforms (high-performance computational simulations) are considered in this industrial sector, mainly thanks to AI and ML's capacity to deal with big data-driven. Pathway to implement AI and

ML follows the same rationale as in any other application, but the difference consists in the type of data sets; in this case, single-nucleotide polymorphisms, gene copy numbers, and gene expression data, including features as genomic variant types such as point mutations, deletions, insertions, and translocations of gene sequences, are possible molecular targets for a new drug (Koromina et al., 2019). The main stages of drug development consist of four steps (Chen et al., 2021): (i) drug discovery by *in-silico* experimentation, where a molecular target is identified, and a set of possible compounds are tested and modified, and in this step, ML can contribute to accelerating the time to assess different novel biomarkers and molecules candidates; (ii) preclinical research, consisting of *in-vitro* (cell cultures) and *in-vivo* (animal models) experimentation, because of better *in-silico* experimentation with ML algorithms, reduces the number of experiments required in this step and is optimized compared with regular workflow; (iii) clinical studies, which involve three phases and correspond to controlled experimentation in humans, and ML could contribute to optimizing the outcome, reducing costs, time, and risk (Mak & Pichika, 2019); Phase I - first time in humans for safety assessment, a reduced number of experiments are carried out, Phase II - proof of concept, assessment of molecule effectiveness for treatment, Phase III - comparison of effectiveness against the available gold standard solution, in which, ML has increased its presence to design and optimize the sample size considering eligible population and prescreening (Harrer et al., 2019); (iv) pharmaco-surveillance (post-marketing studies), drug safety evaluation over time, is also known as Phase IV study, in which, ML could allow the detection of adverse events. This new approach to developing drugs has promoted partnerships between the pharma industry and AI companies such as DeepMind Technologies, a subsidiary of Google. However, up to today, approved results derived from this alternative have not been granted by the authority. The first AI-designed drug candidate to enter clinical trials was reported by Exscientia in early 2020.

10.6 Challenges to Developing and Implementing AI in Medical Areas

Even with the remarkable advances that technologies based on AI have reached and the increased number of applications in medicine, researchers, developers, and companies face a significant challenge. Some authors have proposed a checklist for revising any scientific publication in this area (Mongan et al., 2020). Among the proposed criteria is recommended the correct identification from the title of the methodology being applied, defining that regardless of whether it is an algorithm to be evaluated, it must follow the structure of any other clinical research, including the correct identification of the type of study, as prospective or retrospective. Declaration of the data source and the inclusion and exclusion criteria of the subjects in the clinical trial. In the same way, establish metrics that allow assessing the algorithm's performance in the training, evaluation, and deployment stages.

Large and standardized data sets are standard in many other applications, such as social media analysis. However, that is not the case for the medical area; medical records belong to patients and can be used for research and development only under a rigorous approved clinical protocol; the corresponding permission must be requested from the ethics committee (Kelly et al., 2019). Even when data are available and authorized, there is no standard template or format used by medical devices; the same study recorded in different equipment from a different supplier can result in a different type of information, which adds a preprocessing step to be used. Minimal variations among study conditions, as can be the number of projections in a tomographic image study, could result in non-equivalent information, invalidating the possible outcome; the same algorithm could give different results evaluating the same case. An example of these challenges was the lack of information during the last COVID-19 pandemic, where attempts to create a diagnostic method based on planar X-ray or tomographic image were not achievable (Dong et al., 2021).

AI diagnostic methods' high sensibility and specificity do not assure a generalization on its use; algorithms are frequently trained using a specific local population with a particular ethnicity, representing a severe bias problem with ethical implications. The number of available cases for rare diseases limits obtaining an accurate ML model since the trained model is evaluated by a cross-validation process. This implies that the available data set is split into two sets: one is used for training and the other one for testing the fitted model; this process is repeated, randomizing the elements in each of the two sets each time; with limited elements in a data set, it is impossible to trust on the generalizable result.

AI algorithms, particularly those based on DL, are considered black boxes without the option to be explainable to the medical community. In most ML applications, it is not easy to explain how the outcome is obtained—given a result but without the construction of new knowledge to understanding the physiology and physiopathology events related to the diagnostic.

10.7 Conclusion

Undoubtedly, artificial intelligence is here to stay in medical practice and the entire structure of the national health system, impacting not only diagnosis but the entire care process and even encompassing the new way of creating personalized treatments. However, its relevance will not be such if compliance with the clinical validation requirements is not ensured, such as the strict execution of the protocols necessary for its validation, considering all bioethical implications, additionally looking for methods that are explanatory and understandable for the community.

References

Alami, H., Gagnon, M.-P., & Fortin, J.-P. (2017). Digital health and the challenge of health systems transformation. *MHealth, 3*, 31–31. https://doi.org/10.21037/mhealth.2017.07.02.

Amisha, Malik, P., Pathania, M., & Rathaur, V. (2019). Overview of artificial intelligence in medicine. *Journal of Family Medicine and Primary Care, 8*(7), 2328. https://doi.org/10.4103/jfmpc.jfmpc_440_19.

Bhavnani, S. P., Narula, J., & Sengupta, P. P. (2016). Mobile technology and the digitization of healthcare. *European Heart Journal, 37*(18), 1428–1438. https://doi.org/10.1093/eurheartj/ehv770.

Bitterman, D. S., Aerts, H. J. W. L., & Mak, R. H. (2020). Approaching autonomy in medical artificial intelligence. *The Lancet Digital Health, 2*(9), e447–e449. https://doi.org/10.1016/S2589-7500(20)30187-4.

Chen, Z., Liu, X., Hogan, W., Shenkman, E., & Bian, J. (2021). Applications of artificial intelligence in drug development using real-world data. *Drug Discovery Today, 26*(5), 1256–1264. https://doi.org/10.1016/j.drudis.2020.12.013.

Dong, D., Tang, Z., Wang, S., Hui, H., Gong, L., Lu, Y., Xue, Z., Liao, H., Chen, F., Yang, F., Jin, R., Wang, K., Liu, Z., Wei, J., Mu, W., Zhang, H., Jiang, J., Tian, J., & Li, H. (2021). The role of imaging in the detection and management of COVID-19: A review. *IEEE Reviews in Biomedical Engineering, 14*, 16–29. https://doi.org/10.1109/RBME.2020.2990959.

European Commission. Joint Research Centre. (2020). *AI watch: Defining artificial intelligence: Towards an operational definition and taxonomy of artificial intelligence.* Publications Office. https://data.europa.eu/doi/10.2760/382730.

Farhud, D. D. (2015). Impact of Lifestyle on Health. *Iranian journal of public health, 44*(11), 1442.

Garcia-Gonzalez, A., Fuentes-Aguilar, R. Q., Salgado, I., & Chairez, I. (2022). A review on the application of autonomous and intelligent robotic devices in medical rehabilitation. *Journal of the Brazilian Society of Mechanical Sciences and Engineering, 44*(9), 393. https://doi.org/10.1007/s40430-022-03692-8.

Gerke, S., Babic, B., Evgeniou, T., & Cohen, I. G. (2020). The need for a system view to regulate artificial intelligence/machine learning-based software as medical device. *NPJ Digital Medicine, 3*(1), 53. https://doi.org/10.1038/s41746-020-0262-2.

Harrer, S., Shah, P., Antony, B., & Hu, J. (2019). Artificial intelligence for clinical trial design. *Trends in Pharmacological Sciences, 40*(8), 577–591. https://doi.org/10.1016/j.tips.2019.05.005.

Kelly, C. J., Karthikesalingam, A., Suleyman, M., Corrado, G., & King, D. (2019). Key challenges for delivering clinical impact with artificial intelligence. *BMC Medicine, 17*(1), 195. https://doi.org/10.1186/s12916-019-1426-2.

Koromina, M., Pandi, M.-T., & Patrinos, G. P. (2019). Rethinking drug repositioning and development with artificial intelligence, machine learning, and omics. *OMICS: A Journal of Integrative Biology*, 23(11), 539–548. https://doi.org/10.1089/omi.2019.0151.

Kostkova, P. (2015). Grand challenges in digital health. *Frontiers in Public Health*, 3. https://doi.org/10.3389/fpubh.2015.00134.

Kulkarni, S., Seneviratne, N., Baig, M. S., & Khan, A. H. A. (2020). Artificial intelligence in medicine: Where are we now? *Academic Radiology*, 27(1), 62–70. https://doi.org/10.1016/j.acra.2019.10.001.

Lee, D., & Yoon, S. N. (2021). Application of artificial intelligence-based technologies in the healthcare industry: Opportunities and challenges. *International Journal of Environmental Research and Public Health*, 18(1), 271. https://doi.org/10.3390/ijerph18010271.

Lowery, C. (2020). What is digital health and what do i need to know about it? *Obstetrics and Gynecology Clinics of North America*, 47(2), 215–225. https://doi.org/10.1016/j.ogc.2020.02.011.

Mak, K.-K., & Pichika, M. R. (2019). Artificial intelligence in drug development: Present status and future prospects. *Drug Discovery Today*, 24(3), 773–780. https://doi.org/10.1016/j.drudis.2018.11.014.

Medical devices. (2022). WHO Medical Devices. https://www.who.int/health-topics/medical-devices#tab=tab_1.

Meskó, B., & Görög, M. (2020). A short guide for medical professionals in the era of artificial intelligence. *NPJ Digital Medicine*, 3(1), 126. https://doi.org/10.1038/s41746-020-00333-z.

Mongan, J., Moy, L., & Kahn, C. E. (2020). Checklist for artificial intelligence in medical imaging (CLAIM): A guide for authors and reviewers. *Radiology: Artificial Intelligence*, 2(2), e200029. https://doi.org/10.1148/ryai.2020200029.

Muehlematter, U. J., Daniore, P., & Vokinger, K. N. (2021). Approval of artificial intelligence and machine learning-based medical devices in the USA and Europe (2015–20): A comparative analysis. *The Lancet Digital Health*, 3(3), e195–e203. https://doi.org/10.1016/S2589-7500(20)30292-2.

National Research Council (U.S.) (2011). *Research training in the biomedical, behavioral, and clinical research sciences*.

Perfecto-Avalos, Y., Garcia-Gonzalez, A., Hernandez-Reynoso, A., Sánchez-Ante, G., Ortiz-Hidalgo, C., Scott, S.-P., Fuentes-Aguilar, R. Q., Diaz-Dominguez, R., León-Martínez, G., Velasco-Vales, V., Cárdenas-Escudero, M. A., Hernández-Hernández, J. A., Santos, A., Borbolla-Escoboza, J. R., & Villela, L. (2019). Discriminant analysis and machine learning approach for evaluating and improving the performance of immunohistochemical algorithms for COO classification of DLBCL. *Journal of Translational Medicine*, 17(1), 198. https://doi.org/10.1186/s12967-019-1951-y.

Rizzo, S., Botta, F., Raimondi, S., Origgi, D., Fanciullo, C., Morganti, A. G., & Bellomi, M. (2018). Radiomics: The facts and the challenges of image analysis. *European Radiology Experimental*, 2(1), 36. https://doi.org/10.1186/s41747-018-0068-z.

Semmlow, J. L., & Griffel, B. (2021). *Biosignal and medical image processing*. CRC Press.

Zanca, F., Brusasco, C., Pesapane, F., Kwade, Z., Beckers, R., & Avanzo, M. (2022). Regulatory Aspects of the Use of Artificial Intelligence Medical Software. *Seminars in Radiation Oncology*, 32(4), 432–441. https://doi.org/10.1016/j.semradonc.2022.06.012.

11

The Dark Side of Smart Cities

Karen Hinojosa-Hinojosa and Talía González-Cacho

Tecnológico de Monterrey

CONTENTS

11.1 Introduction

Cities are growing, multiplying, increasing in population and complexity (Bettencourt, 2021). Urban planning, in its attempt to manage this ever-increasing complexity, has been integrating artificial intelligence (AI) in its practice for decades. Since urban systems exhibit predictable patterns, AI is attractive to city planners because it enables large data processing capacities in a significantly lower time than it would take human experts to process (Sanchez et al., 2022). But just as cities are growing in complexity, so are AI models, resulting in processes that are concealed from users and decision-makers, making them difficult to comprehend (Linardatos et al., 2020; Mhlanga, 2022).

Artificial intelligence is hard to define, because the boundary of what is considered AI is constantly pushed forward as it is incorporated into mainstream use. Within urban planning, Geographic Information Systems (GIS) and the multiple algorithms that enable these programs are such basic tools that they probably would not be considered AI by most of its users. One of the most widely accepted definitions is that AI is "a system's ability to interpret external data correctly, to learn from such data, and to use those learnings to achieve specific goals and tasks through flexible adaptation." (Haenlein & Kaplan, 2019). By itself, AI is neither ethical nor evil, but this does not mean that AI's nature is neutral: It is as unbiased as the data we input, as the rules we assign for its operation.

Using AI in decision-making can produce discriminatory results mainly two ways: either because of the biased data or instructions as humans' input, commonly described as a case of "garbage in, garbage out" data (Mellin, 1957), or because we do not know what the algorithm is doing or how it's arriving at decisions, described as the "black box" problem (A. Rai, 2020). Therefore, social inequities can be repeated or even amplified through a lack of diversity in the datasets and in the programmers, and we may be unable to detect them.

Artificial intelligence, big data, the internet of things (IoT), and machine learning coalesced into an ambiguous narrative of Smart Cities that gained popularity in the past decades, peaking around 2015, as Figure 11.1 shows. That year, the United States launched a Smart Cities Initiative, which consisted of $160 million dollars in federal funds for communities to tackle local challenges together through technological collaborations with industry (Secretary, 2015). Smart Cities have captivated adopters with the promise of economic development through efficient resource management, all while contributing to the quality of life of urban dwellers (Allam, 2018). More than 20 years after the introduction of the first Smart Cities, the achievement of it in terms of improving quality of life or economic prosperity is unclear (Cavada et al., 2014; Neirotti et al., 2014) except drastically expanding technology deployment in their self-styled cities (Buyanova et al., 2020) and improving urban cyber-physical integration (Anthopoulos, 2017).

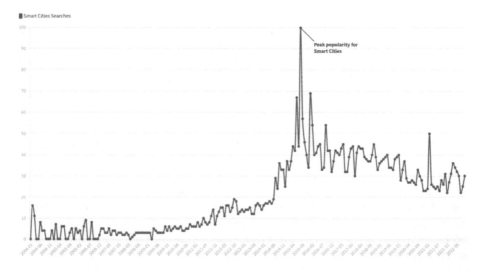

FIGURE 11.1
Smart Cities Google popularity search from 2004 to 2022 (Google Trends, 2022). Note: Popularity gained by the narrative of Smart Cities in the last decades.

As much as the term Smart City became popular as an urban branding and marketing strategy (Anttiroiko, 2015), it also generated criticism because the technologies used within them are typically "envisaged by corporate firms that push them ubiquitously into cities" (Kummitha, 2018). There is also increasing skepticism of the smart city's overestimated potential to generate well-being and wealth (Anthopoulos, 2017; Datta, 2015), especially for already underprivileged social groups (Masucci et al., 2020). These authors join a larger group of researchers outside of the urban planning discipline questioning how AI can reproduce systems of race (Buolamwini, 2017) gender (D'ignazio & Klein, 2020) and socioeconomic inequality (Joyce et al., 2021). Presently, smart cities seem incapable of confronting the complexity, diversity, and uncertainty of contemporary cities (Fernández-Güell et al., 2016). However, it might be that the Smart City's "deferred potentiality," the promise of what it could yet become, is its fundamental mode of existence (Krivý, 2018), and what ultimately drives more sales.

After the Habitat III conference in Quito in 2016 and the subsequent release of the New Urban Agenda (UN Habitat, 2016), the smart city narrative in urban planning has lost ground to terms such as inclusive, sustainable, resilient, and vibrant cities, toward smarter citizens and communities (De Filippi et al., 2019), although the technology still retains an important enabling role in these urban future visions (Toli & Murtagh, 2020). Artificial intelligence will be increasingly present in companies, governments, and cities worldwide, and early reports forecast that the timeline of AI adoption could widen economic development gaps between countries (Bughin et al., 2018). Considering this scenario, urban planners, geographers, sociologists, and allied professionals need a strong understanding of the externalities generated by AI on cities and societies (Andrews et al., 2022).

The present chapter aimed to discover how AI has been integrated into urban planning, and what resistances are there to the implementation of AI in this disciplinary practice. To that aim, we posit the following research questions: What kinds of urban planning problems is AI being used on? What resistances have been called out in academic literature to the implementation of AI, particularly in urban planning? What can we learn from the AI that is already implemented in urban planning? The chapter is sectioned as answers to each of those questions, closing with limitations of this study and takeaways for researchers, practitioners, and city stakeholders.

A mixed methods approach from a qualitative perspective was used to address this research. A critical systematic literature review was conducted, following the methods proposed by Xiao and Watson (2019) to answer the first two research questions. The inclusion criterion was research published in the SCOPUS database from 2012 to 2022 with two search strings considered, the first one including "Artificial Intelligence" AND "urban planning" AND "problems", and the second one "Artificial intelligence" AND "urban planning" AND "bias". A case study methodology gave insight into how funding, data-gathering, implementation, data use, and regulation are currently working within the urban planning discipline. The information gathered was triangulated to understand how AI bias risk can be mitigated to truly benefit cities and their citizens.

11.2 What Kinds of Urban Planning Problems Is AI Being Used on?

The search revealed 126 research items, including peer-reviewed articles, conference proceedings, scoping reviews, and book chapters that linked urban planning, AI, and problems. Although proceedings are sometimes excluded from reviews, they were particularly relevant for this question as they were found to focus on more practical implementations of AI. Research items were critically analyzed for content, and through this review, the different purposes for using AI in an urban planning context emerged. The main purposes of AI use were to facilitate nine common urban planning practices: planning, optimization, management, decision-making, analysis, sensing/monitoring, classification, forecasting, and visualization/communication. To render information more accessible for city stakeholders, researchers, and urban planners, this thematic synthesis is presented in Table 11.1, providing examples of use.

From an object perspective, optimization or performance improvement is a common thread in AI that is already acting within urban environments, such as improving arrow traffic light recognition for autonomous vehicles (Cai et al., 2012), easing traffic congestion (Jin & Ma, 2018; Kothai et al., 2021; Mahrez et al., 2022; Xu & You, 2022), or reducing electrical energy waste in vacuum urban waste collection systems (Béjar et al., 2012).

From a subject, human-centered perspective, assisting complex decision-making is a common theme, achieved oftentimes by integration and reduction. When data are not already organized, this might require mining and collating information spread in different formats into usable knowledge bases (Babu Saheer et al., 2020). Reducing multicriteria spatial decision's complexity for planning (Ruiz et al., 2012; Demetriou et al., 2013; Natividade-Jesus et al., 2013; Pleho & Avdagić, 2015) or reducing planning complications for ecological connectivity (Ferretti & Pomarico, 2013) ultimately places AI as a supporting system for an expert decision-maker to review and make the final call.

Some AI uses are very transparent and carry a low risk of reproducing systemic inequities, while others require a careful appraisal of benefits and beneficiaries and potential negative impacts. The problems and resistance that have been called out are described in the following section.

TABLE 11.1

Objectives for Using AI in Urban Planning with Examples

Purpose of AI Use	Examples
Planning	*Where?:* urban (M. Wang et al., 2020), city (Duarte & Álvarez, 2019), regional (Rienow & Stenger, 2014), landscape (Elshafei & Negm, 2017), coastal (Su et al., 2020), and spatial (Ocalir-Akunal, 2015) *What?:* Land use (Mohammady et al., 2013), wells (Shuaibi et al., 2021), electric power system (Çelik, 2020), travel route and stations (Fuhua & Shuiqiang, 2013), highway (Salman et al., 2013), infrastructure (Terashima & Clark, 2021), motion (Rivalcoba et al., 2019), and optimal path (Zhang et al., 2020)
Optimization	Multiobjective (Li & Liu, 2016), sequential minimal (Torija et al., 2014), constrained (Fu et al., 2019), layout (T. Huang, 2022), and traffic congestion (Jin & Ma, 2018)
Management	Waste (Béjar et al., 2012; Nowakowski et al., 2018; Pleho & Avdagić, 2015), water (Sidek et al., 2016; Visescu et al., 2017), energy (Januszkiewicz & Paszkowska, 2016), environment (Liu & Gong, 2021; Murillo et al., 2020), data/information (Mukhamediev et al., 2020), and crisis (Baharmand & Comes, 2015)
Decision-making	Decision support system (Calzada et al., 2012), decision-making (Elshafei & Negm, 2017), multicriteria decision support systems (Demetriou et al., 2013), spatial decision support systems (Terribile et al., 2015), environmental decision support systems (Denzer et al., 2013), collaborative decisions (Aldawood et al., 2014), and prioritization of strategic actions (Natividade-Jesus et al., 2013)
Analysis	Analytical hierarchy process (A. Rai & Jagadeesh Kannan, 2017), object-based image analysis (OBIA; Escamos et al., 2015), independent component analysis (Cai et al., 2012), spatial analysis (Turki, 2010), comparative analysis (Yuan & Wang, 2022), scenario analysis (Goh et al., 2019), and sentiment analysis (S. Wang et al., 2018)
Sensing and Monitoring	Soundscapes (Torija et al., 2014), environmental (S. Huang et al., 2018), detection sensors (Reid et al., 2018), embedded sensors (Ghahramani et al., 2022), remote (Liu & Gong, 2021), and non-intrusive load monitoring (Chui et al., 2018)
Simulation	Crowd simulation (Rivalcoba et al., 2019), data-driven simulation (Zhang et al., 2020), and daylight simulation (Aldawood et al., 2014)
Classification	Classification models (Nogueira et al., 2015), clustering algorithms (Salman et al., 2013), automatic classification (Torija et al., 2014), image classification (Nogueira et al., 2015), classification model (González-Vergara et al., 2021), multiclass classification (Sanna et al., 2021), pixel-based classifications (Escamos et al., 2015), grid pattern recognition (M. Wang et al., 2020), city pattern classification in satellite images through the following: deep learning (Aslantaş et al., 2021), machine learning (Agoub & Kada, 2021), and convolutional networks (Nogueira et al., 2015)
Forecasting	Electricity consumption forecasting (Zhai et al., 2021) and hybrid prediction models (Zhang et al., 2020)
Visualization and Communication	Mixed reality (Schubert et al., 2015), dynamic data visualization (Aldawood et al., 2014), and mobility data communication (Mahrez et al., 2022)

11.3 What Kinds of Problems and Resistances Have Been Called Out to the Implementation of AI in Urban Planning?

Our initial search criteria of research indexed on the Scopus database only resulted in 5 research items for the search string "urban planning", "artificial intelligence", and "bias". The research was reviewed and subsequently expanded to an exploratory scoping review in Google Scholar, which yielded 6,020 results in the same time period of 2012–2022. The screening was done by the possibility of acquiring full-text versions of articles using the author's institutional subscriptions, and answering the question, "Is this study offering relevant information in the context of urban planning?" Recognizing along with other authors that true theoretical saturation is a logical impossibility (Low, 2019; Wray et al., 2007), the process was halted when the redundancy of data was recurrent. The results of this exploratory scope are thematically coded and presented in this section.

The main problems that have been called out to the implementation of AI are as follows:

1. *Incorrectly assumed neutrality*: The supposed smartness or apparent neutrality of AI, when not questioned, can further entrench systemic inequities (Joyce et al., 2021) like sexism (D'ignazio & Klein, 2020), racism (Brandao et al., 2020; Buolamwini, 2017), ableism (Costanza-Chock, 2020; Holmes, 2020; van Leeuwen et al., 2021), and ageism (Díaz, 2020).

2. *Limited benefits and beneficiaries* reproduce systemic inequities (Fatima et al., 2020; Okada, 2021; Safransky, 2020).

3. *The lack of transparency or explainability* can hinder detection of bias and unwanted repercussions. This is doubly concerning because explainability in machine learning models is usually inverse to their prediction accuracy: The higher the prediction accuracy, the lower its transparency and performance as Explainable AI or XAI (Arrieta et al., 2020; Gunning et al., 2019; Xu et al., 2019).

4. *The lack of regulations for AI and data use* can hinder achieving ethical AI (Andrews et al., 2022).

5. The combination of the previous two factors, *explainability and privacy of user data*, can hinder achieving responsible AI (Arrieta et al., 2020).

Regarding assumed neutrality and limited beneficiaries, the question of inputting unbiased data and rules to make the AI function as intended is no small feat. The increasing use of data-driven analytics and algorithmic planning demands attention to geographies of algorithmic violence, understood as "a repetitive and standardized form of violence that contributes to the racialization of space and spatialization of poverty" (Safransky, 2020, p. 1).

To diminish the risk of data being biased, it must follow the 5V big data framework: namely being the best, it can be in terms of Volume, Variety, Velocity, Veracity, and Value (Gupta & Gupta, 2016). However, a well-articulated vision must be developed in a collaborative way and effectively communicated to all key stakeholders for AI to deliver on its promises. Planners need to understand whether the AI was trained on representative data of the object of interest and be trained to locate relevant absences. Planners will ultimately serve their communities better by applying the same inclusion lens that participatory processes require, becoming critical users of tech, and ensuring that the benefits of AI use are those valued by diverse members of society. Additionally, a careful review of inclusion discourses must also be conducted, as they can distract or veil an AI agenda of "surveillance, social sorting, and optimization" that generate harmful, unequal, or violent outcomes (Hoffmann, 2021).

Race and gender discrimination through AI use have impacted individuals' access to the housing market (Asplund et al., 2020) and their safety in the face of autonomous vehicles (Bigman & Gray, 2020). On the other hand, AI use in the form of autonomous vehicles promises benefits for the elderly and people with travel-restrictive medical conditions (Harper et al., 2016), yet the cost of this technology could render it initially inaccessible to lower income low-mobility subjects. This illustrates the need for an intersectional understanding of beneficiaries and negative impacts.

The lack of transparency is a contentious subject because it can be at odds with optimization, one of the main uses of AI in urban planning practice. One way to mitigate black box algorithm use risk is to only use interpretable, explainable models for high-stakes decisions (Rudin, 2019). However, enforcing this recommendation is discretional, because presently there is no regulatory framework for AI use nor any general ethics code for AI development or use. Moreover, there is no governance structure: Codes of ethics are not legally binding or enforceable at a larger scale (Andrews et al., 2022). At the time of writing, the European Commission proposed the first-ever legal framework on AI incorporating a risk-based approach, estimating the second half of 2024 as the earliest time the regulation could become operational and applicable (European Commission, 2022).

All these resistances highlight the same battles for social and environmental justice that have been present in urban planning debates for decades. Our ability to generate data at an unprecedented scale demands Responsible Urban Intelligence (RUI; Cao et al., 2022) and clearly delineates both action frameworks and regulations in using data-driven-enabling technologies for tackling urban problems.

The following section illustrates some of these issues through the case study of a bike-sharing system (BSS) in Mexico City.

11.4 What Can We Learn from the AI That Is Already Implemented in Urban Planning?

Bike-sharing systems started in Amsterdam in the 1960s, as a sustainable alternative for public transport (Kabak et al., 2018). This type of solution allows citizens to cycle in the city without a personal bike, allowing for faster urban cycling adoption rates. Since then, there has been a fast growth of governments and companies studying different approaches to the system. This growth has also been linked to the rise of smart cities (Chiariotti et al., 2018) and their technological enablers, such as the internet of things (IoT) and AI (Ruohomaa et al., 2019). These technological advances seem to improve participants' urban experience and create enough data for the planners to improve the systems and life quality.

For Latin-American countries, this was not an exemption, although systems arrived later, starting with Brazil. For Mexico, BSS adoption started in 2010. Mexico City (CDMX) is the capital of the country. According to the 2020 census, the city has 9,209,944 inhabitants, being the second most populated of the country. With its surface area of 1,494.3 km^2, it also has one of the highest densities (National Institute of Statistics and Geography, 2020). CDMX is comprised of 13 counties. The northern counties have the highest population density, while the smaller urban agglomerations are in the southern counties.

As Figure 11.2 shows, even though the bike network infrastructure goes from north to south, it is mainly concentrated in the northern counties.

In 2010, a public eco-bike system, Ecobici, was launched in CDMX in three counties: Benito Juarez, Cuauhtémoc, and Miguel Hidalgo. Since then, the system has been constantly growing to give service to more inhabitants (Amieva Gálvez et al., 2018) as we can see in Table 11.2. On July 26, 2022, SEMOVI announced that the contract would change from Clear Channel to 5m2-BKT. At the same time, 5m2-BKT announced that they would add 207 bike stations and would pass from 6,500 bikes to 9300 bikes (BKT bici pública, 2022). This new contract will add the stations to other three counties: Coyocan, Álvaro Obregon, and Azcapotzalco.

The cycling infrastructure of CDMX in 2020 included a bike path network with 69 km of bike paths, and some of them shared with other road users, see Table 11.3. In 2022, there are 88 km of bike infrastructure. According to the Government of Mexico City, the aim is to achieve three goals: (i) covering big distance travels by bike, (ii) creating connections with the public transport infrastructure, and (iii) creating a peripheral network. According to data from the latest Origin and Destination survey (National Institute for Statistics and Geography, 2017), this bike infrastructure is being used by 340,000 cyclists who transit frequently in the Mexican capital; nevertheless, they do not feel secure nor that the infrastructure is guaranteed.

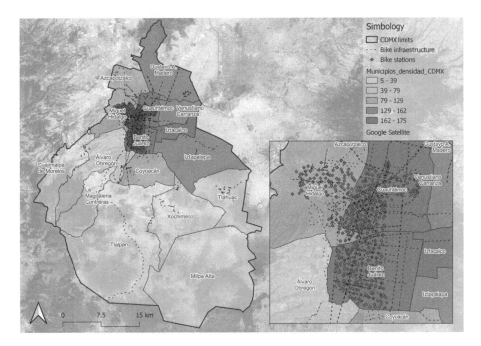

FIGURE 11.2

Bike-sharing system infrastructure of CDMX. (Elaborated by the authors with the information provided by the Government of Mexico City (Mexico City Government, 2022).)

TABLE 11.2

Bike-sharing Service Growth in Mexico City

Year	Action	Company	Bikes	Bike Stations	Counties
2010	Public bike system is launched	Clear channel	114	90	1
2018	Smart bike grows	Clear channel	6,500	480	3
2022	Mibici wins the bidding	5M2-BKT	9,300	687	6

Source: Amieva Gálvez et al. 2018; BKT bici pública 2022.

The BSS in Mexico City has a 12-year story process that has allowed the city to improve the system by adding stations, bikes, and different technology enablers for the user experience and operator security. The users can choose between 4 plans: (i) pay per day $118.00 mxn (5.8 dlls approx.), (ii) pay per 3 days $234.00 mxn (11.6 dlls approx.), (iii) pay per 1 week $391.00 mxn (19.4 dlls approx.), and (iv) annual pay $521.00 mxn (25.9 dlls approx.). There is also an option for hours: (i) 45 to 0 min $25.00 (1.24 dlls. approx.) and (ii) extra hour $50.00 (2.5 dlls. approx.). Bike-sharing system costs have two categories: (i) costs for users and (ii) costs for providers. For the users, the cost seems attractive when comparing the annual payment with the budget to pay for other public transportation systems in the city (Caggiani et al., 2021).

TABLE 11.3

CDMX Bike Infrastructure in 2020

Process	km by Section	km by Way	Investment
Bike infrastructure 2020 (improvements)	20.22 km	27.52 km	50 mdp
Fund for bikers and pedestrians	18.9 km	26.69 km	48 mdp
Metropolitan fund	8.40 km	14.80 km	40 mdp
Total	46.52 km	69 km	18 mdp

Source: Self-elaboration with information from SEMOVI (Mobility Secretariat, 2020).

TABLE 11.4

Data Given to the User by "Mibici"

Bike Station	ID
	Location
	Available bikes in the station
	Available bike sports in the station
	Instruction to arrive from point A to point B

Source: Self-elaboration with information from SEMOVI (Mobility Secretariat, 2020).

In order to use a bike, there is an app for IOS and Android. The app provides you two choices: either select a plan and pay or go to the map. The data demanded and received by the users can be seen in Tables 11.4 and 11.5.

"Mibici" is the last version of BSS adopted by Mexico City. It includes technology enablers that allow the operator to respond to the nearly real-time demand and is approaching being part of a multimodal mobility system. As can be seen in Figure 11.3, the city has 28 multimodal bike stations (6% of the total), 102 bike stations with 4G (21% of the total), and 350 bike stations with 3G technology (73% of the total). Therefore, it can be considered as part of the 4th generation of BSS at this stage.

While in the 1960s Amsterdam was launching the first BSS, Mexico mobility plan was focused on "peseros," named after small buses due to their moderate fare ($1.00 mxn). Since then, they have become the most popular transportation method in CDMX. It was followed by the metro station 1969. At the same time, Europe was learning with this first BSS generation, an innovative mobility option; nevertheless, the systems were deterred due to theft and low-incentivized bicycle return. In the late 1990s, Mexico City was focused on bus stations with a local and regional reach and two trolley cars. During this period, the second generation of BSS came into existence mainly in Europe, especially in France. The second generation is known as the coin-deposit system that allowed the operator to secure the system lifelong term. In 2005, the Government of Mexico City decided to invest in BRT to improve mobility conditions for the citizens. Finally, in 2013 the first BSS system arrived at CDMX as a solution to the pollution and

TABLE 11.5

Personal Data Required to use CDMX "Mibici"

Mail	
Cellphone number	
Password	
Name	
Last name	
Birthday	
ID	INE (Mexican ID)
	Driver's license
	Job title
	Passport
	FM2 (Mexican permanent resident ID)
	FM3 (Mexican temporal resident ID)
Address	Street
	Country
	State
	Zip

Source: Self-elaboration with information from SEMOVI (Mobility Secretariat, 2020).

Bike stations by type in CDMX

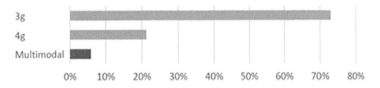

FIGURE 11.3
"Mibici" bike station-type distribution in the city. (Self-elaborated with SEMOVI information.)

traffic problem the city was facing. The system was operated by Clear Channel, a publicity company working with BSS in several countries around the world. During this period, BSS had evolved to the third generation, an IT-based system. This generation incorporates advanced technologies for bicycle reservations, pickup, drop-off, and information tracking. However, technologies evolve faster and are giving opportunities to a fourth generation of BSS that are demand-responsive and with multimodal systems. CDMX in 2022 reassigned the bike system to 5m2-BKT. The system already counts multimodal bike stations, and additionally, the city and operator are pursuing to add more in order to attend to a larger population. Table 11.6 shows the evolution of BSS in the world.

While BSS in Mexico City is growing due to time and technology enablers, it is not getting to most of the population. In the first stage, the system reaches just 16% of the Mexico City population. With the additional counties

TABLE 11.6

BSS Timeline

Generation	Generation Name	Year	Continent	Location	BSS Name
1st generation	White bikes / free bikes system	1965	Europe	Amsterdam	White Bike Plan
		1974	Europe	La Rochelle, France	Vélos jaunes
		1993	Europe	Cambridge, United Kingdom	Green Bike Scheme
		1994	America	Portland, Oregon	Yellow Bike
		1995	America	Boulder, Colorado	Green Bike Program
		1995	America	Wisconsin, United States	Red Bikesin Madison
2nd generation	Coin-Deposit Systems	1995	Europe	Copenhagen, Denmark	Bycyken (City Bike)
		1996	Europe	Sandnes, Norway	Bycykler
		2000	Europe	Helsinki, Finland	City Bikes
		2005	Europe	Arhus, Denmark	Bycykel
		1996	America	Minneapolis, United States	Yellow Bike program
		1996	America	St. Paul, United States	Yellow Bike program
		1996	America	Washington, United States	Olympia bike library
		1997	America	Austin, Texas, United States	Yellow bike
		1998	America	New Jersey, United States	Free-wheels in Princeton
		2002	America	Georgia, United States	Decatur Yellow Bikes

(Continued)

TABLE 11.6 (*Continued*)
BSS Timeline

Generation	Generation Name	Year	Continent	Location	BSS Name
3rd generation	IT based systems	1998	Europe	Rennes, France	LE vélo STAR
		2005	Europe	Lyon France	Velo'v
		2009	Europe	La Rochelle, France	Yélo
		2009	Europe	Paris, France	Vélib'
		2010	Europe	London, England	BIXI
		2008	America	Washington, DC, United States	Smartbike
		2009	America	Montreal, Canada	BIXI
		2009	America	Quebec, Canada	BIXI
		2010	America	Boston, United States	BIXI
		2008	America	São Paulo, Brazil	UseBike
		2008	America	Rio de Janeiro, Brazil	Bikerio
4th generation	Multimodad and demand-responsive	2010	America	Mexico City, Mexico	Smartbike
		2014	America	Guadalajara, Mexico	Mibici
		2015	America	Toluca, Mexico	Huizi
		2022	America	Mexico City, Mexico	Mibici

Source: Self-elaboration with information of Shaheen et al. (2010).

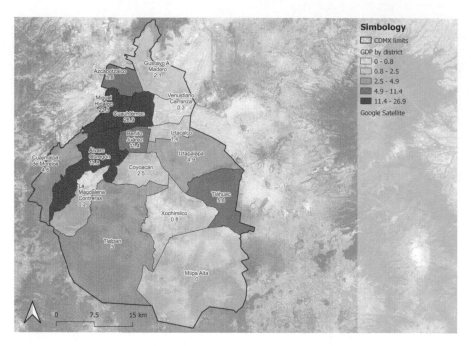

FIGURE 11.4
Local GDP concentración. (Social development evaluation council, 2020)

projected in 2022, it will add another 20% of the population, leading to 36% of the total population. The three counties that have the BSS today, the three new counties that have BSS soon, and the other seven counties that include 64% of the total population in CDMX are highest, middle, and lowest income-generating areas, respectively. On the contrary, Cuauhtémoc, Benito Juarez, and Miguel Hidalgo counties concentrate 60% of the PIB of the CDMX (Social development evaluation council, 2020) as presented in Figure 11.4. There is a positive relation between the areas where the government invests in technology and the areas with highest incomes.

Technology enabled the quality of life of the population in smart cities. Hence, smart cities were initiated only in the areas where per capita income, such as in the CDMX system, could afford to have such technology. The first BSS operator in CDMX, Clear Channel, defined a smart city as a web-connected city that achieved a balance with the environment. The company is aware that data generated from the system allows them to know everything about the bike user behavior and reactions toward different stimuli. Hence, marketing companies like them can impact users because they already have their personal behaviors (Clear channel Mexico, 2019). Today, the only citizen information providers have is of those with the highest incomes.

Although some users in CDMX are improving their quality of life since upgrading to BSS, research shows that some cycling apps can be unfair, racist, and gender-biased (Golub et al., 2016)."

11.5 Conclusions

The potential of AI within urban planning is great, although there are associated risks that should be confronted immediately. Conceptual and operational clarity for AI use in urban environments are needed if we are to achieve the vision of smart, sustainable, inclusive, resilient cities that many contemporary narratives describe, most notably the New Urban Agenda.

The risks of reproducing systemic inequities or increasing social divides could be greater in contexts of socio-economical and spatial inequality, such as those present in many cities of the global south, of which Mexico City is an example. Framing processes in local contexts and communities is not only a question of social justice, failing to do so in territories of great inequities could also mean that the smart city is but an island to which precious few have access.

The clarity and intersectional understanding that is needed to produce engagement with different stakeholders in other urban decision-making processes is also crucial for AI. This demand is synthesized in the *eXplainable* AI (XAI) framework. Transparency for many different actors is key, and opaque decision systems should be avoided.

The undeniable fast growth of AI across many sectors, including urban planning, must be accompanied by regulatory foresight, monitoring, and supervision of AI-based technologies. Without effective regulation, there may be inconsistencies in generating RUI and achieving ethical AI use.

There is no reason, and probably no feasible way, to stop the incorporation of AI in the way we design, manage, and engage with cities. Instead, these critical remarks highlight a series of underlying and yet unresolved questions about AI integration in cities, about the concept of the smart city itself and what it represents. Urban planners, local governments, academics, and city stakeholders have a social responsibility to ground and align AI objectives with the welfare of greater public and the multiplicity of diverse needs of urban dwellers.

References

Agoub, A., & Kada, M. (2021). Exploring the latent manifold of city patterns. *ISPRS International Journal of Geo-Information, 10*(10). https://doi.org/10.3390/ijgi10100683.

Aldawood, S., Aleissa, F., Alnasser, R., Alfaris, A., & Al-Wabil, A. (2014). Interaction design in a tangible collaborative decision support system: The city schema DSS. In *Communications in Computer and Information Science: Vol. 435 PART I.* https://doi.org/10.1007/978-3-319-07854-0_88.

Allam, Z. (2018). Contextualising the smart city for sustainability and inclusivity. *New Design Ideas, 2*(2), 124–127.

Amieva Gálvez, J., Müller García, T., Gómez, R. M., Crotte, A., Garduño, J., Graue, E., Lomelí, L., Oyama, A., Aguilar, R., González, M., Martínez, N., & Suárez, M. (2018). *Mexico City Bike Plan*. https://semovi.cdmx.gob.mx/storage/app/media/PlanBici-baja-sitio-1_2019.pdf

Andrews, C., Cooke, K., Gomez, A., Hurtado, P., Sanchez, T., Shah, S., & Wright, N. (2022). *AI in Planning: Opportunities and Challenges and How to Prepare*. https://planning.org/publications/document/9255930/.

Anthopoulos, L. (2017). Smart utopia VS smart reality: Learning by experience from 10 smart city cases. *Cities, 63*, 128–148.

Anttiroiko, A. (2015). City branding as a response to global intercity competition. *Growth and Change, 46*(2), 233–252.

Arrieta, A. B., Díaz-Rodríguez, N., Del Ser, J., Bennetot, A., Tabik, S., Barbado, A., García, S., Gil-López, S., Molina, D., & Benjamins, R. (2020). Explainable Artificial Intelligence (XAI): Concepts, taxonomies, opportunities and challenges toward responsible AI. *Information Fusion, 58*, 82–115.

Aslantaş, N., Bayram, B., & Bakirman, T. (2021). Building segmentation from VHR aerial imagery using DeepLabv3+ architecture. *42nd Asian Conference on Remote Sensing, ACRS 2021*. Asian Association on Remote Sensing. Can Tho city. https://a-a-r-s.org/proceeding/ACRS2021/4%20Cloud%20Computing_Big%20data%20and%20AI%20in%20Remote%20Sensing/ACRS21_143.pdf

Asplund, J., Eslami, M., Sundaram, H., Sandvig, C., & Karahalios, K. (2020). Auditing race and gender discrimination in online housing markets. *Proceedings of the International AAAI Conference on Web and Social Media, 14*, 24–35.

Babu Saheer, L., Shahawy, M., & Zarrin, J. (2020). Mining and analysis of air quality data to aid climate change. In *IFIP Advances in Information and Communication Technology* (Vol. 585 IFIP). https://doi.org/10.1007/978-3-030-49190-1_21.

Baharmand, H., & Comes, T. (2015). A framework for shelter location decisions by ant colony optimization. *ISCRAM 2015 Conference Proceedings -12th International Conference on Information Systems for Crisis Response and Management*, 2015-Janua.

Béjar, R., Fernández, C., Mateu, C., Manyà, F., Sole-Mauri, F., & Vidal, D. (2012). The automated vacuum waste collection optimization problem. *Proceedings of the National Conference on Artificial Intelligence, 1*, 264–266.

Bettencourt, L. M. A. (2021). *Introduction to Urban Science: Evidence and Theory of Cities as Complex Systems*. MIT Press.

BKT Bici pública. (2022). Más bicis mejor ciudad. Mibici. https://mibici.net/

Bigman, Y. E., & Gray, K. (2020). Life and death decisions of autonomous vehicles. *Nature, 579*(7797), E1–E2.

Brandao, M., Jirotka, M., Webb, H., & Luff, P. (2020). Fair navigation planning: A resource for characterizing and designing fairness in mobile robots. *Artificial Intelligence, 282*, 103259.

Bughin, J., Seong, J., Manyika, J., Chui, M., & Joshi, R. (2018). Notes from the AI frontier: Modeling the impact of AI on the world economy [Discussion paper]. *McKinsey Global Institute*https://www.mckinsey.com/~/media/mckinsey/featured%20insights/artificial%20intelligence/tackling%20europes%20gap%20in%20digital%20and%20ai/mgi-tackling-europes-gap-in-digital-and-ai-feb-2019-vf.pdf

Buolamwini, J. A. (2017). *Gender Shades: Intersectional Phenotypic and Demographic Evaluation of Face Datasets and Gender Classifiers*. Massachusetts Institute of Technology. https://dspace.mit.edu/handle/1721.1/114068

Buyanova, M. A., Kalinina, A. A., & Shiro, M. S. (2020). Smart city branding massively expands smart technologies. *Institute of Scientific Communications Conference*, pp. 1063–1069.

Caggiani, L., & Camporeale, R. (2021). Toward sustainability: Bike-sharing systems design, simulation and management. In *Sustainability* (Vol. 13, Issue 14, p. 7519). MDPI.

Cai, Z., Li, Y., & Gu, M. (2012). Real-time recognition system of traffic light in urban environment. *2012 IEEE Symposium on Computational Intelligence for Security and Defence Applications, CISDA 2012*. https://doi.org/10.1109/CISDA.2012.6291516.

Calzada, A., Liu, J., Wang, H., & Kashyap, A. (2012). Uncertainty and incompleteness analysis using the RIMER approach for urban regeneration processes: The case of the greater Belfast region. *Proceedings - International Conference on Machine Learning and Cybernetics*, 3, 928–934. https://doi.org/10.1109/ICMLC.2012.6359478

Cao, R., Gao, Q., & Qiu, G. (2022). Responsible Urban Intelligence: Towards a Research Agenda. *ArXiv Preprint ArXiv:2208.04727*.

Cavada, M., Hunt, D. V., & Rogers, C. D. (2014). Smart cities: Contradicting definitions and unclear measures. *World Sustainability Forum*, pp. 1–12.

Çelik, E. (2020). Improved stochastic fractal search algorithm and modified cost function for automatic generation control of interconnected electric power systems. *Engineering Applications of Artificial Intelligence*, 88, 103407.Chiariotti, F., Pielli, C., Zanella, A., & Zorzi, M. (2018). A dynamic approach to rebalancing bike-sharing systems. *Sensors*, 18(2), 512.

Chui, K. T., Lytras, M. D., & Visvizi, A. (2018). Energy sustainability in smart cities: Artificial intelligence, smart monitoring, and optimization of energy consumption. *Energies*, 11(11). https://doi.org/10.3390/en11112869.

Clear channel Mexico. (2019, May 22). Smart cities of the future. https://clearchannel.com.mx/smart-cities-del-futuro/

Costanza-Chock, S. (2020). *Design Justice: Community-Led Practices to Build the Worlds We Need*. The MIT Press. https://designjustice.mitpress.mit.edu/

D'ignazio, C., & Klein, L. F. (2020). *Data Feminism*. MIT press. https://data-feminism.mitpress.mit.edu/

Datta, A. (2015). A 100 smart cities, a 100 utopias. *Dialogues in Human Geography*, 5(1), 49–53.

De Filippi, F., Coscia, C., & Guido, R. (2019). From smart-cities to smart-communities: How can we evaluate the impacts of innovation and inclusive processes in urban context? *International Journal of E-Planning Research (IJEPR)*, 8(2), 24–44.

Demetriou, D., Stillwell, J., & See, L. (2013). LACONISS: A planning support system for land consolidation. In *Lecture Notes in Geoinformation and Cartography* (Vol. 0, Issue 199649). https://doi.org/10.1007/978-3-642-37533-0_5.

Denzer, R., Schlobinski, S., Gidhagen, L., & Hell, T. (2013). How to build integrated climate change enabled EDSS. In *IFIP Advances in Information and Communication Technology* (Vol. 413). https://doi.org/10.1007/978-3-642-41151-9_43.

Díaz, M. (2020). *Biases as Values: Evaluating Algorithms in Context*. Northwestern University. http://markjdiaz.com/wp-content/uploads/2021/01/Diaz_BiasesAsValues.pdf

Duarte, F., & Álvarez, R. (2019). The data politics of the urban age. *Palgrave Communications*, 5(1). https://doi.org/10.1057/s41599-019-0264-3.

Elshafei, G., & Negm, A. (2017). AI technologies in green architecture field: Statistical comparative analysis. *Procedia Engineering, 181,* 480–488. https://doi.org/10.1016/j.proeng.2017.02.419.

Escamos, I. M. H., Roberto, A. R. C., Abucay, E. R., Inciong, G. K. L., Queliste, M. D., & Hermocilla, J. A. C. (2015). Comparison of different machine learning classifiers for building extraction in LiDAR-derived datasets. *ACRS 2015–36th Asian Conference on Remote Sensing: Fostering Resilient Growth in Asia, Proceedings* (pp. 282–291). Quezon City, Metro Manila Philippines, https://simdos.unud.ac.id/uploads/file_penelitian_1_dir/ada619cc4f0a56fa0effc7e6f75877b4.pdf

European Commission (2022, September 29). Regulatory framework proposal on artificial intelligence. Digital Strategy European Commission. https://digital-strategy.ec.europa.eu/policies/regulatory-framework-ai.

Fatima, S., Desouza, K. C., & Dawson, G. S. (2020). National strategic artificial intelligence plans: A multi-dimensional analysis. *Economic Analysis and Policy, 67,* 178–194.

Fernández-Güell, J.-M., Collado-Lara, M., Guzmán-Arana, S., & Fernández-Anez, V. (2016). Incorporating a systemic and foresight approach into smart city initiatives: The case of Spanish cities. *Journal of Urban Technology, 23*(3), 43–67.

Ferretti, V., & Pomarico, S. (2013). Ecological land suitability analysis through spatial indicators: An application of the Analytic Network Process technique and Ordered Weighted Average approach. *Ecological Indicators, 34,* 507–519. https://doi.org/10.1016/j.ecolind.2013.06.005

Fu, Y., Wang, P., Du, J., Wu, L., & Li, X. (2019). Efficient region embedding with multiview spatial networks: A perspective of locality-constrained spatial autocorrelations. *33rd AAAI Conference on Artificial Intelligence, AAAI 2019, 31st Innovative Applications of Artificial Intelligence Conference, IAAI 2019 and the 9th AAAI Symposium on Educational Advances in Artificial Intelligence, EAAI 2019,* Honolulu, Hawaii, USA. pp. 906–913. https://ojs.aaai.org/index.php/AAAI/article/view/3879/3757

Fuhua, S., & Shuiqiang, Y. (2013). Research on application of improved genetic algorithm in urban full independent tourist route planning. *Proceedings-9th International Conference on Computational Intelligence and Security, CIS 2013,* pp. 200–203. https://doi.org/10.1109/CIS.2013.49.

Ghahramani, M., Zhou, M., Molter, A., & Pilla, F. (2022). IoT-based route recommendation for an intelligent waste management system. *IEEE Internet of Things Journal, 9*(14), 11883–11892. https://doi.org/10.1109/JIOT.2021.3132126.

Goh, S.-T., Liu, S., Yong, T., & Foo, E. (2019). Scenario analysis with facility location optimization. *IEEE Region 10 Annual International Conference, Proceedings/TENCON, 2018-October,* pp. 1091–1096. https://doi.org/10.1109/TENCON.2018.8650398.

Golub, A., Hoffmann, M., Lugo, A., & Sandoval, G. (2016). *Bicycle justice and urban transformation.* New York: Routledge.

González-Vergara, J., Serrano, N., & Iza, C. (2021). A support vector machine implementation for traffic assignment problem. *PE-WASUN 2021-Proceedings of the 18th ACM Symposium on Performance Evaluation of Wireless Ad Hoc, Sensor, and Ubiquitous Networks,* pp. 41–48. https://doi.org/10.1145/3479240.3488502.

Google Trends (2022). Smart cities worldwide popularity search trend 2004–2022. Available at https://trends.google.es/trends/explore?q=smart%20cities.

Gunning, D., Stefik, M., Choi, J., Miller, T., Stumpf, S., & Yang, G.-Z. (2019). XAI—Explainable artificial intelligence. *Science Robotics, 4*(37), eaay7120.

Gupta, U. G., & Gupta, A. (2016). Vision: A missing key dimension in the 5V Big Data framework. *Journal of International Business Research and Marketing*, 1(3), 50–56.

Haenlein, M., & Kaplan, A. (2019). A brief history of artificial intelligence: On the past, present, and future of artificial intelligence. *California Management Review*, 61(4), 5–14.

Harper, C. D., Hendrickson, C. T., Mangones, S., & Samaras, C. (2016). Estimating potential increases in travel with autonomous vehicles for the non-driving, elderly and people with travel-restrictive medical conditions. *Transportation Research Part C: Emerging Technologies*, 72, 1–9.

Hoffmann, A. L. (2021). Terms of inclusion: Data, discourse, violence. *New Media & Society*, 23(12), 3539–3556.

Holmes, K. (2020). *Mismatch: How Inclusion Shapes Design*. MIT Press. https://mit-press.mit.edu/9780262349635/mismatch/

Huang, S., Welch, C., Iliev, Y., & Rossello, V. (2018). The technical and commercial increment of the fusion of big data analysis, artificial intelligence and earth observation. *Proceedings of the International Astronautical Congress, IAC, 2018-October*. Bremen, Germany.

Huang, T. (2022). Facility layout optimization of urban public sports services under the background of deep learning. *Computational Intelligence and Neuroscience*, 2022. https://doi.org/10.1155/2022/1748319.

National Institute for Statistics and Geography (2017). Origin-Destination Survey in Households of the Metropolitan Zone of the Valley of Mexico (EOD) http://en.www.inegi.org.mx/programas/eod/2017/

National Institute for Statistics and Geography (2020). Mexican Census of Population and Housing. https://en.www.inegi.org.mx/programas/ccpv/2020/

Januszkiewicz, K., & Paszkowska, N. (2016). Climate change adopted building envelope for the urban environment. A new approach to architectural design. *International Multidisciplinary Scientific GeoConference Surveying Geology and Mining Ecology Management, SGEM*, 3(BOOK 6), 515–522.

Jin, J., & Ma, X. (2018). Hierarchical multi-agent control of traffic lights based on collective learning. *Engineering Applications of Artificial Intelligence*, 68, 236–248. https://doi.org/10.1016/j.engappai.2017.10.013.

Joyce, K., Smith-Doerr, L., Alegria, S., Bell, S., Cruz, T., Hoffman, S. G., Noble, S. U., & Shestakofsky, B. (2021). Toward a sociology of artificial intelligence: A call for research on inequalities and structural change. *Socius*, 7, 2378023121999581.

Kabak, M., Erbaş, M., Cetinkaya, C., & Özceylan, E. (2018). A GIS-based MCDM approach for the evaluation of bike-share stations. *Journal of Cleaner Production*, 201, 49–60.

Kothai, G., Poovammal, E., Dhiman, G., Ramana, K., Sharma, A., AlZain, M. A., Gaba, G. S., & Masud, M. (2021). A new hybrid deep learning algorithm for prediction of wide traffic congestion in smart cities. *Wireless Communications and Mobile Computing*, 2021. https://doi.org/10.1155/2021/5583874

Krivý, M. (2018). Towards a critique of cybernetic urbanism: The smart city and the society of control. *Planning Theory*, 17(1), 8–30.

Kummitha, R. K. R. (2018). Entrepreneurial urbanism and technological panacea: Why Smart City planning needs to go beyond corporate visioning? *Technological Forecasting and Social Change*, 137, 330–339.

Li, X., & Liu, X. (2016). Spatial information technology for facilitating "three-plan integration" using geographical simulation and optimization. *Yaogan Xuebao/Journal of Remote Sensing*, 20(5), 1308–1318. https://doi.org/10.11834/jrs.20166161.

Linardatos, P., Papastefanopoulos, V., & Kotsiantis, S. (2020). Explainable AI: A review of machine learning interpretability methods. *Entropy, 23*(1), 18.

Liu, H., & Gong, P. (2021). 21st century daily seamless data cube reconstruction and seasonal to annual land cover and land use dynamics mapping-iMap (China) 1.0. *National Remote Sensing Bulletin, 25*(1), 126–147. https://doi.org/10.11834/jrs.20210580.

Low, J. (2019). A pragmatic definition of the concept of theoretical saturation. *Sociological Focus, 52*(2), 131–139.

Mahrez, Z., Sabir, E., Badidi, E., Saad, W., & Sadik, M. (2022). Smart urban mobility: When mobility systems meet smart data. *IEEE Transactions on Intelligent Transportation Systems, 23*(7), 6222–6239. https://doi.org/10.1109/TITS.2021.3084907.

Masucci, M., Pearsall, H., & Wiig, A. (2020). The smart city conundrum for social justice: Youth perspectives on digital technologies and urban transformations. *Annals of the American Association of Geographers, 110*(2), 476–484.

Mellin, W. D. (1957). Work with new electronic 'brains' opens field for army math experts. *The Hammond Times, 10*, 66.

Mexico city government. (2022). Open data portal. *Mobility open data.* https://datos.cdmx.gob.mx/dataset/?res_format=SHP&groups=movilidad

Mhlanga, D. (2022). Human-centered artificial intelligence: The superlative approach to achieve sustainable development goals in the fourth industrial revolution. *Sustainability, 14*(13), 7804.

Mobility Secretariat (2020). Infrastructure and cycling culture. https://semovi.cdmx.gob.mx/storage/app/media/diamundialdelabici.pdf

Mohammady, S., Delavar, M. R., & Pijanowski, B. C. (2013). Urban growth modeling using ANFIS algorithm: A case study for Sanandaj City, Iran. *International Archives of the Photogrammetry, Remote Sensing and Spatial Information Sciences - ISPRS Archives, 40*(1W3), 493–498.

Mukhamediev, R., Kuchin, Y., Yakunin, K., Symagulov, A., Ospanova, M., Assanov, I., & Yelis, M. (2020). Intelligent unmanned aerial vehicle technology in urban environments. In *Communications in Computer and Information Science* (Vol. 1242). https://doi.org/10.1007/978-3-030-65218-0_26.

Murillo, C. Z., Hechavarría Hernández, J. R., & Vázquez, M. L. (2020). Multicriteria analysis in the proposed environmental management regulations for construction in Aurora, Guayas, Ecuador. In *Advances in Intelligent Systems and Computing* (Vol. 965). https://doi.org/10.1007/978-3-030-20454-9_10.

Natividade-Jesus, E., Coutinho-Rodrigues, J., & Tralhão, L. (2013). Housing evaluation with web-SDSS in urban regeneration actions. *Proceedings of the Institution of Civil Engineers: Municipal Engineer, 166*(3), 194–207. https://doi.org/10.1680/muen.12.00022.

Neirotti, P., De Marco, A., Cagliano, A. C., Mangano, G., & Scorrano, F. (2014). Current trends in Smart City initiatives: Some stylised facts. *Cities, 38*, 25–36.

Nogueira, K., Miranda, W. O., & Santos, J. A. D. (2015). Improving spatial feature representation from aerial scenes by using convolutional networks. *Brazilian Symposium of Computer Graphic and Image Processing*, 2015-October, pp. 289–296. https://doi.org/10.1109/SIBGRAPI.2015.39.

Nowakowski, P., Szwarc, K., & Boryczka, U. (2018). Vehicle route planning in e-waste mobile collection on demand supported by artificial intelligence algorithms. *Transportation Research Part D: Transport and Environment, 63*, 1–22. https://doi.org/10.1016/j.trd.2018.04.007.

Ocalir-Akunal, E. V. (2015). Using decision support systems for transportation planning efficiency. In *Using Decision Support Systems for Transportation Planning Efficiency*. https://doi.org/10.4018/978-1-4666-8648-9.

Okada, E. (2021). Artificial Intelligence to Broaden Beneficiaries. In *Management of Science-Intensive Organizations*. Palgrave Macmillan, Cham. https://doi.org/10.1007/978-3-030-64042-2_6.

Okada, E. (2021). Artificial Intelligence to Broaden Beneficiaries. In *Management of Science-Intensive Organizations*. Palgrave Macmillan, Cham. https://doi.org/10.1007/978-3-030-64042-2_6.

Pleho, J., & Avdagić, Z. (2015). Modeling of decision support system for spatial planning based on ANFIS model. *International Multidisciplinary Scientific GeoConference Surveying Geology and Mining Ecology Management, SGEM, 1*(2), 745–752.

Rai, A., & Jagadeesh Kannan, R. (2017). Multi-scale modeling of territorial dynamics of geospatial anthropogenic energy consumption. *Asian Journal of Pharmaceutical and Clinical Research, 10*, 305–311. https://doi.org/10.22159/ajpcr.2017.v10s1.19744.

Rai, A. (2020). Explainable AI: From black box to glass box. *Journal of the Academy of Marketing Science, 48*(1), 137–141.

Reid, A. R., Pérez, C. R. C., & Rodríguez, D. M. (2018). Inference of vehicular traffic in smart cities using machine learning with the internet of things. *International Journal on Interactive Design and Manufacturing, 12*(2), 459–472. https://doi.org/10.1007/s12008-017-0404-1.

Rienow, A., & Stenger, D. (2014). Geosimulation of urban growth and demographic decline in the Ruhr: A case study for 2025 using the artificial intelligence of cells and agents. *Journal of Geographical Systems, 16*(3), 311–342. https://doi.org/10.1007/s10109-014-0196-9.

Rivalcoba, I., Toledo, L., & Rudomín, I. (2019). Towards urban crowd visualization. *Scientific Visualization, 11*(2), 39–55. https://doi.org/10.26583/sv.11.2.04.

Rudin, C. (2019). Stop explaining black box machine learning models for high stakes decisions and use interpretable models instead. *Nature Machine Intelligence, 1*(5), 206–215.

Ruiz, M. C., Romero, E., Pérez, M. A., & Fernández, I. (2012). Development and application of a multi-criteria spatial decision support system for planning sustainable industrial areas in Northern Spain. *Automation in Construction, 22*, 320–333. https://doi.org/10.1016/j.autcon.2011.09.009.

Ruohomaa, H., Salminen, V., Kunttu, I., & Drucker, P. (2019). TIMReview_September 2019 - final-A. *Technology Innovation Management Review, 9*(9), 5–14. https://timreview.ca/article/1264.

Safransky, S. (2020). Geographies of algorithmic violence: Redlining the smart city. *International Journal of Urban and Regional Research, 44*(2), 200–218.

Shaheen, S. A., Guzman, S., & Zhang, H. (2010). Bikesharing in Europe, the Americas, and Asia: past, present, and future. *Transportation Research Record, 2143*(1), 159–167.

Salman, H. A., Ibrahim, L. F., & Fayed, Z. (2013). Enhancing clustering technique with knowledge-based system to plan the social infrastructure services. *ICAART 2013-Proceedings of the 5th International Conference on Agents and Artificial Intelligence, 2*, Barcelona, Spain. pp. 401–408.

Sanchez, T. W., Shumway, H., Gordner, T., & Lim, T. (2022). The prospects of artificial intelligence in urban planning. *International Journal of Urban Sciences*. https://doi.org/10.1080/12265934.2022.2102538.

Sanna, G., Godio, S., & Guglieri, G. (2021). Neural network based algorithm for multi-UAV coverage path planning. *2021 International Conference on Unmanned Aircraft Systems, ICUAS 2021*, pp. 1210–1217. https://doi.org/10.1109/ICUAS51884.2021.9476864.

Schubert, G., Schattel, D., Tönnis, M., Klinker, G., & Petzold, F. (2015). Tangible mixed reality on-site: Interactive augmented visualisations from architectural working models in urban design. In *Communications in Computer and Information Science* (Vol. 527). https://doi.org/10.1007/978-3-662-47386-3_4.

Secretary, O. of the P. (2015). *Administration Announces New "Smart Cities" Initiative to Help Communities Tackle Local Challenges and Improve City Services*. Office of the Press Secretary. https://obamawhitehouse.archives.gov/the-press-office/2015/09/14/fact-sheet-administration-announces-new-smart-cities-initiative-help.

Shuaibi, F., Harthi, M., Large, S., Obilaja, J.-F., Senani, M., Gomez, C. M., Mahrazy, K., Hussain, M., Al Busaidi, M., Savels, T., Savels, T., & Dolle, N. (2021). Leveraging game AI to transform integrated brownfield well planning. *Society of Petroleum Engineers - Abu Dhabi International Petroleum Exhibition and Conference, ADIP 2021*. https://doi.org/10.2118/207947-MS.

Sidek, L. M., Mohiyaden, H. A., Haris, H., Basri, H., Muda, Z. C., Roseli, Z. A., & Norlida, M. D. (2016). Decision Support System (DSS) for MSMA integrated stormwater management ecohydrology for sustainable green infrastructure. *IOP Conference Series: Earth and Environmental Science*, 32(1). https://doi.org/10.1088/1755-1315/32/1/012068.

Social Development Evaluation Council of Mexico City. (2020). A diagnosis of socio-territorial inequality. https://www.evalua.cdmx.gob.mx/storage/app/media/DIES20/ciudad-de-mexico-2020-un-diagnostico-de-la-desigualdad-socio-territorial.pdf

Su, L., Fan, J., & Fu, L. (2020). Exploration of smart city construction under new urbanization: A case study of Jinzhou-Huludao Coastal Area. *Sustainable Computing: Informatics and Systems*, 27. https://doi.org/10.1016/j.suscom.2020.100403.

Terashima, M., & Clark, K. (2021). The precarious absence of disability perspectives in planning research. *Urban Planning*, 6(1), 120–132. https://doi.org/10.17645/up.v6i1.3612.

Terribile, F., Agrillo, A., Bonfante, A., Buscemi, G., Colandrea, M., D'Antonio, A., De Mascellis, R., De Michele, C., Langella, G., Manna, P., Vingiani, S., & Basile, A. (2015). A Web-based spatial decision supporting system for land management and soil conservation. *Solid Earth*, 6(3), 903–928. https://doi.org/10.5194/se-6-903-2015.

Toli, A. M., & Murtagh, N. (2020). The concept of sustainability in smart city definitions. *Frontiers in Built Environment*, 6, 77.

Torija, A. J., Ruiz, D. P., & Ramos-Ridao, T. F. (2014). A tool for urban soundscape evaluation applying Support Vector Machines for developing a soundscape classification model. *Science of the Total Environment*, 482–483(1), 440–451. https://doi.org/10.1016/j.scitotenv.2013.07.108.

Turki, S. Y. (2010). What education in GIS for town planners? A Tunisian experience. *URISA Journal*, 22(2), 15–19.

UN Habitat. (2016). *New Urban Agenda*. https://habitat3.org/the-new-urban-agenda/.

van Leeuwen, C., Smets, A., Jacobs, A., & Ballon, P. (2021). Blind spots in AI: The role of serendipity and equity in algorithm-based decision-making. *ACM SIGKDD Explorations Newsletter*, 23(1), 42–49.

Visescu, M., Beilicci, E., & Beilicci, R. (2017). Integrated hydrographical basin management. Study case - Crasna river basin. *IOP Conference Series: Materials Science and Engineering*, 245(3). https://doi.org/10.1088/1757-899X/245/3/032038.

Wang, M., Ai, T., Yan, X., & Xiao, Y. (2020). Grid pattern recognition in road networks based on graph convolution network model | 图卷积网络模型识别道路正交网格模式. *Wuhan Daxue Xuebao (Xinxi Kexue Ban)/Geomatics and Information Science of Wuhan University*, 45(12), 1960–1969. https://doi.org/10.13203/j.whugis20200022.

Wang, S., Cao, D., Lin, D., & Chao, F. (2018). Traffic condition analysis based on users emotion tendency of microblog. In *Advances in Intelligent Systems and Computing* (Vol. 650). https://doi.org/10.1007/978-3-319-66939-7_26.

Wray, N., Markovic, M., & Manderson, L. (2007). "Researcher saturation": The impact of data triangulation and intensive-research practices on the researcher and qualitative research process. *Qualitative Health Research*, 17(10), 1392–1402.

Xiao, Y., & Watson, M. (2019). Guidance on conducting a systematic literature review. *Journal of Planning Education and Research*, 39(1), 93–112. https://doi.org/10.1177/0739456x17723971.

Xu, F., Uszkoreit, H., Du, Y., Fan, W., Zhao, D., & Zhu, J. (2019, October). Explainable AI: A brief survey on history, research areas, approaches and challenges. In *CCF International Conference on Natural Language Processing and Chinese Computing* (pp. 563–574). Springer, Cham.

Xu, X., & You, H. (2022). Hybrid fusion technology of transportation big data based on deep learning. In *Lecture Notes in Electrical Engineering* (Vol. 791). https://doi.org/10.1007/978-981-16-4258-6_58.

Yuan, Q., & Wang, N. (2022). Buildings change detection using high-resolution remote sensing images with self-attention knowledge distillation and multi-scale change-aware module. *International Archives of the Photogrammetry, Remote Sensing and Spatial Information Sciences - ISPRS Archives*, 46(M-2-2022), 225–231. https://doi.org/10.5194/isprs-Archives-XLVI-M-2-2022-225-2022.

Zhai, Q., Zhang, X., & Cheng, J. (2021). Enterprise electricity consumption forecasting method based on federated learning. In *Communications in Computer and Information Science* (Vol. 1423). https://doi.org/10.1007/978-3-030-78618-2_46.

Zhang, C., Dong, M., Luan, T. H., & Ota, K. (2020). Battery maintenance of pedelec sharing system: Big data based usage prediction and replenishment scheduling. *IEEE Transactions on Network Science and Engineering*, 7(1), 127–138. https://doi.org/10.1109/TNSE.2019.2901833.

12

The Control of Violence between the Machine and the Human

Juliana Vivar-Vera and Francisco Díaz-Estrada

Tecnológico de Monterrey

CONTENTS

12.1 Introduction

The accelerated technological advance has permeated the state function of violence control in such a way that predictive software is being used as legal aid for the dictation of punitive decisions (Parliamentary Assembly, 2020). These decisions present complexities that derive from the legal application, the context of the case, and the attention to the specialized legal-criminal theoretical approach, in tune with the argumentation for a final decision of the criminal process. Expert systems help judges by processing input data and offering results appropriate to the specific case efficiently and quickly (Corvalan, 2018). However, in its application, the existence of biases in the algorithmic combination has been evidenced, as well as "black boxes," that is, moments of loss of human control derived from automatic machine learning (Parliamentary Assembly, 2020). This added to the fact that the main creator of the software is the private sector, which does not make the algorithmic combination process transparent due to business secrecy (Corvalan, 2018). This hinders the sense of material justice with human reasoning even though its result is mathematically correct. Even so, the automatic learning of the processor glimpses a panorama of substitution of the human decision-making function by the characteristics of accuracy and objectivity offered by the technification of artificial intelligence (AI).

Software development companies have a globalizing objective far from the inclusion of vulnerable sectors and therefore from the objectives of sustainable development to meet the 2030 agenda (Corvalán, 2018). In this way, a panorama of the transfer of the power of state control of violence and of the data of the people in the hands of individuals, used for technological advancement by humans and without the intention of human benefit, is glimpsed. This reasoning is at work also in law-enforcement strategies. Thanks to technological progress in the field of network analysis, and thanks to the metadata collected by mass surveillance techniques from different means of communications, the web of relationships that surrounds a given person is now easy to reconstitute. This information is instrumental when it comes to evaluating the danger an individual poses for the state or society and to determine whether further actions should be conducted to monitor their behavior.

The objective of this chapter was to carry out a critical analysis of the intervention of AI in the state function of control of violence in the criminal decision to show the differentiating characteristics between the machine and the human, the globalizing control of creative companies of AI, and the exercise of the freedom of the professional who builds, feeds, operates, and interprets the algorithmic results of AI programs. To achieve this goal, the first section analyzes the scope of the algorithmic prediction of the criminal court decision and its human and mechanical implications in the debate on digital humanism; subsequently, the intervention of the creative agencies of predictive software with a globalizing vision is examined and finally the exercise of the freedom of the human creator and interpreter of algorithmic results is observed.

Different methods will be used; however, mainly the hermeneutic work for qualitative analysis will be highlighted.

12.2 Human and Algorithmic Implications for the State Decision to Control Violence

A state decision to control violence is a function that corresponds to humans at the service of the State and society, they are the judges whose role is combined with their daily experiences, their subjectivity, their sensitivity, and their creativity as characteristics that cannot be detached when deciding the future of other people.

The criminal judicial decision is adjusted to the theoretical legal-criminal positions that adapt to the regulatory framework with evaluative parameters of the act and its sanction, whose application is directed to a person subject to a process of the state order that awaits a decision of criminal responsibility for the violation of the "protected legal interest" (Zaffaroni, et al., 2006). Despite the act sanctioned by law, it is a person subject to the decision of someone else's life.

In this way, there is a theoretical evolution between rigid positivism and criticism that help us understand that the human judicial decision has a necessarily human orientation not toward the act but toward the person on whom it will dictate its decision. Accordingly, it is possible to understand how technology looks at the subject to dictate a decision that is already beyond the reach of the human but directed at people, either as a means of experimentation or as a sensitive and unattainable being in the creative capacity to commit behaviors that cause damage to another person and that is sometimes irreparable.

The dogmatic formal logical reasoning ensures the principle of exact application of criminal law under the apothegm *nullum crime, nulla poena,* and *sine lege,* which establishes that its source of production determines the behaviors to be punished according to the general description that best fits the cases of reality; in this scheme, by delimiting the individualization of the sentence, the context and the particular characteristics of the active subject, clinical criminology (Manzanera, 2005) proposed a type of criminal subject: an asocial being subject to experimentation to "convert" him into a criminal being able to coexist socially.

In contrast, critical criminology (Baratta, 1982) put into debate the criminal justice system where crimes are behaviors selected by the State under the principle of fragmentation, but at the same time, it selects people to be charged with. This is what Zaffaroni (2006) calls the criminalization process. In such a way that the *ius puniendi* that blames and punishes a person is not the one that represents the rational justification of the State, and that makes the difference with private revenge. The initial classification scheme of the criminal justice system constitutes the basis for understanding the biases in the algorithmic combination in the input data of predictive software. This knowledge is what is harmonized with the design of an expert prediction system.

Continuing with the justiciable subject, the criminal decision becomes the reasoning of human rights that implied recognizing the ductility of the law between the law, rights, and justice (Zagrebelsky, 1992). This is between opposing criminal theories of the criminal law of the enemy (Jakobs, 2006) and the minimum criminal law (Ferrajoli, 1985). In this context, the predictive software in the judicial decision works with network data selection algorithms that allow filtering of the necessary and applicable case. The so-called "artificial intelligence" also deals with deep learning, a specific subfield of machine learning that consists of learning successive layers with more and more significant representations (Chollet, 2018) and that is already working on changing its techniques to reach the reasoning and abstraction from the intuition that achieves a logical argumentation with mining and analysis of legal texts, where models of reasoning and representation of legal knowledge are taken according to legal theory, legal interpretation, and the incidence of the dimension epistemic of politics that are seen as advanced help or even as a substitute for the human judge for a correct and fair decision.

The process that the predictive software follows in the decision starts from the reading of previous resolutions posted online; in its panning, it only uses preloaded keywords to search with the file number and, depending on the case, ask the user a few questions and then gives a prediction in seconds. Then, it shows the user or rapporteur on the screen the model that they should use (Corvalan, 2018).

In terms of detection of human rights violations, acquiring data from the European Court of Human Rights, in experimental studies of the use of natural language processing tools to analyze texts of judicial procedures to automatically predict (future) judicial decisions, a 75% accuracy in predicting the violation of 9 articles of the European Convention on Human Rights, which highlights the potential of machine learning approaches in the legal field; however, predicting decisions for future cases based on the cases of the past negatively impacts performance (average accuracy range of 58%–68%) (Medvedeva et al., 2020).

Currently, AI works on a trial-and-error basis for the improvement free of subjective biases (Sourdin, 2018) comparable to the so-called "judicial error" identifiable in the head of the sentence, in the factual grounds, in the legal grounds, and in the ruling (Malem et al., 2012). Contrary to this, the experiences of "algorithmic bias" reveal that there is no significant difference in the hope of justice with the intervention of predictive software in relation to the work of the human judge, for example, the case of the COMPAS processor (Correctional Offender Management Profiling for Alternative Sanctions), was evidenced by ProPublica (Angwin et al., 2016) in cases such as that of Brisha Borden and Vernon Prater for alleged racial and gender biases that it contained and generated a risk rating corresponding to the sentence that should be imposed, in addition to the appeal sentence for the imposition of an incorrect sentence on the recommendation of the algorithm. In certain regional contexts, biases may involve a specific type of discrimination (race, color, sex, religion, language, or social origin) that violates the principle of equality before the law. That is why, in the aforementioned case, the company Northpointe Suite, on which the COMPAS computer system depends, responded to the accusations by focusing on the validity of the software, which could only be interpreted with solid scientific knowledge of the techniques and common methodological nuances toward the subject.[1] This

---

[1] The company presented technical data from the processor: the General Recurrence Risk Scale (GRRS), which includes the criminal involvement scale, the drug problems subscale, age at the time of evaluation, age at the first adjudication, number of previous arrests, arrest rate, and vocational education scale; and the Violent Recurrence Risk Scale (VRRS), age at assessment, age at first adjudication, and history of violence scale. The risk scales are divided into low, medium, and high levels, and the main types of classification statistics are "model errors" and "target population errors". The company accepted the error in the processor, although not the discriminatory bias, since it affirmed that these were the results of percentages of entry in the target population (Dieterich et al., 2016).

argument reinforces the selectivity in the design, creation, operation, and evaluation of the expert system, contrary to the ease of the operative language to understand the algorithmic combinations that should be regulated in the process of determining the accreditation of criminal responsibility. But even in the same disciplinary field that created software, the reduction in human engineers can be glimpsed thanks to the application of techniques such as backpropagation for the development of the capacity for formal reasoning. On the contrary, the company did not give details of the data combination methodology under the argument of business secret protection, but the cause was probably that the monitoring of what was processed was left out of human control by machine learning, what is commonly called "black boxes." This implication is delicate for a transparent penal system where the legal interpretation in relation to the criminal act concludes in a judicial reasoning that gives the defendants access to the legal mechanisms that may be appropriate.

With this comparison, it is possible to reflect on some implications in the construction of the criminal sentence between the human judge and the predictive algorithm:

The input data for the labeling of the level of risk imputed to a person, seem to be transferred to the clinical positivist method of criminal intervention to determine criminal responsibility, that is, to the scientific system of "statistical prognosis" to measure dangerousness – which began by being intuitive and then became technical – and which consists of studying a more or less numerous group of repeat offenders and quantifying the "causes," but it did not prosper because at the beginning they did not know how to grasp the stereotype well and, furthermore, they did not take into account that the result was the condition of the process and state institutionalization (Zaffaroni, 1998) that implies a regression and contrasts with the current state of the constitutional State of law and criminal law, a criminal law of the author and not of an act classified for a clinical study. Likewise, the scientific knowledge that a programming language invites must be solid to determine the result of justice; otherwise, it limits and opposes a right by and for society, understandable, understood, and recognized, even doing the job of the most overwhelming human judge if the software works as an auxiliary, which questions the perfect human-machine judicial binomial.

In procedural matters, the suspicion of discriminatory algorithmic bias would be related to the standard of reasonable doubt referred to in the principle of presumption of innocence, due to the lack of conviction or certainty of the existence of the criminal act and the criminal responsibility of the accused as a result of the assessment of the wealth of evidence practiced in oral proceedings (Suárez, 2011) and that the theory of crime dogmatically systematizes positive and negative elements for such an exercise. Professor Mario Caterini (2020) proposes starting from the in dubio pro reo interpretation that expert systems should implement based on factual precedents that offer interpretive options that choose the most favorable to the accused and not the statistically most frequent. With this, if it is about assisting the judge in the criminal decision, it would be in an acquittal or milder sense, which could be refuted by the judge with a more rigorous motivational charge capable of destroying the argument of the expert system. This proposal highlights the reasoning capacity of the judge, and the penal system ensures the pro-personae principle in the result of the sentence. This is because a convicting criminal decision should start from accurate and sufficient input data for due justification; In addition, for the individualization of the sentence, it would even be worth considering the incidence of the victim (Montano, 2016) and not only that of the person to whom the criminal responsibility is adjudicated.

It is difficult for the algorithmic bias to be eliminated based on the premise that obtaining the input data should start from an effective penal system that is key in the state control of violence; however, the figures show the opposite (World Justice Project, 2021). The mathematical algorithm depends on the input data to make combinations and produce an exact prediction, but if they are insufficient for a prediction that contemplates the material approach, that is, the content of truth and correctness of the premises that lead to a conclusion logically and deductively valid that requires an internal justification (Atienza, 1994), then its justice would be questioned for yielding erroneous results. It should be noted that the issue becomes more complex in an automatic learning process of the algorithm since it will be more difficult to identify the element that triggers the error in the prediction and there is a risk of continuing its operation with the permanence of such bias and feeding on the subsequent cases that resolve dictates in the same sense. This leads to consider the transparency that criminal systems try to enforce, which is diminished by black box systems since the moment of twilight in the step-by-step follow-up in the processing, prevents giving certainty of the fair result and, therefore, the motivation of the prediction could not consider in the external justification the problems of relevance, interpretation, evidence, and qualification that constitute the problems of the case and that as such are called to be a methodological guide for jurisdictional praxis (Zavaleta, 2014).

12.3 The Colonizing Private Sector of Violence Control

The idea that the control of violence is a governing function of the State is a belief that has kept us as citizens in a subordinate relationship with it, having it as a legitimate symbol of authority and representativeness to decide whether a person is innocent or responsible of what he himself has called a crime and absolve or subject him to a painful restriction of his existence, a "reasoned" violation of his human rights. The capitalist dynamic consists of economic competitiveness, and therefore, the use of technology is seen as a necessary investment. It bets on a political hypothesis, a new fable that, since the Second World War, has definitively supplanted the liberal hypothesis: "The cybernetic hypothesis," which should not be criticized but rather fought and defeated (Tiqqun, 2015). The colonizing private sector, the creator of technological developers, displaces the State in its guiding function of violence control, instrumentalizing it so that it continues selecting people and behaviors to "play with them" and make calculations to maximize profit. Technology, with economic and mathematical language, reifies people and codifies their reality, even deciding what is rational and what is not. Thus, whoever uses a computer and learns the mathematical and economic language is a winner (Schirrmacher, 2014).

In the field of criminal judicial decision, the accelerated advance of technology has gone beyond state regulatory control, giving priority to "being able to do" before "duty to do," so it is important that companies also consider the limited panorama contemplated by the law in relation to social facts at the time of the predictive work of the criminal decision and that the values that the most sophisticated algorithmic combinations can achieve are achieved from norms and principles positive in the national and international legal order, but they could not interpret those facts that are not legally identified as crimes by the principle of fragmentation, according to which only the most serious behaviors are selected as punishable within the multiplicity of the total universe, which is known as primary criminalization (Zaffaroni, 2006) or by the interpretation of multiplicity of cosmovision such as those of indigenous peoples where the uses and customs are not express and are currently part of the complex uncertainty of justice.

The AI processor represents an additional symbol of the supremacy of the criminal justice system before the recipient to legitimize the function of social control of violence, an artifact that combines with the imposing visual majesty of the courts of justice, a sacred temple where justice has a secular character and gives life to public power, which extends beyond those involved in the case that follows (Barrera, 2012) and that invites the imperfect and layperson of the technicality and algorithmic translation not to question the criminal decision before the exact decision of the processor. It is thus a strange process, alien to the moral duty to judge and the interaction between people (Batista et al., 2019). Now the case may be about technology solved by technology: a digital tribunal of a higher order to which the majority of the excluded do not belong, only those who use the technology because they have access and, of course, the small groups that control it; differentiated justice in favor of powerful minorities that symbolize the economic power of a country.

The AI offer would seem more attractive taking into account the efficiency and speed of predictive software even when the technological imperative indicates "if it can be done, it should be done," unlike the ethical imperative that precisely questioned this human technicality (González, 2004) and that seems to approach the normative context of law in which, since the creation of the concept of computer law in the sixties by Wilhelm Steinmülle, the necessary relationship between law and technology in order to regulate it was glimpsed, that is, from legislation that would prevent the abuse of technicality to the point of enabling human destruction itself.

In this sense, the risk of decisional control by technology that countries and international assemblies in the world, mainly in Europe and Latin America, speak out for a technological advance of artificial intelligence with ethics and respect for human rights is to maintain trust, government, integrity, privacy, responsibility, and reliable use so that technological development is not only technical but with a view to social welfare. The Parliamentary Assembly of Europe considers that it is necessary to create a transversal regulatory framework for AI, with specific principles based on the protection of human rights, democracy, and the rule of law (Parliamentary Assembly). Resolution 2342 identified some ways in which the use of artificial intelligence in the criminal justice system is incompatible with ethical principles (Parliamentary Assembly) and made recommendations to the countries in the legal and public policy regulatory framework to ensure state control. The UN (United Nations Organization), for its part, advocates for the so-called "digital well-being" and compliance with the 17 Sustainable Development Goals (SDGs) of the United Nations and warns about social neglect and the creation of software with a vision biased by the global north, which is why the General Assembly of the United Nations (in 2019) proposed intense scrutiny of the practices on which the creation, auditing, and maintenance of data are based as a way to counter these biases. In this way, software developers are currently making efforts to adapt them to the complex theoretical approach of applying justice respectful of human rights so that the terms "dignity," "freedom," and "transparency" are values that are measured in the algorithms. Corvalan (2018) states that for AI to be compatible with a "human rights model," it is necessary to incorporate principles into the legal framework, for a redesign of algorithm development that contemplates the new relationships between humans and machines where we can quantify and certify the robot's capacity for intuition, intelligibility, adaptability, and adequacy of objectives.

12.4 Human Function and Mechanical Function in the Penal Decision and in the Algorithmic Interpretation

The predictive algorithm was not designed to decide criminally, nor does human existence make that decision or interpret algorithms. Technology came from the infinite scientific possibilities of people who are experts in the exact sciences and, in turn, humanity is still striving to explain the origin of life beyond what science itself grants, but it is a fact that design human as a whole – mind, body, and soul – is a perfect system with a multiplicity of functions, among them the one assigned by the State, that is, the *ius puniendi* as "maintenance of the norm and as a model of the orientation of social contacts" (Jakobs, 1995). This action represents a great responsibility directly to another person and indirectly to the State and society. Ronald Dworkin's human "Judge Hercules" would seem to finally become a reality now with the "omni-artificial judge," with unrivaled wisdom and intelligence, whose reason overcomes error and human atavism, perfect and perfectible with logical and objective data that combine mathematical algorithms to approach the neuronal combination, distant from preconceptions and contaminations of lived experiences, that is, the hope of a technician justice. But in criminal matters, the function of sentencing is diverse when the conduct or omission sanctioned by criminal laws has a negative impact on another person in the form of damage, and therefore, a punishment must be attributed that requires specific characteristics that complicate the work in a preventive sense. general at the time of the consequence in objective criminal law, as well as a general preventive and special preventive at the time of sentencing and in resocialization as an effect of execution (Roxin, 1997).

The technique does not determine the judge if he recognizes his autonomy and freedom to rationalize and make decisions and only then could he coexist with the creative technique of the software. Subjectivity and human fallibility are not signs of inferiority but of definition complemented by reason and values. In the criminal decision, the human responsibility that gives meaning to the function should then cross the formal border to transform the lives of the defendants, closely because care allows us to recognize ourselves in the other and take responsibility for the simple greeting (Levinas, 2015).

A revolutionary judge is a free person to help his fellow men and put himself before the State and the company. The work of the judge goes far beyond that imposed by an abstract entity for purposes of legitimation; it consists of analyzing the case in particular and in the collective so that with the help of the automatic processor, it obtains panoramas of realities outside the situational context of the judge, mainly those that represent inequalities and violence, with the aim of applying a differentiated justice with a predictor simulator of improved realities that the judge can manipulate and invite his empathy in the understanding of the justice of the parties, even to accompany them in that their derived human rights of the crime are observed and, where appropriate, put before the State those that led to criminal conduct and verify compliance. It will also require updates as new realities become necessary.

Based on introspection, admiring contemplation, and wondering curiosity, Heidegger's thaumazein (Held, 2020) will make an abstraction and will be in a position to create a fair and transformative sentence, a work of art just as life is, as an aesthetic, original and authentic experience (Gadamer, 2006). This authenticity is defined as putting the harmful behavior in its fair dimension, highlighting all the behaviors that harm the individual and the collective, as is the case where the State itself is a participant in the damage to the citizenry or those committed by large companies. to extinguish social identity elements that indigenous peoples try to maintain and generate macrocrime where the States are only instruments of non-state mediate authors of the current financial totalitarianism (Zaffaroni, 2019). Only with sensitivity can we recognize the true damage.

In this context, the judicial function and the replacement of the human judge are a utopia, since the judge would no longer be a mere applicator of rules or an expert in argumentative techniques, but rather a person whose difference with the software is his ability to feel, understand, and understand the other from interpreting their reality as members of a community by fulfilling a single purpose, and that can be updated within the limits of society and not of the State or the company. The machine can detect feelings and emotions only by the textuality of the words. The human has ancestral supremacy, but the bond with the material surpassed the spiritual and forgot the past that is part of its historical construction and therefore otherness; this is true evolution. Human beings have a new opportunity to put into practice their ability to move from negative action to positive action, so we should not be skeptical that creativity and compassion, as exclusively human characteristics, can be applied to a reconfiguration of the human judge.

The transformations and innovations in the criminal justice systems have as their purpose an improvement in the administration of justice; that is why the citizen evaluation of the institutional service works as direct evidence of the objective of the system, which would be to understand before judging and condemning, the latter that the State is not only weak but also empty, since crime as a social problem pales before the privatization of a good part of the territory of justice and social policies (Tenorio, 2014). With a vision for and by the people in the dictation of criminal decisions, the judge himself would dignify his role with freedom and autonomy, contrary to what the mechanical performance of his current punitive function against people would seem.

Subjectivity, as a characteristic of the human being and therefore of the judge, is biological and feeds on experiential learning from the social groups with which it interacts, which biases its vision of the problems with ethical canons. Thus, the technical capacity of the human improves criminal decisions both in the design of the expert system and in the translation of direct data from the defendant himself. The algorithmic result as the reasoning of the case is the same as part of humans in charge of human fraud conflicts. The thematic classification is part of the objectivity expected by the software's claim to exactness, as well as the impartiality that is required of the human judge, even though he tries to imitate the inductive and deductive processes of the human brain from electronic circuits and advanced computer programs that they simulate human neural networks (such as Deep Blue, Watson, Project Debater, and DeepMind, or in the classic Turing test), which would seem to preserve human subjectivity.

Free reasoning and conviction, whose discretion is controlled with decision parameters for the application of the principles of conventionality and extensive interpretation, can coexist with justice and well-being that machine learning aims at, since it can mathematically "polish" the social bias, since the essence of this layered learning for the criminal decision deals with exclusively human experiences, violence between humans, something difficult to understand even for the human himself. However, deep learning runs the risk of being overestimated from attempts to anthropomorphize by misinterpreting techniques, to the point of believing that the predictive machine understands and understands its own decisions just like the human who does it with sensorimotor experiences, for the opposite is simple (Chollet, 2018) and requires a very wide dimensional space that captures the scope of the original data, that is, that manages to see reality as interpreted by the human being. That is why the so-called "neural networks" of AI resemble the human brain but cannot compete with it and is easily tricked, so the correct term should be "layered representation learning or hierarchical representation learning, or something like that, sometimes even deep differentiable models or chained geometric transformations" (Chollet, 2018), since the results they yield only depend on the input data, which in the case of criminal sentences, would be examples of previous sentences and those data are only adapted to the new case; however, it is different from what the human has developed during his life; the moments lived are fixed in his mind and are part of his perception and preconception.

Freedom and autonomy of will do not exist in a processor, because it is the human who enters the initial data for the algorithmic combination process to occur and then its update. However, machine learning, which becomes autonomous, exercises its technical freedom and decision-making without human intervention. In this objective, it hinders the fact that the context of the case is based on the law, that it is changing according to social evolution in human relations, and that engineering literally moves the processor, contrary to the infinite capacity of the human brain and mindto interpret the law with the understanding and imagination of the pain of the conduct described in law and of the people involved.

For programmers, it represents a challenge to emulate the crime based on the law and the history of violent behavior resolved by the authorities so that the machine can learn and make decisions; however, if the behaviors that define the crimes vary from case to case and therefore its resolution, the input data do little to help the accuracy of the algorithmic decision. So, the software must identify the natural capacity of the human for decision-making and then differentiate them when it comes to social harm, just as a novelist does (Posner, 2011), with life experiences, involved in a culture and processes of resilience that expose the judge to contexts that are alien to their own reality that, although they are unknown to them, have the possibility of provoking their awareness to the point of adapting their critical thinking beyond the legal mandate; that is, they have the possibility of an adaptation more fair than correct.

Recognizing that the function of sentencing is a human capacity used as a means to the end of justice, or at least that is the intention, means that human technique has an infinite possibility of simplifying the complex only with a material and secondarily formal effort. "Technique is freedom when we remember that we are facing a medium, a very complex medium, but only that. The ends must be decided by us. In other words, our freedom depends on our being able to assert the ends of each one, as human beings, through technique, art, sports, or daily life" (Lefranc, 2016).

However, without realizing it and before thinking about it, a technology from a market objective dazzles us with surprising products that try to lighten the mechanical load of the criminal judge. This technology is considered valuable because it provides data and laws that the human judge probably had not contemplated, however it could also help to legitimize the criminal justice system through a redefinition of the role of the criminal judge to make it more human, otherwise, the technique of the exact sciences would have an imperialist meaning "Dominating nature, the imperialists teach, is the meaning of all technique" (Agamben, 2006), but the technique does not determine the human being and the law is not the limit of the technique neither.

It has been necessary to rethink ourselves individually and collectively under a logic of care rather than obligations and duties to understand that the positivity of human rights is only a guide that recalls how cruel we have been as humanity, but at the same time, the ability that we have to rediscover ourselves in the essence of dignity, expanding our vision to the past – that which gave us life – with the present and with people who have the same culture and past, who know the sufferings of the social being in inequalities; this identity would seem to be eliminated by a new conquest: the corporate indoctrination that emulates the justification of the Holocaust, as Bauman explains "Victims of violence are dehumanized as a consequence of ideological definitions and indoctrination" (Bauman, 1989).

12.5 Conclusions

State control of violence is defined mainly by the exemplary nature of its theoretically rational decisions in its construction. The complexity it represents is due to the adjustment of the specific case with the law, the context of the procedural parts, and the criminal act, coupled with a changing legal-criminal theoretical approach to the application of *ius puniendi*. The criminal judicial decision-making function also represents a responsibility with the Constitutional State of Law, in which, the control of conventionality is inserted in the principle of legality so that the norm is interpreted extensively to guarantee the pro-personae principle. The coexistence between norms and principles causes, in certain cases, the lack of definition of these in the legal framework, since one of their characteristics is that they are not necessarily expressed, which is sometimes used by judges to expand their discretion and appeal to the called "judicial activism" as a form of legal disobedience.

However, the international regulatory framework of human rights indicates that the ultimate purpose of applying the law is respect for the human rights of people who, in a criminal decision of conviction, can be analyzed from the reasoned application of the amount of punishment. Predictive software is currently used in judicial decision-making contexts, given the above scenario, an effective help to the human judge for the correct dictation of his sentence could be observed, especially due to the application of deep learning, which shows effectiveness and speed. However, errors and biases have been evidenced in the results of the processes that start from the input data of the processor, which may be insufficient or biased. Although the modernization of cutting-edge investigations provides that, in addition to laws and reference cases, software translates complex arguments, emotions, and human rights violations, which will be of great help if the operators have control of the process, especially with the application of machine learning.

International assemblies have detected that AI has a market objective and that it does not contribute to a social improvement for the reduction of inequalities. In this sense, the human consequences of data management, "dark boxes" and technical misunderstanding on the part of citizens, would be negative considering that the control of the design, creation, and implementation is held by a small group of power, mainly business with an own vision according to its context, which is different from the reality of most of society. In such a way that the results not only have a threshold of inaccuracy, but that threshold would represent an injustice, even more so with the risk of the judge depending on the machine to facilitate and speed up his work, causing the machine to perpetuate the bias with incorrect input data.

The effective relationship between the criminal judge and the machine requires a national and international regulatory framework that ensures a correct and fair operation of the software and represents a true auxiliary and never human substitution since the advance of AI has been detected, regardless of the legality derived from the legislative omission on the subject. In addition to this, the State must retake control of the violence by allowing algorithmic intervention in the criminal sentence, if it coincides with the legal and contextual elements of the people and the crime in the specific case because otherwise, judicial activism will now continue through the machines.

The judge, for his part, must reconfigure his judicial role from the assumption as a human being with dignity that identifies him and makes him responsible for the people who are the recipients of his criminal decision, with a commitment to a correct and fair application of justice with the situation of the case and of the people involved and that the normative framework and principles serve this purpose and, therefore, avoid the so-called "judicial error." In this way, not only the judicial authority is legitimized but also the institution it represents, which must also modify its evaluation schemes to add to the technical, the human, that is, an effective application of justice that looks for the improvement of life of people for a decision that is not only punitive but remedial.

That is why the international claim that technology redirects its efforts to social aid will only be possible if the actions of the State are directed outside the logic of the market so that technology is transparent, understood, and understood for the construction of an equal world. The software development companies, for their part, must modify their capitalist sense for a social one for the fulfillment of collective welfare as a preventive faculty of the State.

References

Agamben, G. (2006). *Lo abierto. El hombre y el animal*, Buenos Aires, Adriana Hidalgo.

Angwin, J., Larson, J., Mattu, S., y Lauren, K. (2016). Machine Bias, ProPublica, disponible en https://bit.ly/3pUoLPX.

Atienza, M. (1994). Las razones del derecho. Sobre la justificación de las decisiones judiciales, en *Isonomía*, vol. 1, pp. 51–68.

Baratta, V. (1982). Criminología crítica y crítica del derecho penal. Introducción a la sociología jurídico-penal, trad. Álvaro Búnster, Siglo XXI Editores.

Barrera, L. (2012). *La Corte Suprema en escena: Una etnografía del mundo judicial*, Buenos Aires, Siglo XXI (ed.)

Batista, N. Navarrete. E. C., León, C., Real, M., Chiriboga, J., y Estupiñán, J. (2019). La toma de decisiones en la informática jurídica basado en el uso de los Sistemas Expertos, en *Revista de Investigación Operacional*, vol. 40 (1), pp. 131–139.

Bauman, Z. (1989). *Modernidad y Holocausto*, Madrid, Sequitur.

Chollet, F. (2018). *Deep Learning with Python*, Nueva York, Manning Publications.

Caterini, M. (2020). Il giudice penale robot, en *Legislazione penale, Giustizia penale e nuove tecnologie*, disponible en chrome-extension://efaidnbmnnnibpcajpcglclefindmkaj/https://www.lalegislazionepenale.eu/wp-content/uploads/2020/12/Caterini-Il-giudice-penale-robot.pdf.

Corvalán, J. (2018). Inteligencia Artificial: retos, desafíos y oportunidades – Prometea: La primera Inteligencia Artificial de Latinoamérica al servicio de la Justicia, en *Revista de Investigações Constitucionais*, vol. 5 (1), pp. 295–316.

Dieterich, W., Mendoza, C., y Brennan, T. (2016). *COMPAS Risk Scales: Demonstrating Accuracy Equity and Predictive Parity*, Northpointe Inc., vol. 7, disponible en https://bit.ly/2S4HRCC.

Ferrajoli, L. (1985). *Derecho y Razón. Teoría del Garantismo Penal*, trad. Perfecto Andrés Ibáñez, Alfonso Ruiz Miguel, Juan Carlos Bayón Mohino, Juan Terradillos Basoco y Rocío Cantarero Bandrés, (1989), Madrid, Trotta.

Gadamer, H-G. (2006). *Estética y Hermenéutica*, Madrid, Tecnos-Alianza.

González, G. (2004). El imperativo tecnológico: una alternativa desde el humanismo, en *Cuadernos de bioética*, vol. 15 (53), pp. 37–58.

Held, K. (2020). Asombro, tiempo, idealización. Sobre el comienzo griego de la filosofía, en *Estudios de Filosofía*, vol. 26, pp. 63–74.

Jakobs. G. y Cancio, M.(2006). *Derecho penal del enemigo*, Madrid, Civitas.

Jakobs, G. (1995). *Derecho penal: parte general. Fundamentos y teoría de la imputación*, Madrid, Marcial Pons, Ediciones Jurídicas

Lefranc, F. (2016). Los sujetos de la SIC, en E. Téllez (coord.), *Derecho y TIC. Vertientes actuales*, Instituto de Investigaciones Jurídicas, Ciudad de México, UNAM.

Levinas, E. (2015). Ética e infinito, Madrid, Machado Libros.

Malem, J. F., Ezquiaga, J., y Andrés, P. (2012). *El error judicial. La formación de los jueces.* Ciudad de México, Fontamara.

Manzanera, L. R. (2005). Criminología Clínica, quinta ed., México, Porrúa.

Medvedeva, M., Vols, M., y Wieling, M. (2020). Using machine learning to predict decisions of the European Court of Human Rights, en *Artificial Intelligence and Law*, vol. 28, pp. 237–266.

Montano, P. (2016). *Incidencia de la víctima para la individualización de la pena*, Montevideo, Instituto de Derecho Penal, vol. 5.

Parliamentary Assembly. (2020). *Artificial Intelligence: Ensuring respect for democracy, human rights and the rule of law*, disponible en: https://pace.coe.int/en/pages/artificial-intelligence.

Posner, R. (2011). *Cómo deciden los jueces*, Madrid, Marcial Pons.

Roxin, C. (1997). *Derecho penal. Parte general, I. Fundamentos. La estructura de la teoría del delito*, Madrid, Civitas.

Schirrmacher, F. (2014). *Ego: Las trampas del juego capitalista*, Barcelona, Ariel.

Sourdin, T. (2018). Judge v robot? Artificial intelligence and judicial decision-making, en *UNSW Law Journal*, vol. 41 (4), pp. 1114–1133.

Suárez, J. L. (2011). Inferencia razonable, probabilidad de verdad y conocimiento más allá de toda duda razonable, en *Principia Iuris*, vol. 16, pp. 307–329.

Tenorio, F. (2014). El delito y el control del delito en la modernidad avanzada, Alemania, Publicia.

Tiqqun. (2015). *La hipótesis cibernética*, Madrid, Machado Libros.

World Justice Project (2021). Disponible en: https://worldjusticeproject.org/index.

Zaffaroni, E. R. (1998). Criminología. Aproximación desde un margen, Bogotá, Temis

Zaffaroni, E. R., Alagia, A., y Slokar, A. (2006). *Manual de Derecho Penal. Parte General*, segunda ed., Argentina, Ediar.

Zaffaroni, E. R. (2019). *La nueva crítica criminológica. Criminología en tiempos de totalitarismo financiero*, Buenos Aires, Ediar

Zagrebelsky, G. (1992). *El derecho dúctil: ley, derechos, justicia*, quinta (ed.), trad. Marina Gascón, 1995, Madrid, Trotta, (2003).

Zavaleta, R. (2014). Los problemas de justificación externa como problemas del caso, en M. García Y R. Moreno (coords.), *Argumentación jurídica. Fisonomía desde una óptica forense*, pp. 127–150, Ciudad de México, UNAM.

13

AI in Music: Implications and Consequences of Technology Supporting Creativity

Carolina Sacristán-Ramírez
Tecnológico de Monterrey

Flavio Everardo
Tecnológico de Monterrey
University of Potsdam

Yamil Burguete
Tecnológico de Monterrey

Brecht De Man
PXL University of Applied Sciences and Arts

CONTENTS

DOI: 10.1201/b23345-15

13.1 Why Puccini for Artificial Intelligence? An Introduction

At first glance, artificial intelligence (AI), early 20th-century opera, history, human decision-making, and creativity may seem like unrelated fields. Our goal is to string the fine line uniting all of these concepts, making this work a starting point for scholars and researchers interested in solving relevant technology-related problems, including diverse perspectives or considerations surrounding art music, and AI. As a motivating example, we will start with one of the most brilliant and paradigmatic cases of musical composition tragically left incomplete; the last opera of the Italian composer Giacomo Puccini (1858–1924).

The compositions of Giacomo Puccini usually cause a fantastic first impression. They can be appreciated by almost anyone regardless of their familiarity with art music, especially with opera. Who was Puccini? Born in Lucca in 1858 into a family of organists and choirmasters, Puccini is one of Italy's most famous composers. He studied at the Milan Conservatory with Antonio Bazzini (1818–1897), who was a violinist and composer, and Amilcare Ponchielli (1834–1886), the author of the opera *La Gioconda* (1876). Puccini's first work for the stage was *Le villi* (1883), which was submitted to a contest. Even though it lost, Puccini's supporters funded the opera production, and its debut had great success. *Manon Lescaut* (1893) brought Puccini international recognition and kept him on the track of opera composition. His mature works include *La Bohème* (1896), *Tosca* (1900), *Madama Butterfly* (1904), *La fanciulla del West* (1910), and the unfinished *Turandot* (1924) (Berger, 2005).

At the age of 60, Puccini studied the developments in contemporary music and set out to write an opera that was different from his typical style. He based the new work on the libretto by Giuseppe Adami and Renato Simoni after Count Carlo Gozzi's drama about the cruel Chinese princess Turandot. The story focuses on an enthralling princess who has all her suitors beheaded unless they can solve three riddles that she poses to them. One day, she meets her match: a mysterious prince named Calaf who not only defeats her but pledges to win her heart. The completion of the opera was truncated due to Puccini's ill health. He died of a heart attack on November 29, 1924, after an operation for throat cancer in Brussels, Belgium (Marchese-Ragona et al., 2004). The conductor Arturo Toscanini entrusted Turandot's completion to Franco Alfano (1875–1954), a composer whose opera *La leggenda di Sakuntala* (1921) had recently succeeded. The American composer Janet Maguire states that Puccini had the music for the duet and the final scene of Turandot clearly in mind before his death. If Puccini had survived the operation, he would have had the opera ready for the publisher Ricordi by the end of 1924. The key to the completion of Turandot lies in relatively few pages of musical ideas written by Puccini in shorthand. These 36 sketches were the only notes Puccini left to indicate what he wanted for the completion of his opera (Maguire, 1990).

When Alfano started working on the finale, Ricordi emphasized that his work should be based exclusively on Puccini's sketches. But he did not use all the sketches, even though some were quite clear. Instead, Alfano resorted to his own music style to complete the finale. Moreover, he looked at the orchestration of the completed parts of the opera only twenty days before he had finished his work. As Maguire points out, Alfano could not incorporate Puccini's orchestral thinking in such a short period. As a result, when Toscanini listened to Alfano's composition, the differences in style and conception stood out clearly to him. He tried to bring Alfano's composition closer to the finale Puccini played for him before his death. Toscanini restored music from the sketches and cut Alfano's version, creating a second version of the finale (Maguire, 1990). However, when Toscanini conducted the world premiere of Turandot at La Scala Opera on April 25, 1926, after the death of Liù—a slave girl who was secretly in love with prince Calaf—, he put down his baton, turned to the public, and said, "Here is where the opera ends, because at this point the Maestro died" and left the stage (Maguire, 1990, as cited in Gara, 1975, p. 170).

Other authors attempted to create a fresh completion for Turandot in the 20th and 21st centuries. From 1976 to 1988, Janet Maguire (1990) worked on a new ending based exclusively on the sketches left by Puccini. In 2001, the Italian composer Luciano Berio wrote a second finale using Puccini's sketches and seeking to expand his musical language (Uvietta, 2004). In 2008, Hao Weiya, a Chinese composer trained in Italy, created a third version of the finale, incorporating the traditional Chinese tune "Jasmine Flower" and avoiding post-romantic harmonies (Siegel, 2008). Regardless of professional criticism and public reception of the different versions of Turandot's ending—Maguire's finale has never been performed on the stage—it is evident that the history of Puccini's Turandot is full of unexpected circumstances, individual musical styles, ideas and approaches, human interpretation, and decisions. These were made under different contexts and creative biases and applied to solve one problem: how to understand Puccini's musical thinking more deeply, including its continuities and changes over time, to decipher better the code—paraphrasing the title by Fisher (2013)—implied by Puccini's last musical sketches. For that reason, it would be worthwhile to wonder about technology as a tool that might help find new and different solutions to create more music, but especially to discover new paths in the development of AI. It is time to further explore the possibility that art music from over a century ago might nurture what has been hailed as the ultimate future of technology. For that purpose, we will set an essential context about what AI is and what is its role in the music domain.

13.2 These Hopeful and Intelligent Machines

As a general statement, we might say that Alfano, Maguire, Berio, and Weiya resort to their knowledge of the musical language, composition techniques, and personal analysis of Puccini's oeuvres. Some of them could have closely studied the finale for Turandot created by other colleagues. This assertion can be reframed as composers using their intelligence to solve the problem of an unfinished opera. However, this simple assertion raises one question: What is intelligence?

Within psychology, intelligence is a difficult concept to define and measure. Its complexity is due to the evolving nature of its features. Sternberg (2018) has defined intelligence as the capacity to learn from previous experiences using cognitive strategies to adapt to and shape environmental inputs but attending to sociocultural implications related to the response to external stimuli. Moreover, the adaptive nature of intelligence results from the interaction between the human being, a specific task, and a given situation (Sternberg, 2021).

Due to the difficulty of explaining human intelligence, there is not a unified characterization of AI either. Russell and Norvig (2010) have classified four definitions from various authors, which share the idea of fidelity to human performance. In 1978, Bellman defined AI as "the automation of activities that we associate with human thinking, activities such as decision-making, problem-solving, learning" (Russell and Norvig, 2010, p. 2). Almost a decade later, in 1985, Haugeland described it as "the exciting new effort to make computers think [...] machines with minds, in the full and literal sense" (Russell and Norvig, 2010, p. 2). In 1990, Kurzweil said that AI "is the art of creating machines that perform functions that require intelligence when performed by people" (Russell and Norvig, 2010, p. 2). A year later in 1991, Rich and Knight characterized AI as "the study of how to make computers do things at which, at the moment, people are better" (Russell and Norvig, 2010, p. 2).

These definitions follow the principles of John McCarthy, who, in 1955, coined the term AI as the science and engineering of making intelligent machines (McCarthy, 2004). Since then, the promise of AI has been to develop machines with intelligence comparable to the one that humans possess (Chollet, 2019). *These hopeful machines*[1] (as the 2011 Grammy-nominated album of the American electronic musician and producer BT) are gifted with the intelligence to think and act like human beings with abilities such as learning, reasoning, and perceiving (Russell and Norvig, 2020). These features have laid the foundation of the test proposed by Alan Turing in 1950 in his article Computing machinery and intelligence (Turing, 1950) with the purpose of answering "can a machine think?". A computer is deemed to have passed when the human interrogator cannot tell if the answers come from a person or a machine.[2] It is worth noting that for a computer to pass the

---

[1] Electronic composer BT talks about his new album These Hopeful Machines, about his love for math and his passion for musical technology. https://www.wired.com/2010/02/bt-these-hopeful-machines/.

[2] The Turing Test was designed to provide a satisfactory operational definition of intelligence. However, it is not a test that considers the measure of intelligence but more a philosophical discussion and far from being the ultimate goal of AI.

Turing, it must have different capabilities like knowledge representation and reasoning (KRR) and machine learning (ML) (Russell and Norvig, 2020).[3]

13.2.1 Knowledge Representation and Reasoning

KRR is the foundation of AI, declarative programming, and the design of knowledge-intensive systems capable of performing intelligent tasks (Baral, 2003; Van Harmelen et al., 2008; Gelfond and Kahl 2014). To solve these, sufficient information of the world or environment must be represented in the form of knowledge (Baral and Gelfond, 1994; Russell and Norvig, 2020; Gelfond and Kahl, 2014). This knowledge is used to solve complex tasks through decision-making, through reason and inference of outcomes (Gelfond and Kahl 2014; Russell and Norvig, 2020).

Answer set programming (ASP) is many things in one. On the one hand, it is a declarative problem-solving approach, initially tailored to modeling problems in the area of KRR (Gelfond and Lifschitz, 1988; Gelfond, 2008, Lifschitz, 2008; Lifschitz, 2019) with roots in logic programming (Baral and Gelfond, 1994). On the other hand, ASP is a rule-based formalism for modeling and solving knowledge-intense combinatorial (optimization) problems (Gebser et al., 2012; Schaub and Woltran, 2018; Eiter et al., 2009; Gelfond and Kahl 2014) with a well-developed theory, efficient reasoning systems, methodology of use, and a growing number of applications (Erdem et al., 2016; Falkner et al., 2018).

ASP is well-known for its combination of a declarative modeling language with highly effective solving engines such as *clingo* (Gebser et al., 2014, Kaminski et al., 2017), *dlv* (Leone et al., 2006), or *wasp* (Alviano et al. 2013) enabling the user to concentrate on specifying—rather than programming the algorithm for solving—a specific problem. The solutions to a problem are referred to as *answer sets*. ASP has proven successful in both academics and industry in areas like planning, scheduling, configuration design, biology, logistics (Erdem et al., 2016), and music (Everardo, 2020).

13.2.2 Machine Learning

ML is the ability of a computer to learn from previous experiences (training) and adapt to new circumstances following either a supervised or an unsupervised approach (Zhou, 2021; Bishop and Nasrabadi, 2006). ML learns from a massive amount of data without being explicitly programmed, oriented to make predictions, classify new information, and detect patterns following an unsupervised learning approach. On the other hand, following a supervised

---

[3] Russell and Norvig (2020) include other fields in which AI has been developed as part of the Turing Test, such as natural language processing, computer vision, or robotics. In this work, we are only focusing our attention on areas that are relevant to the subject we are addressing.

learning strategy, training focuses on deducing the general rules by mapping inputs and outputs (Russell and Norvig, 2020).

Taking a step further, deep learning (DL, also known as neural networks) refers to a subset of ML using multiple layers of computing elements (Russell and Norvig, 2020; Goodfellow et al., 2016). Over the years, with the growth and development of graphics processing units (GPUs) (Krizhevsky et al., 2017), DL has become more popular and relevant to solve increasingly complicated problems like Generative Pre-trained Transformer 3 (GPT-3; Floridi and Chiriatti, 2020). This is aligned with an increase in available training data, allowing learning models expand in size without being a concern to the computer resources (Goodfellow et al., 2016). The benefits of applying DL to a vast quantity of applications (Deng, 2014; Deng and Yu, 2014) make evident its relevance in diverse fields, including but not limited to digital signal processing, audio processing, and music (Briot et al., 2017; Gioti, 2021; Miranda, 2021).

13.3 Music-Making with AI

Music and its production have a rich and broad history, with milestones like the invention of stereophonic sound in the early 30s by Alan Blumlein, different formats for storage and playback, including tape cassettes (1963 by Philips) and compact discs (1982 by Philips and Sony), the move from analog magnetic tape to digital recordings and playing music on mobile devices (Burgess, 2014). However, the technology behind music production over the past decades has made few advancements, mainly focusing on digital modeling of analog equipment with the inclusion of cognification in the music-making process.

Cognification is the process of making things intelligent by using AI algorithms, enabling the development of fields such as intelligent music production (IMP; Moffat and Sandler 2019, De Man et al., 2019), which introduce some level of AI into the music-making process (Palladini, 2018; Moffat and Sandler, 2019; Miranda, 2021).[4] In the words of David Moffat and Mark B. Sandler (2019), IMP has the premise to fundamentally change the manner in which engineers and consumers interact with music by not only allowing collaboration between humans and intelligent systems but also exploring and understanding new dimensions and ideas within the space of music production. These intelligent systems are required to observe or perceive the environment in order to act and make decisions to accomplish desired goals.

---

[4] We refer to the term music-making in the broadest sense, from composition to mastering. Moffat and Sandler direct their work only in the context of post-production processes such as mixing and mastering.

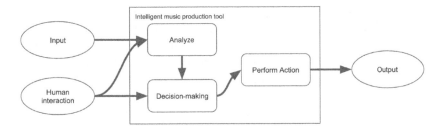

FIGURE 13.1
Generalized flow diagram of an IMP tool. IMP, intelligent music production. (Adapted from Moffat and Sandler, 2019.)

This is congruent with concepts from Russell and Norvig (2020) about a computer able to automatically make decisions, think and reason about those decisions, draw conclusions, and learn from the experience while accumulating knowledge. Figure 13.1 shows a modified diagram from Moffat and Sandler (2019) depicting a general workflow involving interaction with an IMP system. Input is given to the system, and after the input analysis, decisions are made to perform actions before the output is returned to the user. This general diagram relaxes the idea of providing audio input, expecting modified audio as output in contrast to what Moffat and Sandler stated. The input can be audio, music scores, or written text. Similarly, the output could be audio generated from scratch or the modification of an audio source, as well as music scores, visuals, graphics, or text.

Following the principles from Moffat and Sandler, an IMP system must have a predefined scope of action to work. This is materialized in degrees of participation or levels of control of AI in the music-making process. In other words, the realm of possibilities of what AI can (or cannot) do is connected with a specific scope of responsibilities. This appointee of accountabilities is represented in levels of automation for AI, usually consisting of 4 or 5, defining the amount of control we are willing to delegate. We list below the integration of all levels characterized by different authors, namely, Brecht De Man (2017), David Moffat, Mark B. Sandler (2019), and Alessandro Paladini (2018):

1. **Automatic:** It is a fully automated approach where the system is seen as a black box and no human interaction is required. In other words, the system takes complete control of the inner processes and the entire outcome. This approach is desirable for non-experts who count on a system to do the job completely and autonomously.

2. **Independent:** It is an assistant where specific tasks are delegated, providing a close-to-finished result. The user acts as a supervisor and can overrule decisions made by the system.

3. **Suggestive**: The system participation is limited to the analysis and recommendations proposition, while the user has complete control of the outcome. It is functional to set starting points for the user to continue or suggest alternatives if issues in the input are found.

4. **Insightful**: The user has the greatest level of control over the system. This level provides additional insights, textual data, visualizations, among other things to advise the user of information or knowledge, resulting in faster or better decisions to make. This is the ideal solution for professionals and industry experts who are used to having full control over their work.

IMP also tackles user experience and interface design. De Man (2017) described intelligent systems offering more intuitive controls rather than the daily-used digitized analog controls with semantic descriptors that entail inner processes hidden from the user. The user can tell the system to perform filtering or equalization to reduce the *boxiness* of a track or ask the system to make the drums sound *punchy* (De Man, 2017; Stables et al., 2014).

13.4 Solving Puccini

There are several ways to solve the last part of Turandot in the context of AI. For instance, Artificial Intelligence Virtual Artist (AIVA, 2016), the first virtual composer, is capable of composing emotional soundtracks for films, video games, commercials, and any type of entertainment content. It can compose based on specific references. In this regard, we could upload scores from Puccini, Turandot, and the finales from Alfano's, Maguire's, Benio's, and Weiya's to have new possible solutions for what could be the end. However, this chapter is not intended to propose another closure for Turandot or discuss what Puccini could have done, but to focus on the uses of conjunction specific areas of AI with certain levels of control.

By using KRR and ML, possible outcomes might emerge from using each AI field on its own with a given level of control. Table 13.1 shows distinct solutions matching each level of control with each AI field. Each cell has specific inputs considering the requirements for the AI to perform, followed by expected output.

13.4.1 Solving with KRR

The nature of KRR forces the acquisition of more knowledge when less human involvement is required. The idea of a black box system needs an entire body of knowledge to perform with minimum or no input from the user. The

TABLE 13.1

Implications and Consequences of Performing KRR and ML to Each Level of Control

Levels of Control	Fields of AI	
	KRR	**ML**
Automatic	**Input**. The entire knowledge with rules and constraints of Puccini's compositions, styles, preferred instrumentation, the tonality of Turandot, among others. **Output**. Fully automated music composition for the ending of Turandot.	**Input**. A vast and considerable amount of data for intense analysis and feature extraction. **Output**. Fully automated unsupervised music composition for the ending of Turandot.
Independent	**Input**. Partial knowledge from Turandot and Puccini depending on specific areas to solve. **Output**. Specific tasks are solved to complete the user's (composer) work without intervention. The user is free to use the proposed solution, modify it or discard it.	**Input**. A vast and considerable amount of data for intense analysis and specific feature extraction. **Output**. Specific parts of the completion of Turandot following a semi-unsupervised learning approach. The user is free to use the proposed solution, modify it or discard it.
Suggestive	**Input**. Specific knowledge from Turandot and Puccini depending on areas to solve. It may include an initial composition from the user with blanks. **Output**. Complete starting points pronouncing ideas for further development, or completion proposals for the input piece. Recommendations of specific attributes with supporting reasons.	**Input**. A considerable amount of data with a non-exhaustive analysis for specific feature extraction for composing. **Output**. Complete starting points or ideas or completion proposals for the input piece. Specific parts of the completion of Turandot following an unsupervised learning approach. Recommendations of dedicated attributes with supporting reasons learned from the training data set.
Insightive	**Input**. More general knowledge from music theory capturing rules and constraints with not necessarily any from Puccini and Turandot. **Output**. Additional broad insights to correct errors or advise the user.	**Input**. Prior works from Puccini, including Turandot, followed by analysis for specific pattern extraction. It must include the work in progress from the user for analysis. **Output**. Additional broad insights to correct errors or advise the user.

KRR, knowledge representation and reasoning; ML, machine learning.
Notes: The amount of knowledge or data AI needs to perform varies depending on the level of control, revealing a natural correlation that the more participation of the user, the less knowledge or training may be needed.

system is equipped with the necessary representation of the world (in this case, music composition and neighbor areas) so that the computer can use this knowledge to solve challenging tasks through reasoning and decision-making. Composing music is an extremely complex task involving several elements, including harmony, texture, rhythm, and structure. Providing all the rules and constraints to a computer to compose means providing it with enough information that can lead it later to propose new musical solutions.

Everardo (2020) presented the state-of-the-art of ASP across the multiple stages of the music production process. Initial works on automatic music composition with ASP (Boenn et al., 2008; Everardo and Aguilera, 2011) presented this approach where few parameters were given, including tonality, length or duration, and tempo. The ASP solver is responsible for generating the entire composition and delivering score sheets or audio files. The given knowledge is usually genre-specific, consisting of explicit rules and constraints for composing such as melodic and harmonic rules, tonality constraints, or rhythmic progressions. In fact, these systems performed compositions in the style of Palestrina (Boenn et al., 2008) and for Trance music (Everardo and Aguilera, 2011), both of which are packed with predefined rules and constraints to compose within the limits of the genre. To solve the last part of Turandot, we need to capture absolute knowledge of what Puccini did in the first part of Turandot, as well as in other pieces.

As the field started to develop, other works proposed specific tasks to solve relaxing the black box concept and being more of an assistant. Opolka et al. (2015) presented an automatic music composition system that creates simple accompanying pieces of different genres focusing on chord progressions and harmony rules from the early 19th to the late 20th century, with a strong influence from Arnold Schonberg's Harmonielehre.

Releasing more control from AI to perform, other works attacked specific areas such as chord cadences (Eppe et al., 2015) or completing a given score (Boenn et al. 2008; Cabalar and Martín, 2017). In practice, all the works shown in Everardo (2020) serve as starting points rather than a complete and definitive solution. The works from Boenn et al. (2008) and Everardo et al. (2020) can be categorized as the closest to an *insightive* approach. Both provide options to accept compositions as input, allowing the system return information (as text) with detected anomalies or errors such as tonality, structure, or harmony errors.

Lastly, an intermediate work between KRR and ML was presented in Pinetzki (2020) but all performed within ASP. The system takes musical pieces converted to MIDI files and extracts different patterns. These patterns form essential blocks to compose music ideally close to the input data. This proposal is still a black box approach. However, the system does not need to have a vast number of rules and constraints representing individual music styles because the resultant compositions are made from the input observations. This made it a genre-independent system.

13.4.2 Solving with ML

Taking the music of a specific composer as the starting point for creating AI implies the use of ML methods. With the ability of a computer to learn from previous experiences while training to adapt to new circumstances,

the machine becomes capable of making predictions, classifying information, and detecting patterns. These abilities are relevant to music because of
its inherent complexity. First, it carries a set of codes and patterns within a
hierarchical structure with dependencies across time. Second, music consists
of several sounds produced by human voices or musical instruments that
are interdependent and unfold across time. Third, sounds are grouped into
melodies and chords in combinations that at each time step may have multiple outputs. The execution of the composer with these elements is unique.
It gives the music what a more-experienced audience identifies as the composer's personal seal. A computer learning the distinctive style developed by
a musician through the years may be able to propose diverse solutions to create more music similar to what the human composer could have produced.
A machine capable of understanding emotions implied in text such as the
script or libretto might provide relevant clues to other plausible solutions.

It is imperative that more training is required when a computer is expected
to do the entire solution without human involvement. When exploring creativity and solutions in art music, it is unlikely to apply a (fully) supervised
learning approach. The reason is we cannot provide outputs from specific
inputs for the machine to learn. In unsupervised learning, it is possible to
give additional data such as other endings of Turandot. In this manner, the
computer with a given level of control can provide alternatives using the
entire Puccini history or even mix with Alfano's, Maguire's, Benio's, and
Weiya's finales.

Definitely, this area is the one for trying to understand Puccini's musical
thinking more deeply. The algorithms or systems in this domain can learn
musical knowledge without the need to sit a machine in a music conservatory. But not everything is related to a score. The interpretation or performance of a musician is key for this area to learn. Following the score sheet is
similar to following a map. The nuances and expressiveness of each instrument performer are unique and vary every time. A computer learning the
distinctive style developed by a musician through the years may be able to
propose diverse solutions to create more music similar to what the human
composer could have produced.

AIVA (2016) uses DL to look for patterns and rules in compositions and
uses this information to learn the basics of style and music. During the composition process, AIVA predicts what should come next on the track. The
system's main purpose is to support the rapid ideation of potential and, at
the same time, create a producer–session–musician relationship (Gioti, 2021)
taking into account the extensive learning process from a large collection
of music written by Mozart, Beethoven, Bach, etc., to create a mathematical
representation of what music is (Cope, 2021).

Solving Turandot requires fulfilling certain conditions that are not so simple to achieve due to data volume. This is noticed in Sterne and Razlogova
(2021) with the example of LANDR as a response for automated mastering
service. A specific group of mastering practices that mimic a certain "taste,"

an aesthetic that is reshaped or mutated over time, as well as technical aspects of the mastering process, needs to be considered to train the system.

An overview of DL applications in music including challenges and further directions are listed in Briot et al. (2017) and Briot and Pachet (2017).

13.5 Implications and Consequences for Art Music Endeavors

Besides selecting your AI field of choice and the level of control to initiate your journey with AI and music, there are questions to address about several surrounding aspects of AI in the music-making process. These characteristics apply either for composition or for mixing, and they are acceptance, trustworthiness, efficiency, and search space.

13.5.1 Acceptance

Recent years have seen a marked increase in acceptance of technologies, like AI, in creative settings, like music production (De Man et al., 2019). The initial hesitance is neither new nor surprising. Even in more prosaic areas like manufacturing or transportation, progress is often met with resistance as dystopian scenarios go from displaced jobs to any societal, economical, or health-related havoc. These fears are, at times, misguided or exaggerated and, at times, legitimate and prophetic with typical examples of controversial inventions in textile machinery (spurring the Luddite movement in 1811, which became a derisive synonym for a person resisting any type of technological progress), transportation, and communication devices.

Creative disciplines add another obstacle as the idea of a machine having artistic agency is deemed laughable or offensive by some. In music, strong reactions were evoked by the likes of drum machines in the eighties (the concern that they would replace human drummers has not quite materialized), Auto-Tune in the nineties, and the laptop musician's toolset in the noughties. While the aesthetic impact such technology had and continues to have is certainly subjective, it is hard to argue that the history of music and the history of technology are not intertwined. The availability of novel manufacturing processes, electronic devices, and computing power gives rise to new instruments, recording and amplification, and entirely new genres. The converse is also true to some extent: artistic developments and desires inform innovation in music making tools.

As to why the prevalence and the acceptance are rapidly increasing at the time of writing, we can look to increase in computational power, inspiration from lucrative applications including natural language and image processing, and hunger for the "next big thing" after the digital revolution. There is little doubt that time itself helps the industry and audiences getting used to the idea of AI. The ongoing discussion of computers as creators of art has been thoroughly discussed by Hertzmann (2018), who argues that tools cannot be acknowledged as creators. But he also recognizes that modern art creators use elements like AI to bring forth their creations.

13.5.2 Efficiency

More palatable applications of AI in music are those which are menial, time-consuming, tedious, and technical in nature: this will have little effect on the outcome but save time and energy for the musician or producer who wants to focus on creative developments. These are the areas where most creative specialists prefer the implementation of AI solutions. On the other hand, they are very opposed to unsupervised proposals of creative tasks (Pfeiffer, 2018). In other words, most professional practitioners prefer the use of AI as an assistant that improves their workflow in order to expedite their creative process.

Allowing human input or correction further ensures that AI plays an assistive and net positive role, unless the application dictates a "black box" solution—for instance, when the operator has no ability or interest to influence the result.

By extension, while the ostensible goal of musical AI can be to fully automate some aspect of the music production process, this does not preclude unintended uses where the AI serves as a source of inspiration or a co-creator.

13.5.3 Trustworthy AI

In this text, like virtually everywhere else, the term AI is used with little consideration for what it actually means. A strict definition is bound to have serious limitations as the concept is context-dependent. Furthermore, the use of the term evolves as new technologies gain ground, and others are taken for granted. Many forms of computer-aided and automated processes (Roads, 1996) have been classified as AI as they replace part of a human's cognitive or administrative task. The fact that it frees up some time and energy that would be spent thinking by a person, yet limited, makes it AI. However, to call a system intelligent nowadays, the bar is typically higher. Many will call a system intelligent only if it actually learns from new input—be it in the development stage or during use.

There has been an increment in requirements to use AI responsibly, like benchmarks that establish the rules for building trustworthy AI (Floridi, 2019) by means of formal verification to provide provable guarantees and therefore an increment of trust that the system will behave as desired (Wing, 2021). We can have the same outcome being a composition or a mix from a black box AI system, a computer-aided tool, or a stochastic approach. In either case, the three approaches may not give feedback or information to the user about what is happening behind the scenes. In this regard, the user must be able to ask for the trace of the performed decisions.

People do not trust things that they do not understand (Palladini, 2018). Trust in AI comes with the quality of the results. If the result is good (because of trust), the rest becomes unessential. To trust an autonomous system, the user must be acknowledged by AI and its decisions. This is where explainable AI (XAI; Gunning et al., 2019) is relevant as a field of AI that promotes a set of tools, techniques, and algorithms and can generate high-quality interpretable, intuitive, human-understandable explanations of AI decisions (Das and Rad, 2020). In other words, XAI supports the need of conveying safety and trust to users in the "how" and "why" of automated decision-making in different applications (Confalonieri et al., 2021).

The implications of XAI varies in the AI field. For instance, in ML, evidence exists that training conditions the outcome (Barocas and Selbst, 2016). ML computes probabilities of symbolic sequences, and it is trained to give such sequences from its training dataset with high probability, with concrete implications of creative ownership of works (Sturm et al., 2019). With enough data available, ML algorithms may obtain highly accurate predictions, but many ML techniques act as black boxes without the capabilities to provide a verifiable explanation for their results (Cabalar et al., 2021).

In the KRR domain, an ASP program does not contain information about the method to obtain the answer sets, being this task fully delegated to the ASP solver, and the information provided by answer sets may be insufficient to explain the decisions (Cabalar et al., 2020; Fandinno and Schulz, 2019; Trieu et al., 2022).

This justification of a system that makes automatic decisions is relevant to the transparency with different applications, including the applications addressed in this chapter. With a human-readable explanation coupled with a specific level of control, it becomes feasible to start a dialogue between the user and the computer.

13.5.4 Search Space

Once we have accepted, trusted, and adopted the use of AI in our workflow either as assistance or as a recommendation tool to efficiently our creative process, with built-in capabilities to provide reasons or explanations to understand the effects in our compositions or in the generated music, we need to discuss what AI can do in terms of the realm of possibilities. Asking for a solution to an AI system or a fully random process might be sufficient for what we need. However, what if the answer is not what we want or need? Can we ask for another one and get the result we expect? How many results do we need to ask for? To answer these questions, we need to be sure that the selected AI approach and its implementation do not follow monotonic reasoning. This refers to having the same conclusion even if new information is given to the current knowledge. We can call our AI to generate several solutions for Turandot. However, every time, the result will be the same. Even if we know that there exist other interpretations. By contrast, using a non-monotonic reasoning (NMR) engine within a given AI approach will explore different alternatives.

This is AI traversing the search space and returning solutions from specific points. A search space is defined as the feasible region defining the set of all possible solutions. AI with such qualifications is capable of producing distinct outcomes to a single problem. In Everardo (2020), a toy music composition example problem encoded in ASP syntax is depicted to demonstrate that a simple problem has a vast and impossible-to-explore search space. The example consists of a number of timed steps without repeating notes at consecutive times. For the case of eight-time steps, the number of solutions blows up to almost two million solutions computed in two seconds. With existing computation resources, it is noticeable that this is not a hard job for the machine. However, for a human, it is impossible to look at all the scores or listen to millions of solutions searching for one or for a subset of preferred compositions. This is analogous to the amount of solutions that the Mozart Dice Game offers.[5] With a number around 1.29×10^{29}, it is impossible to trace the search space. A similar behavior is present in ML with the curse of dimensionality (Bishop and Nasrabadi, 2006), where input data points may comprise a n-dimensional input vector, making the analysis of all the values complicated.

One possible alternative is to invest time and effort to take a good solution as a starting point for further development, instead of navigating over an unknown sea of solutions. Lastly, transparent AI must be able to give information about how similar or diverse one solution is to another. How close or how far is the first solution to the next five solutions? How are these solutions classified in terms of repeated patterns, bars, progressions, etc.?

With this vision in mind, we can give meaning to the search space of composing music and not ask for individual solutions without any apparent relationship. We can request five distant culminations for Turandot and be sure that we are dealing with representative solutions from all possible to start working with.

13.6 Encore: Another Muse for Music Creation

We will never know the true ending of Puccini's Turandot, even with the integration of AI. That does not mean that there is no use for this technology; we can actually expand our creativity for composition with more information about the search space and the overview of different solutions. As we have described, there are a plenty of implications that must be understood to design and implement this technology. The quality and usefulness of the solutions are dependent on the proper analysis of the nature of the problem, the selection of the field of AI, and the level of autonomy.

There are many facets and possible avenues of human–computer interaction within the music. Different forms of collaboration can be achieved since

---

[5] Mozart Dice Game http://www.cs.cornell.edu/courses/cs1110/2008fa/assignments/a6/a6.html.

AI does not need to be a fully automatic black box process. It can be used to reduce the number of monotonous tasks for the users in order to allow them to have more resources dedicated to their creative endeavors. The value of these solutions is laden on the subjective evaluation by the users in the same way different endings of Turandot are preferred. Their importance and usefulness will have lasting effects if it is asserted as such. There is a caveat; even with all the recent advancements and growth within the field, it is improbable to assume there will be a near future where AI will dominate human-centric creative practices (Anantrasirichai and Bull, 2022).

Additionally, the power of computation and the advanced algorithms driving AI fields enable us with tools that never existed before. It is naive to propose another composition for Turandot, but rather reflect on the possibilities of what AI can do to help us better understand the past and propose solutions that can be applied in other areas of music.

Acknowledgments

We would like to thank Christian Steinmetz for making us aware of relevant research and for all the enriching discussions about our shared vision of what AI can do in the music-making process.

References

AIVA. 2016. AIVA - Artificial Intelligence Virtual Artist. Accessed September 17, 2022. https://www.aiva.ai/about#about.

Alviano, M., Dodaro, C., Faber, W., Leone, N., & Ricca, F. (2013, September). WASP: A native ASP solver based on constraint learning. In *International Conference on Logic Programming and Nonmonotonic Reasoning* (pp. 54–66). Springer, Berlin, Heidelberg.

Anantrasirichai, N., & Bull, D. (2022). Artificial intelligence in the creative industries: A review. *Artificial Intelligence Review*, 55(1), 589–656. https://doi.org/10.1007/s10462-021-10039-7.

[Applied Psychoacoustics Lab Huddersfield]. Palladini, A. (2018, November 26). Alessandro Palladini - Intelligent Audio Machines. 4th Workshop on Intelligent Music Production 2018 (WIMP 2018) [Video]. YouTube. https://www.youtube.com/watch?v=IUz-eHgwDWU

Baral, C. (2003). *Knowledge Representation, Reasoning and Declarative Problem Solving*. Cambridge University Press, Cambridge.

Baral, C., & Gelfond, M. (1994). Logic programming and knowledge representation. *The Journal of Logic Programming*, 19, 73–148.

Barocas, S., & Selbst, A. D. (2016). Big data's disparate impact. *California Law Review*, 671–732.

Berger, W. (2005). *Puccini without Excuses: A Refreshing Reassessment of the World's Most Popular Composer*. New York: Vintage.

Bishop, C. M., & Nasrabadi, N. M. (2006). *Pattern Recognition and Machine Learning* (Vol. 4, No. 4, p. 738). New York: Springer.

Boenn, G., Brain, M., De Vos, M., & Ffitch, J. (2008). Automatic composition of melodic and harmonic music by answer set programming. In *International Conference on Logic Programming, ICLP08*. Lecture Notes in Computer Science (Vol. 4386, pp. 160–174). Springer, Berlin, Heidelberg.

Briot, J. P., Hadjeres, G., & Pachet, F. D. (2017). Deep learning techniques for music generation--a survey. arXiv preprint arXiv:1709.01620.

Briot, J. P., & Pachet, F. (2017). Music generation by deep learning-challenges and directions. arXiv preprint arXiv:1712.04371.

Burgess, R. J. (2014). *The History of Music Production*. Oxford University Press, Oxford.

Cabalar, P., Fandinno, J., & Muñiz, B. (2020). A system for explainable answer set programming. arXiv preprint arXiv:2009.10242.

Cabalar, P., and Martín, R. (2017). Haspie-A musical harmonisation tool based on ASP. In *EPIA Conference on Artificial Intelligence* (pp. 637–642). Springer, Cham.

Cabalar, P., Muñiz, B., Pérez, G., & Suárez, F. (2021). Explainable Machine Learning for liver transplantation. arXiv preprint arXiv:2109.13893.

Chollet, F. (2019). On the measure of intelligence. arXiv preprint arXiv:1911.01547.

Confalonieri, R., Coba, L., Wagner, B., & Besold, T. R. (2021). A historical perspective of explainable Artificial Intelligence. *Wiley Interdisciplinary Reviews: Data Mining and Knowledge Discovery*, 11(1), e1391.

Cope, D. (2021). Music, artificial intelligence and neuroscience. In: Miranda, E.R. (Ed.) *Handbook of Artificial Intelligence for Music* (pp. 163–194). Springer, Cham.

Das, A., & Rad, P. (2020). Opportunities and challenges in explainable artificial intelligence (xai): A survey. arXiv preprint arXiv:2006.11371.

De Man, B. (2017, November 20). Brecht De Man - Audio effects 2.0: Rethinking the music production workflow. Audio Developer Conference 2017 (ADC'17) [Video]. YouTube. https://www.youtube.com/watch?v=yJcqNR7by9o

De Man, B., Stables, R., & Reiss, J. D. (2019). *Intelligent Music Production*. Routledge, Oxfordshire.

Deng, L. (2014). A tutorial survey of architectures, algorithms, and applications for deep learning. *APSIPA Transactions on Signal and Information Processing*, p. 3.

Deng, L., & Yu, D. (2014). Deep learning: Methods and applications. *Foundations and Trends® in Signal Processing*, 7(3–4), 197–387.

Eiter, T., Ianni, G., & Krennwallner, T. (2009, August). Answer set programming: A primer. In Tessaris S., Franconi E., Eiter, T., Gutierrez C., Handschuh S., Rousset M., & Schmidt R., (Eds.), *Fifth International Reasoning Web Summer School (RW'09)*, volume 5689 of Lecture Notes in Computer Science, (pp. 40–110).. Springer-Verlag, Berlin, Heidelberg.

Eppe, M., Confalonieri, R., Maclean, E., Kaliakatsos, M., Cambouropoulos, E., Schorlemmer, M., Codescu, M., and Kuhnberger, K. U. (2015). In Yang Q., & Wooldridge M. J., (Eds.) *Twenty-Fourth International Joint Conference on Artificial Intelligence, IJCAI 2015*, Buenos Aires, Argentina, 25–31 July 2015.

Erdem, E., Gelfond, M., & Leone, N. (2016). Applications of answer set programming. *AI Magazine*, 37(3), 53–68.

Everardo, F., and Aguilera, A. (2011). Armin: Automatic trance music composition using answer set programming. In Osorio M., & Marek V. (Eds.) *Fundamenta Informaticae* (Vol. 113, No. 1, pp. 79–96). Special Issue: Latin American Workshop on Logic Languages, Algorithms and New Methods of Reasoning (LANMR).

Everardo, F. (2020). Over a Decade of Producing Music with Answer Set Programming. A Survey. At *Trends and Applications of Answer Set Programming (TAASP) Workshop*. Klagenfurt, Austria. November 2020.

Everardo, F., Gil, G. R., Pérez Alcalá, O. B., & Silva Ballesteros, G. (2020). Using answer set programming to detect and compose music in the structure of twelve bar blues. *Proceedings of the Thirteenth Latin American Workshop on Logic/Languages, Algorithms and New Methods of Reasoning. Virtual conference at Universidad Nacional Autónoma de México (UNAM)*, December 10–11, 2020.

Falkner, A., Friedrich, G., Schekotihin, K., Taupe, R., & Teppan, E. C. (2018). Industrial applications of answer set programming. *KI-Künstliche Intelligenz*, 32(2), 165–176.

Fandinno, J., & Schulz, C. (2019). Answering the "why" in answer set programming– A survey of explanation approaches. *Theory and Practice of Logic Programming*, 19(2), 114–203.

Floridi, L. (2019). Establishing the rules for building trustworthy AI. *Nature Machine Intelligence*, 1(6), 261–262.

Floridi, L., & Chiriatti, M. (2020). GPT-3: Its nature, scope, limits, and consequences. *Minds and Machines*, 30(4), 681–694.

Gebser, M., Kaminski, R., Kaufmann, B., & Schaub, T. (2012). Answer set solving in practice. *Synthesis Lectures on Artificial Intelligence and Machine Learning*, 6(3), 1–238.

Gebser, M., Kaminski, R., Kaufmann, B., & Schaub, T. (2014). Clingo= ASP+ control: Preliminary report. arXiv preprint arXiv:1405.3694.

Gelfond, M. (2008). Answer sets. In F. van Harmelen, V. Lifschitz, & B. Porter (Eds.), *Handbook of Knowledge Representation*. Foundations of Artificial Intelligence (vol. 3, pp. 285–316). Elsevier.

Gelfond, M., & Kahl, Y. (2014). *Knowledge Representation, Reasoning, and the Design of Intelligent Agents: The Answer-Set Programming Approach*. Cambridge University Press, Cambridge.

Gelfond, M., & Lifschitz, V. (1988). The stable model semantics for logic programming. In Kowalski, R.A. & Bowen, K.A. (Eds.), *Logic Programming, Proceedings of the Fifth International Conference and Symposium on Logic Programming (ICLP/ SLP)*, Seattle, Washington, USA, August 15–19, 1988 (2 Volumes, pp. 1070–1080). Cambridge: MIT Press.

Gioti, A. M. (2021). Artificial intelligence for music composition. In: Miranda, E.R. (Ed.) *Handbook of Artificial Intelligence for Music* (pp. 53–73). Springer, Cham.

Goodfellow, I., Bengio, Y., & Courville, A. (2016). *Deep Learning*. MIT Press, Cambridge.

Gunning, D., Stefik, M., Choi, J., Miller, T., Stumpf, S., & Yang, G. Z. (2019). XAI— Explainable artificial intelligence. *Science Robotics*, 4(37), eaay7120.

Hertzmann, A. (2018, May). Can computers create art? In *Arts* (Vol. 7, No. 2, p. 18). Multidisciplinary Digital Publishing Institute.

Kaminski, R., Schaub, T., & Wanko, P. (2017). A tutorial on hybrid answer set solving with clingo. *Reasoning Web International Summer School*, 167–203.

Krizhevsky, A., Sutskever, I., & Hinton, G. E. (2017). Imagenet classification with deep convolutional neural networks. *Communications of the ACM*, 60(6), 84–90.

Leone, N., Pfeifer, G., Faber, W., Eiter, T., Gottlob, G., Perri, S., & Scarcello, F. (2006). The DLV system for knowledge representation and reasoning. *ACM Transactions on Computational Logic (TOCL)*, 7(3), 499–562.

Lifschitz, V. (2008). What is answer set programming. In *Proceedings of the Twenty-third National Conference on Artificial Intelligence (AAAI'08)* (Vol. 8, pp. 1594–1597) AAAI Press.

Lifschitz, V. (2019). *Answer Set Programming* (pp. 1–147). Springer, Heidelberg.

Maguire, J. (1990). Puccini's version of the duet and final scene of Turandot. *The Musical Quarterly*, 74(3), 319–359.

Marchese-Ragona, R., Marioni, G., & Staffieri, A. (2004). The unfinished Turandot and Puccini's laryngeal cancer. *The Laryngoscope*, 114(5), 911–914.

McCarthy, J. (2004). What is artificial intelligence. http://www-formal.stanford.edu/jmc/whatisai.html.

Miranda, E. R. (Ed.). (2021). *Handbook of Artificial Intelligence for Music: Foundations, Advanced Approaches, and Developments for Creativity*. Springer International Publishing, Cham.

Moffat, D., & Sandler, M. B. (2019, September). Approaches in intelligent music production. In *Arts* (vol. 8, p. 125). Multidisciplinary Digital Publishing Institute..

Opolka, S., Obermeier, P., and Schaub, T. (2015). Automatic genre-dependent composition using answer set programming. *Proceedings of the 21st International Symposium on Electronic Art*. Vancouver, Canada.

Palladini, A. (2018). Intelligent audio machines. In *Keynote Talk at 4th Workshop on Intelligent Music Production (WIMP-18)* (Vol. 14), September. Huddersfield, UK.

Pfeiffer, A. (2018). Creativity and technology in the age of AI: Research Report (USA, Europe and Japan). Pfeiffer Report. https://www.pfeifferreport.com/essays/creativity-and-technology-in-the-age-of-ai/.

Pinetzki, L. (2020). Using Pattern Mining and Answer Set Programming to Compose Music. *Bachelor Thesis*. University of Potsdam, Germany.

Roads, C. (1996). *The Computer Music Tutorial*. MIT press, Cambridge.

Russell, P. N., & Norvig, P. (2010). *Artificial Intelligence: A Modern Approach*. Pearson Education, Inc., London.

Russell, S., & Norvig, P. (2020). *Artificial Intelligence: A Modern Approach* (Pearson Series in Artifical Intelligence) (4th ed.). Pearson, London.

Schaub, T., & Woltran, S. (2018). Answer set programming unleashed! *KI-Künstliche Intelligenz*, 32(2), 105–108.

Siegel, R. (2008, April 29). *Chinese Composer Gives 'Turandot' a Fresh Finale*. NPR, July 28, 2022. https://www.npr.org/transcripts/90037060

Stables, R., Enderby, S., De Man, B., Fazekas, G., & Reiss, J. D. (2014). SAFE: A system for extraction and retrieval of semantic audio descriptors. In Wang, H.-M., Yang, Y.-H., & Lee, J. H. (Eds.). *Proceedings of the 15th International Society for Music Information Retrieval Conference, ISMIR 2014*, Taipei, Taiwan, October 27–31, 2014. http://www.terasoft.com.tw/conf/ismir2014/.

Sterne, J., & Razlogova, E. (2021). Tuning sound for infrastructures: Artificial intelligence, automation, and the cultural politics of audio mastering. *Cultural Studies*, 35(4–5), 750–770. https://doi.org/10.1080/09502386.2021.1895247.

Sternberg, R. J. (2018). Theories of intelligence. In S. I. Pfeiffer, E. Shaunessy-Dedrick, & M. Foley-Nicpon (Eds.), *APA Handbook of Giftedness and Talent* (pp. 145–161). American Psychological Association. https://doi.org/10.1037/0000038-010.

Sternberg, R. J. (2021). Adaptive intelligence: Intelligence is not a personal trait but rather a person × task × situation interaction. *Journal of Intelligence*, 9(4), 58.

Sturm, B. L., Iglesias, M., Ben-Tal, O., Miron, M., & Gómez, E. (2019, September). Artificial intelligence and music: Open questions of copyright law and engineering praxis. In *Arts* (Vol. 8, No. 3, p. 115). Multidisciplinary Digital Publishing Institute..

Fisher B. (2013). Tosca and Il tabarro. In *Puccini's Il Trittico*. Miami: Opera Journeys Pub, 20, 71–72.

Trieu, L. L., Son, T. C., & Balduccini, M. (2022). An explanation generation system for answer set programming. In *International Conference on Logic Programming and Nonmonotonic Reasoning* (pp. 363–369). Springer, Cham.

Turing, A. M. (1950). Can a machine think. *Mind*, 59(236), 433–460.

Uvietta, M. (2004). 'È l'ora della prova': Berio's finale for Puccini's Turandot. *Cambridge Opera Journal*, 16(2), 187–238. doi:10.1017/S0954586704001843

Van Harmelen, F., Lifschitz, V., & Porter, B. (Eds.). (2008). *Handbook of Knowledge Representation*. Elsevier, Amsterdam.

Wing, J. M. (2021). Trustworthy AI. *Communications of the ACM*, 64(10), 64–71.

Zhou, Z. H. (2021). *Machine Learning*. Springer Nature, London.

Section III

Future Scenarios and Implications for the Application of AI

Section three of this book focuses on scenarios that could be achieved by using Artificial Intelligence (AI). The section is composed of eight contributions, which touch on aspects of Industry 4.0, AI impacting our lifestyle or influencing well-being, as well as the implications of either robotics or AI in psychology. It also includes an analysis of one of the most relevant topics in AI today, which is called robustness.

Regarding Industry 4.0, we have two contributions. The first one is by Juan Roberto López and co-authors. In this work, the authors focus on proposing a method based on machine learning to develop new energy management systems, which have become increasingly necessary as new sources of energy generation are being incorporated. These alternative sources are usually intermittent, and therefore, they require planning, operation, control, and management tasks that are already well studied in a traditional system. The implications of these systems are so relevant that in the future they could be immersed in smart cities.

The second chapter in this topic is written by Pedro Fonseca et al., in which the authors talk about the preponderant role that construction has played in the development of entire civilizations. In addition to the evolution of materials and techniques, there is a growing trend towards the development of intelligent processes in this activity. This chapter explains some of the most relevant applications, ranging from the intelligent surveillance of construction sites, to the use of intelligent robots, to new management systems that make

DOI: 10.1201/b23345-16

construction processes much more efficient. We will see the implications of AI here in the following decades, when it will not be strange to see robots building spaces that have been designed and optimized by intelligent systems.

Then, we move to collaborations that focus on AI changing our lifestyle. We have three contributions in this topic. The first being the chapter developed by Omar Mata and co-authors deals with the application of one of the most recent and controversial methods of AI, called Deep Learning (DL). DL is having a number of impressive applications, especially in image analysis and processing, as well as in aspects of natural language processing. In this specific collaboration, the authors describe how convolutional neural networks can be used to solve tasks such as designing clothing that meets certain characteristics such as durability, thermal comfort, and other aspects. Undoubtedly, the incursion of these techniques will revolutionize even tasks that are currently unimaginable. Subsequently, we have two contributions by Gildardo Sanchez and co-authors. Both are related to health issues. The first one talks about how machine learning is supporting the development of new antibiotic drugs. This is relevant because more and more bacteria are being detected that have been developing resistance to certain drugs, and this has triggered research into the use of new nanomaterials with the capacity to inhibit the development of these bacteria. Machine learning, especially neural networks, has been of value in making a process that can be long and costly more efficient. The following contribution describes how by using computer vision tools and machine learning, it is possible to train a system to detect anomalies in retinal images. This would allow the semi-automatic analysis of images from hundreds of patients with either vascular or diabetic conditions to monitor their evolution. Tools such as these will make it possible for more processes to be developed in a hybrid way between an intelligent computer and a doctor, saving the latter's time and minimizing errors.

From its origins, AI has been considered with connections to disciplines such as psychology and neurosciences. The section includes two interesting contributions related to those topics.

The first one by Gerardo Castañeda and his co-authors discusses the limitations of health systems in general in situations such as the COVID-19 pandemic, in which conditions such as depression and anxiety are manifested in a large proportion of the population. Faced with this situation, AI is considered as an alternative, and we are presented with a very interesting analysis of how it could be used through various technologies, particularly for cognitive-behavioral therapy. The chapter does not limit itself to describing these interactions, but also analyzes ethical, privacy, security, and accessibility issues that arise in cases such as the one analyzed. The second one by Nora Aguirre and her co-author talks about some of the possibilities offered by AI for neurosciences, specifically to try to understand how the brain assigns meaning to words. When talking about AI, one of the questions that often arises is how much relation it has or maintains with respect to natural intelligence. Trying to answer this question is extremely complex, partly because there are still many doubts about how humans develop our intelligence. Almost invariably, the possible answers to this question involve analyzing how our brain works and how it internalizes knowledge. In this chapter, the authors lead us to reflect on how we give meaning to words. They discuss the role that perception plays in the acquisition of knowledge in humans and how modeling these phenomena at the computational level offers some insight into how the brain might represent meaning.

Finally, we close the section with a chapter developed by Ivan Reyes and co-authors. This chapter discusses a topic of fundamental relevance today, especially with regard to the application of deep learning in such critical tasks as medicine or the driving of autonomous vehicles. And that topic is focuses on how reliable is an intelligent system. This question has led to the development of AI systems to attack other systems, looking for their vulnerabilities. Hence, the concept of adversarial robustness, which gives place to a field that tries to analyze the behavior of neural networks when deliberately altered information is fed to them. This type of analysis is undoubtedly of great importance, and new techniques will surely be developed and applied to improve the intelligent applications of the future.

14

Classification Machine Learning Applications for Energy Management Systems in Distribution Systems to Diminish CO_2 Emissions

Juan R. Lopez, Pedro Ponce, and Arturo Molina

Tecnológico de Monterrey, Mexico

CONTENTS

14.1 Introduction

It is known that the increment of greenhouse gas emissions is one of the main factors behind the global warming phenomenon. Global warming is a heavy influence to abnormal variations in the earth's temperature, causing irreversible changes in the environment and an unbalanced behavior of the nature. It is also known that CO_2 is one of the most concerning greenhouse

DOI: 10.1201/b23345-17

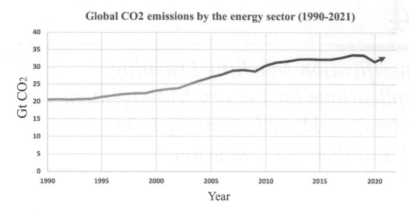

FIGURE 14.1
Global CO_2 emissions by the energy sector [1].

gases and therefore a main contributor to the global warming paradigm. The production share of CO_2 comes from various sectors such as:

(i) the energy industry; (ii) industrial processes; (iii) waste; and (iv) agriculture, where the energy sector is responsible for 73.2% of total CO_2 emissions. According to the International Energy Agency (IEA) [1], CO_2 emissions by the energy sector grew in 5% adding 1.5 gigatons (GTs) of CO_2 in a single year from 2020 to 2021, and this trend is shown in Figure 14.1. For such motives, the global energy landscape has been pushing forward a revolution of critical aspects within the generation, distribution, and consumption areas in the energy industry with the purpose of decarbonizing internal process and alleviate the global warming paradigm.

According to a report by the IEA, the global energy demand grew a 6% in a single year from 2020 to 2021 and it is expected to keep this growing pace for the year 2022 thanks to a rebound effect caused by the world's economic recovery from the 2019 COVID pandemic. As a consequence of this rebound effect, the global CO_2 emissions rose by a 7% in the last year mainly due to an increment in the demand of fossil fuel-based generation that grew by a 9% when compared to 2020 statistics. On the other hand, the installed generation capacity based on renewable sources grew only by a 9% in the last year, going from 2,807,265[MW] in 2020 to 3,063,926[MW] in 2021; in such a scenario, photovoltaic energy generation and wind energy generation are considered as the fastest growing renewable energy sources as shown in Figure 14.2.

In addition to the growing energy demand, the cost of producing energy increased by a 400% when compared to previous yearly averages due to political constraints and a shortage of nonrenewable resources [1]. According to the IEA [1], most consequences of the economic rebound effect should last until 2024, while other factors may never return to a "normal state," such is the case of the global oil and gas industries, which demand is predicted to keep increasing till the year 20206 according to the oil and gas market reports

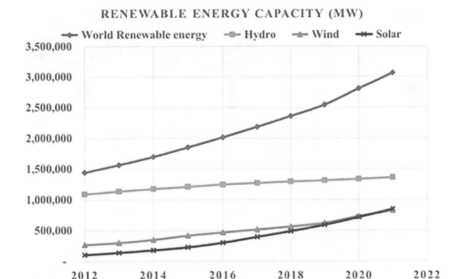

FIGURE 14.2
Renewable energy power generation capacity [2].

by the IEA [3,4]. This would indicate that cost of producing energy can still be compromised for the following years depending on the availability of nonrenewables. This scenario creates new challenges for the energy generation and energy management sectors, where new technologies and strategies need to be developed to counter this effect and regulate the availability of resources and energy generation prices.

As mentioned, the numbers of installed renewable-based generation capacity have been growing on an average of 9% per year on a global scale. This actions not only help to alleviate the dependency on nonrenewable assets but also assist on the expansion of distribution network with a incremental contribution to the global energy generation capacity. In essence, this yearly incremental contribution plays an important role in future regulation of the energy generation market, where, by having a bigger renewable generation share, the dependency on conventional energy generation methods can start decreasing as the demand for nonrenewables becomes less significant. However, as the penetration of these Alternative Generation Units (AGUs) becomes more compelling, other challenges start to emerge, such is the present need to improve on the planning, control, management, and operation strategies that follow the deployment of AGUs.

These paradigms can be addressed at different levels within the distribution system. In example, research and industry efforts have been done in segmenting these different levels and scale down the complexity of each paradigm, from general to more particular level, and these segments can be comprehended as: (i) picogrid; (ii) nanogrid; (iii) microgrid; and

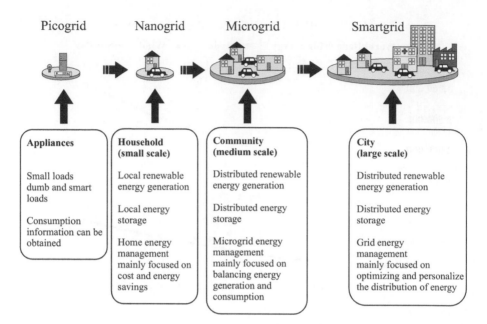

FIGURE 14.3
Partitions of the distribution network.

(iv) smartgrid; these partition levels of the distribution network are illustrated in Figure 14.3. Picogrids are usually deployed within the home level, and these are mainly composed of small loads that do not possess any other connection or interaction with a cloud or information service, such is the case of household appliances and DC loads [5]. Picogrids are mostly managed by the operators, making the application of management strategies more challenging, and these complications can be solved by integrating the old technologies with new devices that allow their connection to a cloud or information service; in essence, making "dumb" electrical loads smart. Nanogrids take a leap forward and introduce a local energy generation and energy storage levels; mostly on renewable assets and Energy Storage Systems (ESS); in addition, these systems are paired with controllable loads, and everything is enclosed within a single system. Usually, nanogrids are considered as entire homes with an energy demand in be tween 15 and 70 KW; in addition, the stochastic consumption and generation behavior at this level require of control and management system to regulate the generation and optimize its use across the entire system; commonly, demand-side and source-side management techniques are preferred at this level [6].

On a bigger scale, microgrids now incorporate multiple loads and multiple distributed generation sources forming a small distribution network; these systems can be studied in many formats, from complete buildings and factories to university campuses, and small communities [7,8]. Microgrids are known in the literature for having the capability of operating with or without a direct connection to the main electrical network, classifying its modes of operation in grid-connected and island systems. Across the literature, microgrids are recognized as systems with challenges in the control, management, and operation areas. For such motives, microgrids have adopted a hierarchical control structure of three levels to alleviate the required efforts for its right operation.

In this hierarchical control structure, the first and primary level is responsible for regulating the output of low-level components in variables such as voltage, frequency, and delivered power. The secondary level deals with the power generation distribution among the many interconnected AGUs, procuring the fulfilment of power sharing levels according to the available generation levels and power demand. Lastly, the tertiary level seeks global system gains by integrating energy management systems that contemplate a various number of factors such as weather conditions, load consumption patterns, and energy generation and consumption prices, just to mention a few. As mentioned, the global gains can be oriented toward sustainable or economical goals depending on the implemented management strategy and desired operational objectives.

Lastly, smartgrids level possesses similar architectures for the interfaced consumption and generation agents than those of microgrids. However, smartgrids differentiate from microgrids by operating more on the digital communication side, by taking advantage of consumers and the generated information through a vast network of sensors and Internet of Things (IoT) interconnected devices to optimize and personalize the distribution of energy in a large scale [9]. At the end, and in spite of the level of application, Energy Management Systems (EMS) operate on available data with specific objectives toward global or particular system gains. Nowadays, consumption and generation data is easier to obtain thanks to smart metering devices; with all the gathered information, it would be appropriate to adopt advanced data analysis strategies to observe, clean, understand, and crate models that allow the discovery of meaningful information to support the decision-making process behind EMS [10].

In the same understanding, having an advanced data analysis strategy is a great solution to confront the stochastic behavior of consumers and generation devices; since power generation is expected to be based on renewables, it is important to understand that their generation profile is dependent on weather conditions and on other physical aspects such as antiquity and material composition [11]. On the other hand, the stochastic user consumption can be handled more effectively with the development of a personalized model according to the gathered consumer information. At the end, the stochastic generation and consumption models can be generated more easily with advanced data analysis strategies; in addition, these can be paired with cost and suitability objectives to obtain effective energy management strategies at the different levels of the distribution network.

Nowadays, advanced data analysis strategies are associated with approaches based on Artificial Intelligence (AI). AI strategies enable computers to mimic human intelligence by following a set of logic rules, Decision Trees (DT), or Machine Learning (ML) models. In recent years, ML has been gaining popularity as one of the most relevant data analysis engines for predictive energy management services to balance the generation forecast with the expected consumption as described by Zhang Yingchen [12].

However, there are many ML methods, and each can be suited for different applications; most common ML methods include the following: (i) regression; (ii) classification; (iii) clustering; (iv) deep learning and neural nets; (v) transfer learning; (vi) reinforcement learning; (vii) dimensionality reduction; and (viii) ensemble forecasting methods [13]. The benefits associated with energy management services that take advantage of some of these ML strategies will play a crucial role in the future of the distribution networks. The distinction falls directly on which data acquisition and analysis strategy is used in the decision-making process within energy management systems; especially since each ML strategy has its own attributes and disadvantages, therefore, special attention may be required to select the most suitable strategy. With all of these in mind, there are clear factors that determine the impact of ML in EMS, such as:

- The chosen ML strategy according to a specific distribution level.
- The preferred ML strategy for specific energy management strategies.
- How the selected global or partial system goal can suggest the application of specific ML strategies
- The differentiating factors between the many ML strategies

It is also essential to consider that the immediate application of ML strategies to EMS is not an automatic benefit in saving costs or improving the sustainability aspects. ML strategies have to be trained and validated according to their environment; that is, the acquired data can be significantly different from one asset to another; therefore, a training or adaptive learning process needs to be taken into consideration for an effective prediction model and circle around the paradigms of a variability and stochastic behavior of the many assets within the distribution network. For instance, nanogrids and microgrids can have different energy management objectives and agents; however, both can seek to maximize the power delivery while reducing the dependency on the fossil generation to improve the sustainability of the system and also reduce the dependency on the connection with the main grid; essentially, the objectives become the same, but the scalability of the environment in which the energy management services operate cannot be done in a straightforward manner since the data from one system to another is vastly different.

It is the above problematic that motivates this work. As mentioned, the known benefits of applying ML to EMS can lead to system gains in costs and sustainability areas. However, these ML methods can be highly dependent on their operation environment, making their applications different at the many levels of distribution systems. This work gives an overview of the many ML strategies and areas of application within the partitions of the distributing network. The remaining of this work is as follows; Section 14.2 gives an introduction to EMS, focusing on the various strategies at the different levels; then, Section 14.3 introduces basic ML theory to answer the question "What is ML?". Formerly, these terms (EMS and ML) are associated in the following sections; first, Section 14.4 starts by highlighting the importance that ML has in the decision-making process of emerging EMS technologies; it clarifies on why ML is being preferred over traditional methods; secondly, Section 14.5 solidifies the role that ML has in developing EMS technologies by presenting a classification of ML applications in the scope of EMS at different levels of distribution network. To finalize, the closing sections start by giving condensed information as regards the future challenges that the implementation of ML will encounter in developing EMS; then, the wrap-up of this is presented as the ending section of this work, where the main ideas and highlights are presented in a succinct tone.

14.2 Energy Management Systems

Distribution networks can be seen as unique systems where each power distribution sector has its own set of generation and consumption agents. This creates a vast number of possible grid topologies and architectures motivating the requirement for generic energy management strategies that can be easily adapted to any given level and topology of the distribution system. Nowadays, generic management strategies can be included as a core part in the operation of EMS, representing a collective solution to the distribution system variability paradigms. In essence, EMS are responsible for regulating the operation of controllable agents to achieve global or system objective in terms of cost and/or sustainability. This is accomplished by adapting energy management strategies to take actions according to a huge amount of data that has been previously gathered and analyzed. These data enclose the vast number of variables that can affect directly or indirectly the performance of these generation and consumption agents. These variables include historical weather information, current weather information, energy market fluctuations, and stochastic consumer profiles just to mention a few.

The operation of EMS can be summarized in five sequential steps: (i) data gathering; (ii) data analysis; (iii) planning operation; (iv) application; and (v) motoring. These steps are also illustrated in Figure 14.4. The initial step is formed by an Advanced Meter Infrastructure (AMI) that gathers consumption and demand information from smart meters and smart devices;

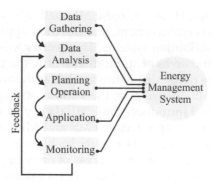

FIGURE 14.4
Energy Management System process tree.

sequentially, the gathered data is passed to a data analysis stage; in this step of the process, the data is used to build and update current prediction and operation models with the purpose of bettering the decision-making process of the following stages. A crucial part of this second stage is selecting the right data analysis strategy and choosing from fixed optimization models or more automated solutions such as ML-based strategies.

As a third stage, planning and operation include the chosen energy management and also undergo through a decision-making process powered by the translated data from the previous stage. This process can also be accommodated with automated solution to improve the performance of the flow of operation of the chosen energy management strategy. The remaining stages include the application and monitoring of the management strategy, where feedback loop is interfaced between the monitoring and data analysis stages to include information as regards the current stage of the system to improve learning and adaptive features.

To carry on with the third stage, it is imperative to study the wide amount of EMS proposed in the literature. One of the most popular EMS strategies is known as Demand-Side Management (DSM); this category of strategies has the advantage of directly regulating the demanded power through the controllable loads that can be found within the target system [14]. These strategies are mostly used to avoid over -- and under-generation scenarios where crucial variables such as system frequency and delivered voltage can stay within nominal parameters. In addition to the controllable loads, DSM also includes the active participation of users in cases where the target system does not include controllable assets. In itself, the EMS that can be categorized as DSM are the following:

- Energy Efficiency
- Demand Response (DR)
- Time of use

- Tariff awareness
- Generation reserve
- Spinning reserve
- Rapid reserve

As mentioned, depending on the chosen energy management strategy, the user can become a crucial variable for the energy management tasks; in example, the energy efficiency strategy aims to educate the consumer about their energy consumption profile and highlight the areas of opportunity where the system can become more energy-efficient by correcting energy consumption habits and/or replacing consumption agents with a more energy-efficient version [15]. Additionally, other system variables can also be contemplated in more localized systems such as houses and buildings; energy efficiency strategies can suggest structural modification to improve thermal comfort and efficiency. This becomes more of a crucial factor in locations where an intense use of HVAC system is required [14].

In a different method, time of use and tariff awareness strategies also seek to educate the consumer on their consumption profiles; however, this is achieved by informing the consumer about the peak-pricing schedules and high consumption times to avoid and expensive use of energy. Nonetheless, this strategy can include dynamic or static tariffs; the former adapts in values according to the target system environment and the value of this tariffs depends on the target system power demand and available generation, while the static version has specific prices according to the time of the day [14].

As a more automated approach, DR strategies are implemented to regulate consumption of controllable assets according to pricing and scheduling strategies with limited consumer interaction. DR strategies also serve the purpose of providing ancillary services to the target system, by balancing the power demand with the available distributed generation [16]. Last on the list, generation reserve strategies deal with unloaded generation in different manners with the same objective of compensating for an unexpected load demand or loss of a generation unit [17]. Spinning reserve takes advantage of the synchronous speed and inertia of synchronous generators to match the system frequency to meet the unexpected rise in the power demand, while rapid reserve operates on ESS, where the fast response and reliability serve the same purpose; at the end, both strategies prevent the degradation and/or interruption of the target system.

With these strategies in mind and by taking into consideration the fact that power systems are moving toward more autonomous systems, it is imperative to start thinking in smarter and adaptive solution that can enhance the performance of current EMS strategies. In such understanding, research efforts in the area of EMS are inclining toward the inclusion of ML as a core part in the decision-making process of these strategies [18]. In fact, the incorporation of ML can provide tailored EMS since such system can learn over their own dynamics instead of just relying on static models or constrained rule-based methods [19]. There is no doubt that ML will become a core part in future iterations of power systems.

14.3 What Is Machine Learning?

Nowadays, ML has become one of the most popular branches of AI and computer science, and ML is mainly focuses on the analysis of gathered data and algorithms to clone a human way of learning by computers with similar adaptive reasoning that allows a gradual knowledge improvement of certain processes. ML has gained its place as one of the most critical components in the field of data science. It is by the implementation of statistical methods that ML algorithms are trained to classify and predict key insights of specific processes that result in an enhanced operation of the target systems. These subsequent insights drive decision-making actions within applications and systems and improve the course of action against certain scenarios [20]. According to the literature, area of ML is composed of three main branches:

- Supervised learning
- Unsupervised learning
- Reinforcement learning

Supervised learning consists in knowing the relationship between the input and the output data, where basically the system is trained to recognize patterns and point out an existing relationship between the input and the output data label. Since these methods are trained on the principles of pattern recognition, the inclusion on new data to the training data sets can further improve the accuracy of the trained model. Supervised learning strategies are a good option for regression and classification problems. However, one big disadvantage of these methods is the quality of the data sets used in the training process, where an inaccurate data set can hinder the performance of the end model [21].

Contrary to supervised learning, unsupervised learning methods do not need information regarding the input and the output data; in fact, in most methods, only the input data is required to train a model that can predict or classify the input of a system. These characteristics give unsupervised methods more flexibility over supervised methods, allowing unsupervised strategies to function on more complex data sets. A big disadvantage of unsupervised ML strategies is the unpredictability of the trained model, where some categories or prediction might be undesired for the end result [22]. Lastly, reinforcement learning strategies are based on a reward and punishment dynamic, where a certain agent is rewarded for desired behaviors and

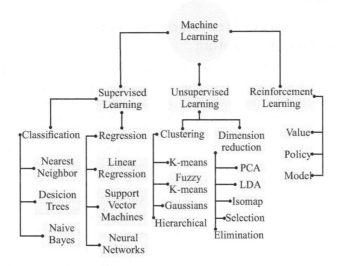

FIGURE 14.5
Classification of ML methods and main strategies [24,25].

punished in if it misconducts. Reinforcement strategies make the decision-making agents learn based on "rules" which can depend on the value, policy, or a model [23]. The main difference between reinforcement learning strategies and supervised learning is that in the latter strategies, and the data set used for training contains the answer to the inputs of the system, whereas in former strategies, the answer is not given, forcing the reinforcement agent to act according to set learning model and learn from experience. On the other hand, reinforcement strategies differ from unsupervised methods in how these intact with its environment; unsupervised methods do not interact with environment, whereas reinforcement strategies are obligated to act upon the agents within the environment and learn based on trial and error providing more insights in regard to the environment dynamics [24]. Figure 14.5 shows a diagram that contains these ML categories and few of the most used strategies for such categories.

It is due to such characteristics of ML strategies that these have been an attractive solution for classification and prediction paradigms in the context of EMS and their stochastic behavior, where in some cases supervised learning fits more the necessities because the data sets of a system are known from the input to the output; however, the complexity of distribution systems can cause certain uncertainty in regards the output of the system; for such cases, unsupervised or reinforcement learning techniques can be implemented. In any of such cases, ML strategies pretend to enhance certain features of EMS and there is no doubt that their implementation will keep pushing the evolution of power distribution network toward smarter, personalized, and autonomous systems.

14.4 Driving Energy Management Systems with Machine Learning

Current high-level EMS gather the necessary system information from a robust AMI with the purpose of monitoring system variables such as demanded energy by the interfaced loads, generated power (local and/or distributed), and State of Charge (SoC) of ESS (local and/or distributed). These actions generate a huge amount of data that can result overwhelming for conventional EMS operators such as in rule-based, schedule-based, static model-based, and manual-based systems. Essentially, the performance of conventional EMS can become insufficient and unreliable thanks to the deficiency, limited, or restricted data processing capabilities for the non-convex data sets produced by the many agents within the target system. Non-convex data sets can possess multiple suitable sectors and multiple optimal points for each sector; this means that only local solution can be obtained if the problematic is constrained to a bounded region of operation. Traditional EMS methods constrain this problematic by applying different set of rules or static models that only satisfy a bounded region; at the end, these constraints can end up having a negative effect in the operation of the target system by having scenarios out of the initial configuration scope of the EMS.

For such motives, ML strategies have gained popularity as data analysis engines; ML strategies possess the capability of processing non-convex data sets with ease; and this characteristic enhances the decision-making processes within EMS and contributes to a more reliable operation [26]. One of the most attractive features for applying ML to EMS is the capability to create and adapt system models for specific domains, and these actions are made based on past data and contribute to a more personalized and faster response of the system, whereas conventional methods require of an external agent to intervene and modify their configuration to perform any changes to the starting model.

In reason, ML is becoming a core part in the development of smart and adaptive features that follow the implementation of the future management of the power grid in its many levels; because of this, it is reasonable to believe that this technological tool can be used to analyze and improve certain aspect or areas within the distribution, consumption, and generation of electrical energy, including environmental topics that follow the sustainability of the different grid partition in regard to energy demand and environmental impacts of their operation. For example, ML can be applied to optimize the EMS within industrial parks to consume less resources without compromising the final output, and such point is also stated in the manufacturing process scenario described by Ref. [27].

In a more precise application, ML can be used to develop a tailored and adaptive profile of the users and generation assets to create accurate multi-dimensional classification of these agents according to the collected data by the AMI and the target system feedback [28]. In example, the work presented in Ref. [26] applies ML techniques to demonstrate how a ML-based EMS can suggest tailored tariff schemes to specific characteristics of a home-level system. Such study is focused on time-of-use and real-time-pricing energy management strategies where a k-means clustering ML strategy is selected to segregate consumers into different groups and learn more about their behavior during the energy management operation and offer optimal tariffs to both; generation and consumption agents. At the end, the applied ML strategy successfully categorizes consumers into five clusters: (i) system size; (ii) home system type; (iii) building system type; (iv) number of occupants; and (v) terrain. Such work concludes by highlighting that thanks to a successful clustering of consumers, tailored dynamic piecing strategies can be applied to consumers according to specific characteristics and result in cost and sustainability saving when compared to the application of a single dynamic pricing strategy without caring for characteristics of the consumer.

However, ML can also be applied to other areas in power grid management such as disaster mitigation. This case is studied in Ref. [29], where a Support Vector Machine (SVM) ML classification algorithm is selected to obtain different operational modes of the power units in case of predicted compromising scenarios. Principal training variables include weather conditions (normal or severe), voltage sags (over/under voltage), and peak-hour schedule; then seven modes of operation are obtained to successfully respond to possible compromising scenarios that include islanding events and short or complete outages. Another application is also presented in Ref. [30], where a DT ML algorithm is implemented to maximize the power output of local solar generation system. The DT algorithm is trained to switch between supplying the local loads with only the local solar generation systems or with the utility grid in cases when the local generation system does not meet the predicted target system requirements.

In recent studies, the ML strategy known as deep reinforcement learning has been gaining popularity as an alternative data analysis engine. Reinforcement learning reshapes the way in which the EMS operator learns about the different dynamics of the target system by rewarding an optimal behavior of the interfaced agent and punishing those actions that lead to a negative impact in the overall target system performance [31]. Deep Reinforcement Learning (DRL) combines the reward-based approach of traditional reinforcement methods with the learning and adaptive features of Artificial Neural Networks (ANN); this combination of strategies improves the analysis of high-dimensional and continuous state spaces, offering additional system information to the DRL method and eliminating the uncertainty of having high-dimensional system. An example of this ML application is exposed in Ref. [32], where the authors use reinforcement learning in combination with a Markov decision process to enable an energy trading interaction between agents in a microgrid system. At the end, the microgrid EMS treats local renewable generation systems as a centralized energy pool to allow a seamless energy trading and optimize the use of available energy produced by renewable assets.

Recent applications also demonstrate that ML strategies can be applied to enhance the reliability and security of the target system [33]. In example, the system reliability paradigm that follows fault diagnosis methods to prevent a hindering operation of EMS is heavily studied in the literature; in example, the work presented by Rahman Fahim [34] studies the application of a ML method based on ANN to classify fault conditions in microgrid structures and provide self-healing capabilities to the target system; this method analyzes voltage and current wave forms of distributed generators and outputs of predicted conditions that can lead to improvements in the transient response of the microgrid; in such test cases, ML strategies are trained offline with previously gathered data from past events. In a different example, a DT ML classification method is used proposed in Ref. [35] to classify power system faults, and DT methods have a rule-based operation, which allows a quick classification based on input parameters.

On the other hand, the paradigms that deal with the security of the target system need to take special care to the amount of data that is being collected by the AMI, where ML strategies are adopted to detect, identify, respond, and mitigate adversary attacks to the decision-making process behind management operations, enhancing classical defense mechanism and the overall security level of the system [36]. Across the literature, three main areas of interest are highlighted in the area cybersecurity microgrids and smartgrids; (i) data gathering; (ii) data processing; and (iii) classification stages [37]. Supervised and unsupervised methods are commonly used to prevent cybercriminals from modifying the collected data and impact the training process of other ML-based processes down the operation line of EMS.

At the end, ML has shown a great involvement in many areas behind or ahead the operation of EMS despite the level of the target system, showing promising result in such areas of application; ML is quickly becoming a core part in the decision-making process to achieve sustainable, reliable, and secure systems. This would further enable the integration of smartgrids and microgrids to the conventional distribution network, improving the smartness of the system toward a completely decentralized, self-healing, personalized, and autonomous distribution network.

14.5 Classification and Trends of Machine Learning Applications in Energy Management Systems

The objective of this section is to give a classification of ML strategies according to the area of application in the last ten years. Subsequently, the following subsections will detail a brief overview of the different ML application and strategies at the different levels of distribution system. For this section, a total of 45 papers were selected and reviewed in scopus for a more detailed

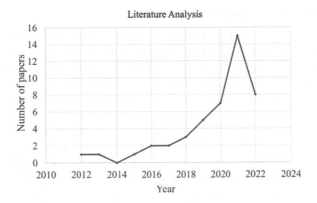

FIGURE 14.6
Reviewed papers in SCOPUS.

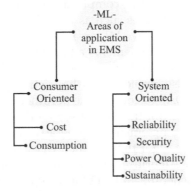

FIGURE 14.7
ML areas of application in Energy Management System.

analysis; such papers date from 2012 to 2022 and were published mainly in subject areas of engineering, computer science, and energy. Figure 14.6 shows the recent growing publication numbers in subjects of cooperative strategies between ML and EMS.

By analyzing the reviewed literature, the areas of application for ML in the context of EMS can be classified into two main groups: (i) consumer-oriented, and (ii) system-oriented. These two groups are also divided according to the objective of the ML strategy within the EMS. As an example, consumer-oriented applications operate mostly on cost reduction and consumption optimization to favor the end consumer; such areas of application can be divided into (i) cost and (ii) consumption. On the other hand, system-oriented applications act over operational aspect of EMS such as: (i) reliability; (ii) security; (iii) power quality; and (iv) sustainability. These categories and their applications are further illustrated in Figure 14.7.

14.5.1 Costumer-Oriented ML Applications

In example, authors in Ref. [38] implemented an ANN along with a DR technique to reduce the overall cost of consumed electricity, where the ANN was in charge of making decisions over controllable appliances depending on the cost of electricity during specific instances of the day. In a different example, authors in Ref. [39] use a DRL strategy to enhance the consumption schedule of controllable appliances and contribute to the cost reduction. On the other hand, authors in Ref. [40] also make use of a DRL strategy to enhance the operation to the EMS; however, such work is more oriented toward the reduction of consumed power by a single-home systems; in such case study, authors pair the DRL with a time of use tariff strategy, to reduce consumption times according to dynamic consumption tariffs. From these examples, it is clear how EMS in nanogrids are more focused on the consumer interests, and make use of various ML strategies to achieve such economic objective.

Unlike in nanogrids, microgrids and smartgrids are systems where the consumer grows in capacity and complexity leading to the application of more generalized EMS strategies such as DSM [41]. Microgrid systems can allocate individual or cooperative ML areas of application. For instance, the work presented in Ref. [42] implements a Support Vector Regression (SVR) to predict the charging demand by the electric vehicles in a microgrid; in addition, the authors also implement a heuristic optimization algorithm (Dragonfly Algorithm (DA)) to counter the nonlinear nature of the data and clean the input parameters for the supervised method. In a similar example, authors in Ref. [43] also implement a cooperative operation of a ML strategy and an optimization algorithm; in such case, an ANN and a Bacterial Foraging Optimization Algorithm (BFOA) are implemented along with a optimized generation schedule and a DR technique to minimize the energy generation cost and enhance the use of renewable energy by also managing the power flow between the renewable assets and the power grid.

It is at the smartgrid level that cooperative actions between ML strategies start to appear. As an example, authors in Ref. [44] proposed a hybrid application of two ML strategies to tackle the load forecasting paradigm. The two strategies are implemented in a series configuration, by starting with a clustering technique to classify two types of consumers based on daily and adjacent consumption patterns; after the classification process, a Back Propagation Neural Network (BPNN) is implemented to perform the load forecast of the users. In such proposal, the BPNN is previously trained with data that describes the load rate-of-change and load value at adjacent times. Overall, the hybrid method demonstrates that a cooperative approach of ML strategies can be an adequate solution to more convoluted systems.

As stated in Ref. [45], another paradigm within smartgrids is the sustainability factor. This area of application is indirectly tackled in Ref. [45] by implementing a Long Short-term Memory (LSTM) neural network to discover any abnormal drifts in the consumption pattern of certain consumers

TABLE 14.1

Costumer-oriented ML Application

ML Area of Application in EMS Consumer-oriented				
System Level	Area of Application	ML Strategy	EMS Strategy	References
Nanogrid	Cost	ANN	DR	[38]
		DRL	DR	[39, 46]
	Consumption	DRL	Time of use	[40,47]
Microgrid	Consumption	DRL	DSM	[48]
		ANN	DSM	[49]
		DA SVR	DSM	[42]
	Cost	BFOA ANN	DR	[43]
	Cost and Consumption	DT	DR	[50]
Smartgrid	Consumption	K-means BPNN	DSM	[44]
		LSTM	DSM	[45]
		SVM	DSM	[51]

and preserve the sustainability of the system as positive after effect. Such work demonstrates that ML can also be implemented to deal with management paradigms related to sustainability objectives by predicting unfavorable scenarios and prepare the most suiting countermeasures or correct certain system conditions to avoid its occurrence.

Generally speaking, consumer-oriented applications are focused on cost and consumption factors at the mentioned levels of the distribution network (nanogrid, microgrid, and smartgrid). This is mainly because the interest of the consumer shows a predominance toward and economic benefit instead of a sustainable one. It is the task of EMS to offer a sustainable-related incentive to consumer to start tackling sustainability applications in consumer-oriented. Table 14.1 summarizes the above-mentioned ML classification and classifies them according to the system level and area of application.

14.5.2 System-Oriented ML Applications

Across the literature, ML applications can also be found with more focus on system characteristics. Such applications defined as system-oriented, where their main objective is to enhance the reliability, security, power quality, and sustainability aspects within the different levels of the distribution system. In the case of nanogrids, the dominant areas are reliability, sustainability, and security; at this level, reliability applications are found for sensor data estimation, to validate the obtained data from the local metering devices by comparing it to a predicted data set; this application is found in Ref. [52]

where the authors employed a Deep Neural Network (DNN) to output the predicted data behavior and compare it to the data set obtained from the metering devices. On the sustainability aspect, nanogrids are more focused on controllable loads and have a better usage of local RES, and such applications are studied in Refs. [53] and [54] where the most popular ML methods for such application are k-nearest neighbors (KNN), Back Propagation Neural Network (BPNN), SVM, ANN. On the area of security, the threat of energy theft is a major concern at the nanogrid level. To avoid the unauthorized consumption of energy, the article in Ref. [55] proposed an energy theft detection system based on four different ML methods, LSTM, Multi-Layer Perceptron (MLP), Recurrent Neural Network (RNN), and Gated Recurrent Unit (GRU).

Under the scope of reliability-focused application, ML strategies are used to predict scenarios that can hinder the performance of the system, and such scenarios include: (i) islanding; (ii) blackouts; (iii) and faults across the distribution system. For instance, authors in Ref. [56] proposed an islanding detection mechanism based on Convolutional Neural Networks (CNN) and LSTM to monitor harmonic anomalies at the point of common coupling between the microgrid and the power grid. Another use of ML strategies at the microgrid level includes the prediction of power grid blackouts; as an example, authors in Ref. [57] utilize a Raining Forest Algorithm (RFA) to perform a regression procedure to predict certain conditions that can trigger such event. Separate areas with the reliable scope are fault localization and fault identification; these areas are commonly interfaced ML classification methods such as DT and SVM as studied in Refs. [58] and [59].

On another scope, security issues are more related to the detection of anomalies in the data stream of metering devices to identify cyberattacks; for such case, authors in Ref. [60] use a supervised learning method based on passed occurrences. On the contrary, the power quality and sustainability areas use supervised classification methods to improve the operation of Energy Storage Systems (ESS) and the performance of RES which operations are bounded to specific environmental conditions; in the latter scenario, ML is used to classify the environment and target system and act over certain system agents to enhance the use of RES and minimize the dependency on the power grid.

Lastly, at the smartgrid level, the reliability and security areas are more inclined toward classification methods such as SVM and LSTM in applications similar to the microgrid level. It is at the power quality area where a new application can be found, that being the use ML to have an improved planning strategy for the placement of metering devices; as an example, authors in Ref. (61) use a Bayesian Network (BN) to solve the metering placement paradigm. Finally, the area of power quality also involves the use of ML to classify power disturbances such as over- and under-generation scenarios and avoid any hindering conditions across the distribution network, in the work in Ref. [62] that compares six different ML strategies for such objective; the work concludes that a combination of a CNN and LSTM is suitable for learning the

TABLE 14.2

System-oriented ML Application

		ML Application in EMS System Oriented		
System Level	Area of Application	ML Strategy	Application	References
Nanogrid	Reliability	DNN	Sensor data estimation	[52]
	Sustainability	KNN BPNN	Load energizing and de-energizing	[53]
		SVM ANN	Increase the use of local RES	[54]
	Security	MLP RNN LSTM GRU	Energy Theft detection system	[55]
Microgrid	Reliability	CNN	Islanding detection	[56]
		LSTM RFA	Blackout prediction	[57]
		DT	Fault localization	[58]
		SVM	Fault identification	[59]
	Security	DNN	Cyberattacks Anomaly detection	[60]
	Power Quality	DT	ESS management	[63]
		SVM	PV performance prediction	[64]
	Sustainability	SVM DT	Minimize consumption of resources	[65]
Smartgrid	Reliability	SVM	Blackout prediction	[66]
	Security	SVM	Detect fraudulent data injection	[67]
		CNN LSTM		[68]
		CNN LSTM	Energy Theft detection system	[69]
	Power Quality	CNN RNN LSTM GRU	Power disturbance classification	[62]
		BN	Planning of AMI	[61]

temporal characteristics of the disturbances and effectively classifying them to perform any corrective or preventive actions. Overall, the classification of these methods for system-oriented applications is given in Table 14.2.

14.5.3 Comparison between ML Strategies in the EMS Context

As shown in the subsections above, ML strategies are employed for different tasks and with various objectives at different levels of the distribution network. Figure 14.8 shows the distribution of implemented ML methods at the different levels according to the reviewed literature. For instance, Figure 14.8 illustrates how NN-based methods predominate at each level; from the nanogrid to the smartgrid, although NN-based methods require a larger set of data when

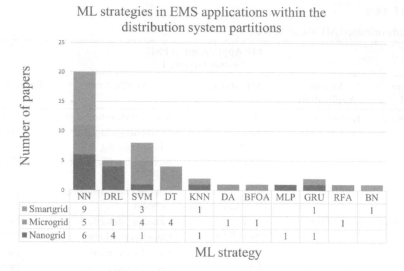

	NN	DRL	SVM	DT	KNN	DA	BFOA	MLP	GRU	RFA	BN
■ Smartgrid	9		3		1				1		1
■ Microgrid	5	1	4	4		1	1			1	
■ Nanogrid	6	4	1		1			1	1		

ML strategy

FIGURE 14.8
Summary of ML strategies used for EMS applications.

compared to SVM, DT, or RFA, this characteristic also provides certain advantages, such a better adaptability as more data is fed into the system, resulting in enhanced categorization and accurate prediction with a low error rate.

On the other hand, ML strategies such as SMV, DT, and RFA are preferred for more trivial task, where the input data can be of a lower magnitude. This results in a minimal or meagerly compelling processing of the input data, ensuring less use of time. Another observation from the reviewed literature is in implementation of DRL algorithms, where their use is more significant at lower levels of the distribution network. DRL requires a large amount of data to be trained and its performance can be enhanced by increasing the input data dimensionality as reviewed in Ref. [70]. Overall, DRL is more suitable for lower distribution levels since this method often requires to wait for the reaction of the system to its decision, so it can learn from its past mistakes and improve in further iterations, whereas in larger systems these initial mistakes can challenge or hinder the operation of other areas in the systems, until the operation is corrected in further iterations of the DRL.

Another area of comparison is how these ML strategies are implemented. For example, the implementation method depends mostly on the level of application; that is, for nanogrid systems the implementation is commonly done on controller boards with the embedded ML algorithm as suggested in Ref. [52]; also, most proposals for the nanogrid level can be validated in low-cost computers and implemented in low-cost embedded systems, whereas most EMS of higher partitions often require of SCADA systems to improve the reliability on connections and assure the quality of service, resulting in a higher investment for these ML applications [71].

14.6 Challenges of Machine Learning Driving Energy Management Systems

Multiple challenges have been popping up as the topic incorporating ML strategies as a core part of EMS keeps on developing. The most mentioned concerns in the literature circle around the data security topic. This is in part with the fact that ML strategies are trained and operated based on data sets gathered through an AMI, where data can be stolen, modified, or used to hinder operational aspects of the target system. In par with data security challenges, also data privacy can be seen a mayor blocker for ML applications, and data privacy is a major concern for the users since personal data can be used for cyberattacks or energy theft, affecting the end consumer and the power service provider. Particularly, privacy and security topics are more sensitive to smaller systems since these are closer to the consumers personal information, and it would be the work of future management systems to find a less invasive approach to acquire such data, or to strengthen the privacy policies of the metering devices at such level.

On a different topic, as reviewed in Table 14.2, most of the ML methods used for system-level applications fall in the classification category, where different algorithms have been proposed to tackle different aspects of the systems; although this approach is effective, the superposition of such methods when working in single system is still a topic to be studied, where these can cooperate together toward a global goal or block the system operation. It is then the challenge of future management systems to find a holistic approach to classify a wider range of scenarios and efficient use and collection of data by having a single or less classification methods.

Essentially, research trends are focusing more on the development of multi-objective energy optimization to solve preset issues that involve any conflicts of interests between optimization goals [72]. This multi-objective paradigm also includes the realization of multi-timescale objectives, where ML strategies need to develop multi-time predictive actions as proposed in Ref. [73], where a multi-timescale irradiance forecasting is constructed to predict inside the hour and ahead of time operation conditions for photovoltaic panels.

Despite the technical challenges, social challenges still remain as a key unexplored area in the implementation of ML in EMS as stated in Refs. [74,75]. These social challenges include the acceptance of the user to these new technologies against traditional energy management architectures, where the user feels more familiarized. It is then the responsibility of future researchers to develop interfaces that help the user to adapt these new technologies, and such is the case of the work in Ref. [76], where the authors propose an interface that monitors the thermal comfort of the user and suggests the energy saving measurements for the user by managing controllable loads such as HVAC systems.

At the end, it is the topic that follow these technical and social challenges that represent significant areas of opportunity for ML within EMS. Pursuing the resolution of these challenges should not only be influenced by only local objectives according to particular goals at each level; this resolution must be motivated to contemplate holistic objectives to achieve global gains that involve all the areas of application for ML strategies within EMS. It is crucial to remember that many of the implemented smart EMS are not fully implemented; most of the results in the literature are obtained through simulation or small-scale representative systems. Thus, the real application of smart EMS is still a crucial task for future researchers and engineers and push a true reshaping of the electrical network.

14.7 Wrap-Up

The decarbonization of the energy sector has been one of the most critical aspects behind the continuous evolution of the power distribution network. This gradual progression has slowly been incorporating advanced technologies that lighten the tasks of traditional grid operations by making long-established systems smarter due to the incorporation of advanced data analysis strategies that allow continuous learning, classification, and adaptive features to the agents in charge of the decision-making actions. In addition, this evolutionary effect is also responsible for the development of new system partitions of the distribution network, where these partitions serve the purpose of simplifying control and management tasks by reducing the order of complexity of the target system. Literature has shown that it is the combination of new advanced technologies in areas control, management, and operation that can lead to the further development of these grid partitions and exploit their potential toward the development of more intelligent, autonomous, sustainable, and overall a more tailored distribution network operation.

This work gives a categorization of the many ML strategies in the context of EMS at different levels of the distribution network. From the reviewed literature, we classified the ML applications for EMS into two main groups: customer- and system-oriented. In particular, it was found that costumer-oriented applications are more inclined toward economic, consumption, and sustainability preferences, while system-oriented applications make use of ML strategies to improve crucial operational areas in the target system, such as reliability, security, power quality, and sustainability. From these applications, it can be derived that certain ML strategies are chosen over others depending on the system level and the area of applications, where clustering and classification techniques are more adequate for higher systems levels such as the smartgrid, while learning and adaptive ML techniques are more frequently adopted at lower systems levels such as the nanogrid. In addition, system-oriented applications are found to be a different application pattern, where the reliability and security areas are more leaned toward learning and adaptive strategies, while the ML strategies in areas of power quality and sustainability are more inclined toward classification and clustering methods. Overall, this work shows that consumer-oriented ML applications depend more on the system level, while system-oriented ML applications are more dependent on the area of application in EMS.

Acknowledgement(s)

This work would not have been possible without the support of the Institute of Advanced Materials and Sustainable Manufacturing, Tecnológico de Monterrey, Mexico.

Disclosure Statement

Authors do not have conflict of interest.

Funding

Funding was provided by Institute of Advanced Materials and Sustainable Manufacturing, Tecnológico de Monterrey, Mexico.

References

1. Agency, I.E. Electricity Market Report - January 2022, IEA, Paris https://www.iea.org/reports/electricity-market-report-january-2022, 2022. Accessed: 2022-28-06.
2. IRENA. *Renewable Capacity Statistics 2022*, International Renewable Energy Agency (IRENA), Abu Dhabi, 2021.
3. Toril, B.; Christophe, B.; Olivier, L.; Peg, M.; Anne, K.; Kristine, P.; Masataka, Y.; Jeremy, M. Oil 2021 Analysis and Forecast to 2026, https://iea.blob.core.windows.net, 2021. Accessed: 2022-07-05.

4. Jean-Baptiste, D.; Louis, C.; Jean-Baptiste, D.; Tetsuro, H.; Akos, L.; Gergely, M. Gas Market Report Q1 2022, https://iea.blob.core.windows.net, 2021. Accessed: 2022-07-05.

5. Quek, Y.T.; Woo, W.L.; Logenthiran, T. Smart Sensing of Loads in an Extra Low Voltage DC Pico-Grid Using Machine Learning Techniques, *IEEE Sensors Journal* **2017**, *17* (23), 7775–7783.

6. Yousaf, S.; Mughees, A.; Khan, M.G.; Amin, A.A.; Adnan, M. A Comparative Analysis of Various Controller Techniques for Optimal Control of Smart Nano-Grid Using GA and PSO Algorithms, *IEEE Access* **2020**, *8*, 205696–205711.

7. Holdmann, G.P.; Wies, R.W.; Vandermeer, J.B. Renewable Energy Integration in Alaska's Remote Islanded Microgrids: Economic Drivers, Technical Strategies, Technological Niche Development, and Policy Implications, *Proceedings of the IEEE* **2019**, *107* (9), 1820–1837.

8. Washom, B.; Dilliot, J.; Weil, D.; Kleissl, J.; Balac, N.; Torre, W.; Richter, C. Ivory Tower of Power: Microgrid Implementation at the University of California, San Diego, *IEEE Power and Energy Magazine* **2013**, *11* (4), 28–32.

9. Fang, X.; Misra, S.; Xue, G.; Yang, D. Smart Grid — The New and Improved Power Grid: A Survey, *IEEE Communications Surveys & Tutorials* **2012**, *14* (4), 944–980.

10. Han, T.; Muhammad, K.; Hussain, T.; Lloret, J.; Baik, S.W. An Efficient Deep Learning Framework for Intelligent Energy Management in IoT Networks, *IEEE Internet of Things Journal* **2021**, *8* (5), 3170–3179.

11. Ibarra, L.; Lopez, J.R.; Ponce, P.; Molina, A. Empowering Energy Saving Management and Microgrid Topology to Diminish Climate Challenge. In *Handbook of Climate Change Mitigation and Adaptation*; Lackner, M., Sajjadi, B., Chen, W.Y., Eds.; Springer New York: New York, 2020; pp. 1–31.

12. Zhang, Y.; Yang, R.; Zhang, K.; Jiang, H.; Zhang, J.J. Consumption Behavior Analytics-Aided Energy Forecasting and Dispatch, *IEEE Intelligent Systems* **2017**, *32* (4), 59–63.

13. Zeki´c-Su˘sac, M.; Mitrovi´c, S.; Has, A. Machine Learning Based System for Managing Energy Efficiency of Public Sector as an Approach Towards Smart Cities, *International Journal of Information Management* **2021**, *58*, 102074. https://www.sciencedirect.com/science/article/pii/S0268401219302968.

14. Palensky, P.; Dietrich, D. Demand Side Management: Demand Response, Intelligent Energy Systems, and Smart Loads, *IEEE Transactions on Industrial Informatics* **2011**, *7* (3), 381–388.

15. Yang, H.; Shen, W.; Yu, Q.; Liu, J.; Jiang, Y.; Ackom, E.; Dong, Z.Y. Coordinated Demand Response of Rail Transit Load and Energy Storage System Considering Driving Comfort, *CSEE Journal of Power and Energy Systems* **2020**, *6* (4), 749–759.

16. Xie, J.; Zheng, Y.; Pan, X.; Zheng, Y.; Zhang, L.; Zhan, Y. A Short-Term Optimal Scheduling Model for Wind-Solar-Hydro Hybrid Generation System With Cascade Hydropower Considering Regulation Reserve and Spinning Reserve Requirements, *IEEE Access* **2021**, *9*, 10765–10777.

17. Parker, C. Applications – Stationary — Energy Storage Systems: Batteries. In *Encyclopedia of Electrochemical Power Sources*; Garche, J., Ed.; Elsevier: Amsterdam, 2009; pp. 53–64. https://www.sciencedirect.com/science/article/pii/B9780444527455003828.

18. Parisio, A.; Wiezorek, C.; Kyntäjä, T.; Elo, J.; Strunz, K.; Johansson, K.H. Cooperative MPC-Based Energy Management for Networked Microgrids, *IEEE Transactions on Smart Grid* **2017**, *8* (6), 3066–3074.

19. Alnejaili, T.; Drid, S.; Mehdi, D.; Chrifi-Alaoui, L.; Belarbi, R.; Hamdouni, A. Dynamic Control and Advanced Load Management of a Stand-Alone Hybrid Renewable Power System for Remote Housing, *Energy Conversion and Management* **2015**, *105*, 377–392. http://doi.org/10.1016/j.enconman.2015.07.080.
20. Shalev-Shwartz, S.; Ben-David, S. *Understanding Machine Learning: From Theory to Algorithms*, 1st ed.; Cambridge University Press, 2014. https://www.cambridge.org/core/product/identifier/9781107298019/type/book.
21. Gan, T.H.; Kanfoud, J.; Nedunuri, H.; Amini, A.; Feng, G. Industry 4.0: Why Machine Learning Matters? In *Advances in Condition Monitoring and Structural Health Monitoring*, Singapore; Gelman, L., Martin, N., Malcolm, A.A., (Edmund) Liew, C.K., Eds.; Springer: Singapore, 2021; pp. 397–404.
22. Sathya, R.; Abraham, A. Comparison of Supervised and Unsupervised Learning Algorithms for Pattern Classification, *International Journal of Advanced Research in Artificial Intelligence* **2013**, *2*, 34–38.
23. Zhang, Z.; Wang, D.; Gao, J. Learning Automata-Based Multiagent Reinforcement Learning for Optimization of Cooperative Tasks, *IEEE Transactions on Neural Networks and Learning Systems* **2021**, *32* (10), 4639–4652.
24. Morales, E.F.; Escalante, H.J. Chapter 6: A Brief Introduction to Supervised, Unsupervised, and Reinforcement Learning. In *Biosignal Processing and Classification Using Computational Learning and Intelligence*; Torres-García, A.A., Reyes-García, C.A., Villaseñor-Pineda, L., Mendoza-Montoya, O., Eds.; Academic Press, 2022; pp. 111–129. https://www.sciencedirect.com/science/article/pii/B9780128201251000178.
25. Hastie, T.; Tibshirani, R.; Friedman, J. *The Elements of Statistical Learning*; Springer Series in Statistics; Springer New York: New York, 2009. http://link.springer.com/10.1007/978-0-387-84858-7.
26. Koolen, D.; Sadat-Razavi, N.; Ketter, W. Machine Learning for Identifying Demand Patterns of Home Energy Management Systems with Dynamic Electricity Pricing, *Applied Sciences* **2017**, *7* (11), 1160. http://www.mdpi.com/2076-3417/7/11/1160.
27. Rai, R.; Tiwari, M. K.; Ivanov, D.; Dolgui, A. Machine learning in manufacturing and industry 4.0 applications, *International Journal of Production Research* **2021**, *59* (16), 4773–4778. https://doi.org/10.1080/00207543.2021.1956675
28. Cınar, Z.M.; Abdussalam Nuhu, A.; Zeeshan, Q.; Korhan, O.; Asmael, M.; Safaei, B. Machine Learning in Predictive Maintenance towards Sustainable Smart Manufacturing in Industry 4.0, *Sustainability* **2020**, *12* (19), 8211. https://www.mdpi.com/2071-1050/12/19/8211.
29. Maharjan, L.; Ditsworth, M.; Niraula, M.; Caicedo Narvaez, C.; Fahimi, B. Machine Learning Based Energy Management System for Grid Disaster Mitigation, *IET Smart Grid* **2019**, *2* (2), 172–182. https://onlinelibrary.wiley.com/doi/10.1049/iet-stg.2018.0043.
30. Gautam, M.; Raviteja, S.; Mahalakshmi, R. Household Energy Management Model to Maximize Solar Power Utilization Using Machine Learning, *Procedia Computer Science* **2019**, *165*, 90–96. https://linkinghub.elsevier.com/retrieve/pii/S1877050920300831.
31. Nakabi, T.A.; Toivanen, P. Deep Reinforcement Learning for Energy Management in a Microgrid with Flexible Demand, *Sustainable Energy, Grids and Networks* **2021**, *25*, 100413. https://linkinghub.elsevier.com/retrieve/pii/S2352467720303441.

32. Zhou, S.; Hu, Z.; Gu, W.; Jiang, M.; Zhang, X.P. Artificial Intelligence Based Smart Energy Community Management: A Reinforcement Learning Approach, *CSEE Journal of Power and Energy Systems* **2019**, *5* (1), 1–10.

33. Jamil, N.; Qassim, Q.S.; Bohani, F.A.; Mansor, M.; Ramachandaramurthy, V.K. Cybersecurity of Microgrid: State-of-the-Art Review and Possible Directions of Future Research, *Applied Sciences* **2021**, *11* (21). https://www.mdpi.com/2076-3417/11/21/9812.

34. Rahman Fahim, S.; K. Sarker, S.; Muyeen, S.M.; Sheikh, M.R.I.; Das, S.K. Microgrid Fault Detection and Classification: Machine Learning Based Approach, Comparison, and Reviews, *Energies* **2020**, *13* (13). https://www.mdpi.com/1996-1073/13/13/3460.

35. Kar, S.; Samantaray, S.R.; Zadeh, M.D. Data-Mining Model Based Intelligent Differential Microgrid Protection Scheme, *IEEE Systems Journal* **2017**, *11* (2), 1161–1169.

36. Berghout, T.; Benbouzid, M.; Muyeen, S. Machine Learning for Cybersecurity in Smart Grids: A Comprehensive Review-Based Study on Methods, Solutions, and Prospects, *International Journal of Critical Infrastructure Protection* **2022**, *38*, 100547. https://www.sciencedirect.com/science/article/pii/S1874548222000348.

37. Alimi, O.A.; Ouahada, K.; Abu-Mahfouz, A.M. A Review of Machine Learning Approaches to Power System Security and Stability, *IEEE Access* **2020**, *8*, 113512–113531.

38. Zhang, D.; Li, S.; Sun, M.; O'Neill, Z. An Optimal and Learning-Based Demand Response and Home Energy Management System, *IEEE Transactions on Smart Grid* **2016**, *7* (4), 1790–1801.

39. Liu, Y.; Zhang, D.; Gooi, H.B. Optimization Strategy Based on Deep Reinforcement Learning for Home Energy Management, *CSEE Journal of Power and Energy Systems* **2020**, *6* (3), 572–582.

40. Haq, E.U.; Lyu, C.; Xie, P.; Yan, S.; Ahmad, F.; Jia, Y. Implementation of Home Energy Management System Based on Reinforcement Learning, *Energy Reports* **2022**, *8*, 560–566. 2021 The 8th International Conference on Power and Energy Systems Engineering. https://www.sciencedirect.com/science/article/pii/S2352484721013172.

41. Puranik, S. Demand Side Management Potential in Swedish Households: A Case Study of Dishwasher, Laundry and Water Heating Loads. *Ph.D. Thesis*, Chalmers University of Technology, 01, 2014.

42. Lan, T.; Jermsittiparsert, K.; T. Alrashood, S.; Rezaei, M.; Al-Ghussain, L.; A. Mohamed, M. An Advanced Machine Learning Based Energy Management of Renewable Microgrids Considering Hybrid Electric Vehicles' Charging Demand, *Energies* **2021**, *14* (3). https://www.mdpi.com/1996-1073/14/3/569.

43. Roy, K.; Krishna Mandal, K.; Chandra Mandal, A.; Narayan Patra, S. Analysis of Energy Management in Micro Grid – A Hybrid BFOA and ANN Approach, *Renewable and Sustainable Energy Reviews* **2018**, *82*, 4296–4308. https://www.sciencedirect.com/science/article/pii/S1364032117310997.

44. Bian, H.; Zhong, Y.; Sun, J.; Shi, F. Study on Power Consumption Load Forecast Based on K-Means Clustering and FCM–BP Model, *Energy Reports* **2020**, *6*, 693–700. 2020 The 7th International Conference on Power and Energy Systems Engineering. https://www.sciencedirect.com/science/article/pii/S2352484720315730.

45. Fenza, G.; Gallo, M.; Loia, V. Drift-Aware Methodology for Anomaly Detection in Smart Grid, *IEEE Access* **2019**, 7, 9645–9657.
46. Forootani, A.; Rastegar, M.; Jooshaki, M. An Advanced Satisfaction-Based Home Energy Management System Using Deep Reinforcement Learning, *IEEE Access* **2022**, *10*, 47896–47905.
47. Lissa, P.; Deane, C.; Schukat, M.; Seri, F.; Keane, M.; Barrett, E. Deep Reinforcement Learning for Home Energy Management System Control, *Energy and AI* **2021**, *3*, 100043. https://www.sciencedirect.com/science/article/pii/S2666546820300434.
48. Dridi, A.; Afifi, H.; Moungla, H.; Badosa, J. A Novel Deep Reinforcement Approach for IIoT Microgrid Energy Management Systems, *IEEE Transactions on Green Communications and Networking* **2022**, *6* (1), 148–159.
49. Abdolrasol, M.G.M.; Mohamed, R.; Hannan, M.A.; Al-Shetwi, A.Q.; Mansor, M.; Blaabjerg, F. Artificial Neural Network Based Particle Swarm Optimization for Microgrid Optimal Energy Scheduling, *IEEE Transactions on Power Electronics* **2021**, *36* (11), 12151–12157.
50. Zhang, P.; Lu, X.; Li, K. Achievable Energy Flexibility Forecasting of Buildings Equipped With Integrated Energy Management System, *IEEE Access* **2021**, *9*, 122589–122599.
51. Ayub, N.; Javaid, N.; Mujeeb, S.; Zahid, M.; Khan, W.Z.; Khattak, M.U. Electricity Load Forecasting in Smart Grids Using Support Vector Machine, In *Advanced Information Networking and Applications*, Cham; Barolli, L., Takizawa, M., Xhafa, F., Enokido, T., Eds.; Springer International Publishing: Switzerland, 2020; pp. 1–13.
52. Akpolat, A.N.; Dursun, E.; Kuzucuoğlu, A.E. Deep Learning-Aided Sensorless Control Approach for PV Converters in DC Nanogrids, *IEEE Access* **2021**, *9*, 106641–106654.
53. Tsai, M.S.; Lin, Y.H. Modern Development of an Adaptive Non-Intrusive Appliance Load Monitoring System in Electricity Energy Conservation, *Applied Energy* **2012**, *96*, 55–73. https://www.sciencedirect.com/science/article/pii/S0306261911007240.
54. Natarajan, K.P.; Vadana, P. Machine Learning Based Residential Energy Management System, In *Proceedings of the 2017 IEEE International Conference on Computational Intelligence and Computing Research (ICCIC)*, Tamil Nadu, India, 12, 2017; pp. 1–4.
55. Li, W.; Logenthiran, T.; Phan, V.T.; Woo, W.L. A Novel Smart Energy Theft System (SETS) for IoT-Based Smart Home, *IEEE Internet of Things Journal* **2019**, *6* (3), 5531–5539.
56. Ozcanli, A.K.; Baysal, M. Islanding Detection in Microgrid Using Deep Learning Based on 1D CNN and CNN-LSTM Networks, *Sustainable Energy, Grids and Networks* **2022**, *32*, 100839. https://www.sciencedirect.com/science/article/pii/S2352467722001230.
57. Mbuya, B.; Dimovski, A.; Merlo, M.; Kivevele, T. Very Short-Term Blackout Prediction for Grid-Tied PV Systems Operating in Low Reliability Weak Electric Grids of Developing Countries, *Complexity* **2022**, *2022*, 1–13.
58. Seyedi, Y.; Karimi, H.; Grijalva, S.; Mahseredjian, J.; Sans'o, B. A Supervised Learning Approach for Centralized Fault Localization in Smart Microgrids, *IEEE Systems Journal* **2022**, *16* (3), 4060–4070.

59. Ali, Z.; Terriche, Y.; Hoang, L.Q.N.; Abbas, S.Z.; Hassan, M.A.; Sadiq, M.; Su, C.L.; Guerrero, J.M. Fault Management in DC Microgrids: A Review of Challenges, Countermeasures, and Future Research Trends, *IEEE Access* **2021**, *9*, 128032–128054.
60. Marino, D.L.; Wickramasinghe, C.S.; Singh, V.K.; Gentle, J.; Rieger, C.; Manic, M. The Virtualized Cyber-Physical Testbed for Machine Learning Anomaly Detection: A Wind Powered Grid Case Study, *IEEE Access* **2021**, *9*, 159475–159494.
61. Ali, S.; Wu, K.; Weston, K.; Marinakis, D. A Machine Learning Approach to Meter Placement for Power Quality Estimation in Smart Grid, *IEEE Transactions on Smart Grid* **2016**, *7* (3), 1552–1561.
62. Mohan, N.; Soman, K.P.; Vinayakumar, R. Deep Power: Deep Learning Architectures for Power Quality Disturbances Classification, In *2017 International Conference on Technological Advancements in Power and Energy (TAP Energy)*, Kollam, India, 2017; pp. 1–6.
63. Gutierrez-Rojas, D.; Mashlakov, A.; Brester, C.; Niska, H.; Kolehmainen, M.; Narayanan, A.; Honkapuro, S.; Nardelli, P.H.J. Weather-Driven Predictive Control of a Battery Storage for Improved Microgrid Resilience, *IEEE Access* **2021**, *9*, 163108–163121.
64. Mohamed, M.; Mahmood, F.E.; Abd, M.A.; Chandra, A.; Singh, B. Dynamic Forecasting of Solar Energy Microgrid Systems Using Feature Engineering, *IEEE Transactions on Industry Applications*, **2022**, *58* (6), 7857–7869.
65. Jesus, I.; Pereira, T.; Marques, P.; Sousa, J.; Perdigoto, L.; Coelho, P. End-to-End Management System Framework for Smart Public Buildings, In *2021 IEEE Green Energy and Smart Systems Conference (IGESSC)*, 2021; pp. 1–6.
66. Gupta, S.; Kambli, R.; Wagh, S.; Kazi, F. Support-Vector-Machine-Based Proactive Cascade Prediction in Smart Grid Using Probabilistic Framework, *IEEE Transactions on Industrial Electronics* **2015**, *62* (4), 2478–2486.
67. Aziz, S.; Irshad, M.; Haider, S.A.; Wu, J.; Deng, D.N.; Ahmad, S. Protection of a Smart Grid with the Detection of Cyber-Malware Attacks Using Efficient and Novel Machine Learning Models, *Frontiers in Energy Research* **2022**, *10*, 964305. https://www.frontiersin.org/articles/10.3389/fenrg.2022.964305/full.
68. Mahi-al rashid, A.; Hossain, F.; Anwar, A.; Azam, S. False Data Injection Attack Detection in Smart Grid Using Energy Consumption Forecasting, *Energies* **2022**, *15* (13). https://www.mdpi.com/1996-1073/15/13/4877.
69. Hasan, M.N.; Toma, R.N.; Nahid, A.A.; Islam, M.M.M.; Kim, J.M. Electricity Theft Detection in Smart Grid Systems: A CNN-LSTM Based Approach, *Energies* **2019**, *12* (17). https://www.mdpi.com/1996-1073/12/17/3310.
70. Ota, K.; Oiki, T.; Jha, D.K.; Mariyama, T.; Nikovski, D. Can Increasing Input Dimensionality Improve Deep Reinforcement Learning? June, 2020. ArXiv:2003.01629 [cs, stat], http://arxiv.org/abs/2003.01629.
71. De Almeida, L.F.F.; Santos, j.R.D.; Pereira, L.A.M.; Sodré, A.C.; Mendes, L.L.; Rodrigues, J.J.P.C.; Rabelo, R.A.L.; Alberti, A.M. Control Networks and Smart Grid Teleprotection: Key Aspects, Technologies, Protocols, and Case-Studies, *IEEE Access* **2020**, *8*, 174049–174079.
72. Yu, L.; Qin, S.; Zhang, M.; Shen, C.; Jiang, T.; Guan, X. A Review of Deep Reinforcement Learning for Smart Building Energy Management, *IEEE Internet of Things Journal* **2021**, *8* (15), 12046–12063.
73. Mishra, S.; Palanisamy, P. An Integrated Multi-Time-Scale Modeling for Solar Irradiance Forecasting Using Deep Learning, 19. arXiv preprint arXiv:1905.02616.

74. Méndez, J.I.; Peffer, T.; Ponce, P.; Meier, A.; Molina, A. Empowering Saving Energy at Home through Serious Games on Thermostat Interfaces, *Energy and Buildings* **2022**, *263*, 112026. https://www.sciencedirect.com/science/article/pii/S0378778822001979.

75. Ponce, P.; Meier, A.; Méndez, J.I.; Peffer, T.; Molina, A.; Mata, O. Tailored Gamification and Serious Game Framework Based on Fuzzy Logic for Saving Energy in Connected Thermostats, *Journal of Cleaner Production* **2020**, *262*, 121167. https://www.sciencedirect.com/science/article/pii/S0959652620312142.

76. Medina, A.; Mendez, J.I.; Ponce, P.; Peffer, T.; Meier, A.; Molina, A. Using Deep Learning in Real-Time for Clothing Classification with Connected Thermostats, *Energies* **2022**, *15* (5). https://www.mdpi.com/1996-1073/15/5/1811.

15

Artificial Intelligence for Construction 4.0: Changing the Paradigms of Construction

Pedro Fonseca, Juan Pablo Solís, and Juan Guadalupe Cristerna Villegas
Tecnológico de Monterrey

CONTENTS

15.1 Introduction

Construction is known to be one of the largest industries in human activity. According to the Mckinsey global institute, construction represents 13% of the world's GDP, accounting for about 10 trillion spent in construction-related goods and services a year. It is also true that construction is one of the oldest human activities; according to Cambridge professor Alan Barnard (2011), the oldest evidence of human constructing something is 1.8 million years old. Considering only these two statements, we can safely say that construction is a very important aspect of human civilizations. In fact, construction advancement has been a significant indicator of the development of civilizations. Take for instance, the pyramids in Egypt and pre-Hispanic cultures of the great wall of China. These structures represent human groups development, from simple tribes to advanced civilizations.

DOI: 10.1201/b23345-18

As human species develop, there are certain activities and industries that develop faster than others. For instance, manufacturing has been one of the leading industries in terms of innovation. Nevertheless, construction has advanced at a slow pace. For example, even though there is constant research toward improving construction materials, humans have relied on some basic construction components for centuries, being the Bessemer process for mass production of steel in the 1850s, one of the last significant construction milestones in history. Why does construction have a development delay in comparison with other industries? It is because it is already an advanced industry? Is there no more room for improvement? Maybe the lack of regulations that promote innovation has discouraged the investment in new construction technology; perhaps a solid and hard culture prevents experimenting with new techniques, or it could be that tradition leaves no room for creativity in the construction process. Whatever the case, it is true that there is a large gap between construction and other industries. Mckinsey global institute estimates that there is a $1.6 trillion opportunity to close the gap. In this chapter, we go through a variety of cases of studies in which artificial intelligence promises to be a key concept in reducing this innovation gap.

15.2 Construction 4.0

Historically, the construction industry has had great economic power; however, it presents a large shortage in productivity and technology use in comparison with other product and service industries that transit currently in the state of the art of Industry 4.0 concepts. In the interest of establishing context, Industry 4.0 is stated as the last industrial revolution where automation and data interchange is the tendency. This is with the intention of creating "smart factories' ' improving every aspect of the manufacturing process, where emerging technologies enable each system to adjust production factors in real time in order to enhance productivity.

According to several world studies and statistics, the construction industry suffers great shortages in the generation, management, and control of the information. This data is generated in the whole process of planning, design, execution, safety, quality assurance, and operation of each state. Construction industry can benefit from digital progress, using Industry 4.0 concepts to promote innovation in how construction assets are designed, constructed, and delivered to the final user. Aslam Hossaing and Abid Nadeem (2019) describe how Construction 4.0 would make a great impact in the construction industry through improved value chain of construction projects, productivity improvement, and safe and sustainable construction. This is current state-of-the-art technology associated with Industry 4.0 such as BIM (Building Information Modeling), Virtual and Augmented Reality,

Drone technology, machine learning, which are starting to be implemented in construction projects around the globe. In this development line, artificial intelligence is set to be the core assembler that brings together all these technologies. This is to enable constructors to stop relying on traditional methods and techniques that stagnate productivity. As a result, artificial intelligence under Construction 4.0 methodologies can reduce costs and construction time, by minimizing errors and interferences, and at the same time enhancing accuracy, safety, and efficiency.

15.3 Information Management in Construction 4.0

One of the key components of Construction 4.0 concept is the generation, processing, and delivery of large amounts of information during the whole construction process. Through the different stages of a project life cycle, from its planning to its operation, a great quantity of data is generated. To this date, developers, constructors, management, and specialists have not been able to achieve its traceability, control, and adjustment through performance indicators. This is because there are several key players integrated in the process, each one generating different types of outputs. Even though a construction process generates a huge amount of data, the processing of it, in order to transform data into information, appears to be the challenge.

At the present time, key players of the construction industry make decisions reactively and subjectively. Preventive decisions do not exist in the strategy of information management in order to allow maintaining technical and economic risk in a controllable range. Implementation of information processing techniques could finally allow projects to be constructed in the established time, with high-quality standards and inside the base contracted budget.

Data science algorithms appear as a solution for data processing and analysis in every possible industry. When talking about the construction process, the big challenge in data generation and processing is the lack of digitalization. Most construction methodologies and techniques rely on analog and manual activities to input information generated by another instance and to report construction progress. One example is the interpretation of an architectural drawing document by the construction team. Nevertheless, new technologies like IoT (Internet of things), cloud computing, and mobile devices open the possibility to better and more efficient ways to gather and transfer data. Furthermore, data gathered by sensors and on-site devices can be centralized and analyzed in real time with the help of cloud computing.

One technology that is taking advantage of data analysis by artificial intelligence is BIM modeling. Even though BIM modeling has been present in construction for almost 50 years, being the conception of a computerized

Building Design System (Eastman, 1974) one of the first glance of a technique to manage digitally all the parts of a construction process, BIM modeling has not been able to deploy its full potential. As a matter of fact, Gholizadeh et al. (2018) analyzed 14 BIM functions and found that only three are used consistently in the construction industry: 3D visualization, clash detection, and constructability analysis. Most of the remaining ones have to do with analysis, like code validation, constructability analysis, structural analysis, and energy analysis. Artificial intelligence tools like data analysis can be the answer to take real advantage of these functions. Take for example code validation, which is declared as "checking the compliance of a construction project with regulatory requirements, including analyzing project design, finding related codes, and matching the design against codes." If a technology system can analyze and become "expert" in construction regulations, it will be possible to automate the process of validating every element of the construction process from the very moment of the construction design and constantly provide feedback as the design phase progresses. This can help avoid rework, saving a huge amount of time and money, not just for the construction company but for the authority in charge of code compliance review. A construction permit could be granted or declined the very moment a construction company submits the BIM model of the project to the corresponding authorities.

Through the BIM methodology and inside every process of it, a large amount of data is generated. This information requires to be controlled during design, procurement (the act of obtaining or purchasing goods or services for a construction project), and construction. Through artificial intelligence, this control can be done by means of performance indicators. All of this data, including legal, financial, and resource information, among others, can be projected through different monitoring systems that reflect the "project health" in a very straightforward fashion; in this manner, the directive team can take more accurate and anticipated decisions.

15.4 On-Site Surveillance

One critical part of a construction project is the on-site surveillance. For a construction project, it is important to validate that the planned project is being constructed as it was designed. Furthermore, a site surveillance is needed to verify that the project is on schedule and that the materials and resources used are according to the original plan. Usually, this process is done manually and could take several resources, including time, to perform. Commonly, a site surveillance relies on human visual inspection and often requires specialized equipment and skills. For example, if one wants to measure construction volume, it could take several days for a human party to measure components and calculate its volume. During these days, the project continues advancing, and by the end of it, the measures can be outdated. Furthermore, if a construction error is not detected on time, and the development continues, this could represent extra cost and delays in the delivery times.

One way artificial intelligence is helping to improve on-site surveillance is with the use of autonomous systems that can perform on-demand or even constant measurements. One example of this system is the deployment of drones in construction sites. These devices had the ability to carry several types of sensors which can gather any type of data that can be used to generate a constant and accurate traceability of the whole project. Also, these vehicles can cover great areas and distances in a few minutes, surpassing the limited cover of a human party. For example, a drone can carry a lidar scanner (Laser Imaging Detection and Ranging) which can make millions of measurements that are translated into millions of points with an x, y, x position. At the end of the surveillance, these points are reconstructed using an artificial intelligence algorithm to generate what is called a point cloud. This point cloud is a 3D representation of any visible detail in a construction to a millimeter precision. This digital representation can be used to validate the accuracy of construction versus plan.

Another drone technique widely used in construction is photogrammetry, which is the use of a large set of photographs that contain certain types of metadata (georeference, sensor size, camera angle, orientation, etc.) to generate 3D models, often called "as-built models." These 3D digital representations are made using computer vision algorithms. This is to reconstruct an accurate digital twin of the real construction in which one can perform measurements such as volume, area, and distances as it was done in the real site. Anwar et al. (2018) determined that drone surveillance "can significantly reduce the effort required in traditional construction monitoring and reporting procedures" demonstrating the effectiveness of the technique. As a result of the case of study, it was proven the convenience of drone site supervision and management, resulting in better operations, planning, and effective on-site adjustments.

Another artificial intelligence application in a construction site is the use of machine learning algorithms for safety and health issues. Using IoT techniques, several sensors and cameras can be placed strategically in the construction site, and with the use of machine learning algorithms the systema can detect risk situations. For example, a camera using computer vision algorithms can detect when a worker is performing a risk activity without the necessary equipment or without safety measures. A set of sensors can detect irregularities in the condition of materials and equipment which can represent a risk for human health. A computer algorithm can be trained to detect defects in the execution of an activity, or weathering of materials that can lead to structural damage and therefore a risk situation. Overall, machine learning algorithms can perform risk assessment and improve the accuracy of it in each iteration.

15.5 Quality Control

Construction firms constantly have to cope with quality challenges; this constitutes an important issue regarding the compliance with regulations and international standards. Failing in the fulfillment of regulations can result in economic penalties or even the closure of a project. AI techniques can improve the implementation of quality assurance methodologies. As mentioned by Chakkravarthy (2019), data collected in several projects can be compared to offer valuable learning information for AI applications. These AI programs can deliver precise data and insights to ultimately assist contractors to optimize the whole construction process.

One methodology that is broadly used for quality assurance in many industries is the Lean model. In construction, Lean management is used in order to generate performance indicators in several aspects of the project such as materials, equipment, and processes. Perico and Mattioli (2020) describes Lean systems as the conversion of organizational and theoretical knowledge into practical behavior and business value. It also states that artificial intelligence algorithms can replicate cognitive activities such as perception, understanding, learning, decision, and dialogue, skills needed to implement Lean practices. One application of artificial intelligence in quality management under Lean thinking for construction projects is the use of statistics algorithms that are able to detect tendencies and generate forecasts of construction outcomes. This can be performed by analyzing data retrieved in all of the project stages under the supervision of a human expert. AI algorithms can process the data retrieved by audits and reports making it readily accessible. Furthermore, machine learning algorithms can detect correlation within analysis criteria used to perform quality assessment of an activity or element in the construction process. This analysis can be performed almost in real time with the help of cloud computing, making quality information available at any stage of the project. With this, it is possible to anticipate unfavorable conditions. This can represent economic savings without compromising time and quality of any construction element.

15.6 Design Optimization

In every construction project, the design phase is critical to a project's success. The ability to consider every possible variable into the design of any construction is key to improve efficiency on the use of the project's resources. Optimization as it is has been a constant case of study for artificial intelligence. For example, the use of genetic algorithms is broadly used in the generation of optimization tools for designing complex devices like communication hardware or aerospace components.

In construction, the concept of optimization has been constantly related to the construction phase, for instance the optimization of material used in building different structures, or the optimization of the human resource by reducing death times. Nevertheless, optimizations have not been a key component in the design activities. This can result in situations such as the construction of a structure different from the one in the drawing documents, in order to fit in budget limitations or to fulfill time constraints. Still the implementation of optimization techniques in the design phase can improve the project execution.

One example of how artificial intelligence can benefit the optimization process is mentioned by Gilan and Dilkina (2015) where it is analyzed how the design of a building has a major impact on the energy footprint of the building. The study stated that there are several factors affecting the building's energy efficiency, such as floor plan design, building orientation, construction materials, daylight and solar control measures, and activity-related parameters. For a human team, it could take a great amount of time to map and consider a large quantity of parameters in the design of a building. This can discourage the application of optimizations tools in early stages of the projects due to the factor of time. However, machine learning algorithms can be used to include all types of parameters and simulate a variety of scenarios in order to find the most optimal case. This can be applied to factors such as position, materials, and management of resources used in the construction phase.

Another opportunity of artificial intelligence in the construction industry is in urban planning and zoning. For instance, in highly populated areas, or places with a population growth tendency, it has become complicated to plan assertively the development of any type of infrastructure. One good example of the use of AI for design optimization in urban planning is the one described by Kamal Jain (2011) where artificial neural networks are used to forecast the change of land use. Neural networks can overcome the limitations of human analysis where there is a large amount of interdependent variables. Optimization during the design stage of a construction project can lead to having more energy-efficient buildings, to reduce traffic congestion in high-population-density areas, to reduce construction time and costs, just to mention a few benefits. In this way, the use of optimization algorithms can improve optimization outcomes by being able to take into account more variables than a manual technique. Furthermore, an optimization algorithm can produce a series of possible outcomes, give more accurate forecasts, and integrate more variables and constraints with fewer costs to the project.

15.7 Other Artificial Intelligence Technologies

In the last decade, new technologies like cloud computing, IoT, and block chain have enabled artificial intelligence to increase the possible applications in almost any aspect of human life. In construction, Pan, (2021) performs a scientometric review of keywords in over 4,473 journal articles published in 1997–2020. It was found that the focus on papers shifted in the last 10 years from expert systems to the implementation of technology in the construction industry. It was also found that there are six directions of research in construction science: smart robotics, cloud virtual and augmented reality, Artificial Intelligence of Things (AIoT), digital twins, 4D printing, and blockchains. Much of this can serve from artificial intelligence to improve aspects of construction activities.

15.7.1 Smart Robotics

In the topic of smart robotics, one popular trend is the use of robots for "printing houses." In this technology, the premise is that a robot can be programmed with an architectural drawing document and replicated it with high precision. The robot can be adapted for using many types of materials, and switch between them as the construction is advancing. Furthermore, the robot has a payload of several sensors that can take many variables into account, such as material composition and weather, and constantly adjust its processes to maintain the quality of the construction. Another benefit is that a robot can run for extended periods of time, allowing the mass production of houses in a short amount of time. An additional advantage is the safety issue, and one main challenge in construction is the constant risk of accidents; when using a robot, this risk is minimized. This is because the robot is programmed to fulfill every safety regulation, taking the human factor out of the picture. In case of a robot malfunction, it can be rapidly replaced, and the new robot will be loaded with the same program and take from the point the last robot left in a matter of minutes.

Furthermore, we can consider robots not only during the construction phase, but also introduce them into a pre-fabrication process. Instead of having a robot constructing a house on-site, an array of robots can manufacture a series of components that are later assembled on-site. Although modular construction and pre-fabricated components are already a reality, the introduction of robots in a more active role could finally switch the gap between manufacturing and construction, where components are manufactured using industrial methodologies and quality standards.

Elattar (2008) analyzes the use of robots in several types of constructions like concrete works, construction of roads, and finishing works (window mounting, ceramic tile installation, welding, etc.) and concludes that robots can maintain highly accurate actions and reduce hazardous risks achieving improved control and safety. Also that the automation of process can be achieve with high accuracy, which can lead to several benefits in the construction process

15.7.2 Digital Twins

We discuss the use of digital twins in the information management section of this chapter. Nevertheless, the use of digital twins can serve more opportunities when applied artificial intelligence concepts into it. A digital twin is an information system that can replicate the output of his real twin, when receiving the same inputs. For example, a digital twin can simulate the behavior of a structure in specific ambient conditions. This allows them to detect failures before construction, saving time, and money by doing modifications prior to the development of the project.

We can describe two types of digital twins, the one that is modeled based on drawing documents before construction, usually inside BIM methodology. And the as-built models, which are 3D representations of the actual element already constructed, often generated by a laser scanner of photogrammetry. In both cases, it allows the generation of relevant and well-timed information. This information can serve as input for machine learning algorithms that can provide relevant findings and forecasts before starting the real construction. This allows the validation of the executive project and engineering without the cost of failure and errors during the construction phase.

Alexopoulos et al. (2020) propose an interesting approach of digital twin-driven machine learning technique for manufacturing, which can be easily applied to construction. Starting with the premise that machine learning algorithms rely on a great quantity of data for supervised learning. And that the generation of these data sets must be acquired through manufacturing itself, representing high costs and time spams. A systematic framework is an approach in which a digital twin serves as the data set generator, shifting from real-world data sets to virtual ones. Considering that the digital twin is a one-one representation, the virtual data set should behave in the same way that the real-world data set. The virtual data sets are generated with less resources and enable supervised learning for the machine learning algorithms. The result can later be enriched with real ones.

15.7.3 Artificial Intelligence of Things (AIoT)

AIoT is the combination of concepts around the Internet of things (IoT) and artificial intelligence techniques. Whereas IoT describes the concept around the interchange of information around several physical objects with sensors, the enrichment of artificial intelligence algorithms can take better advantage of the data sense by these objects. In construction, Ghosh et al. (2020) analyze several potential applications of IoT in construction, and identify the level of importance perceived in the principal research areas. The ones identified with more relevance are remote structural health monitoring, construction safety, optimization, and simulation and image processing.

IoT can solve one of the most challenging situations in construction projects: Traceability. This is the ability to be able to verify the progress of each part of the project, having a complete perspective of risks, delays, problems, etc. The construction industry can apply IoT concepts to design a cyber-physical system deployed on-site which is constantly capturing data in every work front. This data is instantly uploaded to the cloud, where data analysis algorithms process the information. The system can be programmed to send push notifications to the decision-making team and the project management when a risk situation is identified by a camera with image processing algorithms, or when the system detects that materials in the warehouse are getting low, or when the resources assigned to an activity are getting out of the planned budget, or whenever a situation that requires attention presents itself. Furthermore, machine learning techniques can use the situation's data set to forecast the possibility of it occurring again.

15.7.4 Blockchain

Even though blockchain is one of the technologies less present in the construction industry nowadays, it is listed as one of the most promising for the industry. Pan and Zhang (2021) identifies three possible applications of blockchain powered by artificial intelligence:

- Integration with BIM, in which blockchain can facilitate the management of assets. This by storing in each block important and sensitive information of each asset.
- The creation of smart contracts. Where blockchain can constitute a secure code document which enhances trust and prevents fraud.
- Supply chain, where trustworthy information is generated, where information like payments, contracts, and agreements can be stored in each block.

15.7.5 Architectural Design

One part of the construction industry that has not been discussed is the architectural design. Even though this is a very creative and human process, there are great examples of how architecture is benefiting from artificial intelligence techniques.

One example in which artificial intelligence can help architecture is parametric and generative design. In parametric design, the user indicates the artificial intelligence the parameters to be solved. In each iteration, the system shows the results and the user is able to alternate the parameters. In generative design, the user proposes the inquiries to be solved to the artificial intelligence and through simulation and iterations, the system offers the best solutions.

The benefits in the use of this artificial intelligence-aided design tools are the reduction of time and costs. The disadvantages of this technique are the loss of the emotional craft that is implicit in the architectural design process. Therefore, the result of this methodology requires further analysis by the architects to adapt the solutions presented by the system to the requirements of the project.

15.7.6 Document Management

One last area of the construction process that can benefit from artificial intelligence is document management and control. A construction project produces a huge amount of different types of documents, for example, drawing documents, construction permits, financial documents, among others. Artificial intelligence could help to develop a more precise trace of the documents and its content. For example, when a list of engineering documents is generated, the system can analyze and calculate the man-hour produced by type of document. It can also calculate the amount of estimations or contracts. Artificial intelligence can help in reducing the document analysis that is made manually.

15.8 Conclusions

Even though construction is one of the largest industries by worth, it still relies on human manual labor. This results in the decrease of productivity and increase of costs and in time spent for correcting mistakes. However, artificial intelligence presents a necessary solution. The use of these technologies will allow the information generated in each stage of the project to be shown as performance indicators, graphics, and tendencies. This will grant management and directors to make better technical and economical preventive decisions. This with the objective of making the adjustments needed to fulfill the goals and original scopes of each project. Artificial intelligence represents the tool that will grant the opportunity to be able to pursue government over the great quantity of data generated in each project. One last thing that is needed to take into account is the regulation part of the industry. As of this day, many countries rely on outdated normativities and construction rules. Governments and regulating entities can benefit from the information analysis capabilities of artificial intelligence to improve and update current regulations. This is to enable innovation and encourage investment in new technologies to improve the construction industry.

References

Alexopoulos, K., Nikolakis, N., & Chryssolouris, G. (2020). Digital twin-driven supervised machine learning for the development of artificial intelligence applications in manufacturing. *International Journal of Computer Integrated Manufacturing*, 33(5), 429–439.

Anwar, N., Izhar, M. A., & Najam, F. A. (2018, July). Construction monitoring and reporting using drones and unmanned aerial vehicles (UAVs). In *The Tenth International Conference on Construction in the 21st Century (CITC-10)*, Colombo (pp. 2–4).

Barnard, A. (2011). *Social Anthropology and Human Origins*. Cambridge: Cambridge University Press. doi:10.1017/CBO9780511974502.

Chakkravarthy, R. (2019). Artificial intelligence for construction safety. *Professional Safety*, 64(1), 46. Retrieved from https://0-search-proquest-com.biblioteca-ils.tec.mx/scholarly-journals/artificial-intelligence-construction-safety/docview/2165604383/se-2.

Eastman, C. M. (1974). An outline of the building description system. Research Report No. 50.

Elattar, S. M. S. (2008). Automation and robotics in construction: Opportunities and challenges. *Emirates Journal for Engineering Research*, 13(2), 21–26.

Gholizadeh, P., Esmaeili, B., & Goodrum, P. (2018). Diffusion of building information modeling functions in the construction industry. *Journal of Management in Engineering*, 34(2), 04017060.

Ghosh, A., Edwards, D. J., & Hosseini, M. R. (2020). Patterns and trends in Internet of Things (IoT) research: Future applications in the construction industry. *Engineering, Construction and Architectural Management*. doi: 10.1108/ECAM-04-2020-0271.

Gilan, S. S., & Dilkina, B. (2015, April). Sustainable building design: A challenge at the intersection of machine learning and design optimization. In *Workshops at the Twenty-Ninth AAAI Conference on Artificial Intelligence*, Austin, TX.

Hossan, M. A & Nadeem, A. (2019). Towards digitizing the construction industry: State of the art of construction 4.0. *Proceedings of International Structural Engineering and Construction*. 6. doi:10.14455/ISEC.res.2019.184.

Kamal Jain, P. (2011). A review study on urban planning & artificial intelligence. *International Journal of Soft Computing & Engineering*, 1(5), 101–104.

Mckinsey Global Institute. (n.d.). Reinventing construction executive summary - Mckinsey & company. Retrieved October 31, 2022, from https://www.mckinsey.com/~/media/McKinsey/Business%20Functions/Operations/Our%20Insights/Reinventing%20construction%20through%20a%20productivity%20revolution/MGI-Reinventing-Construction-Executive-summary.pdf.

Pan, Y., & Zhang, L. (2021). Roles of artificial intelligence in construction engineering and management: A critical review and future trends. *Automation in Construction*, 122, 103517.

Perico, P., & Mattioli, J. (2020, September). Empowering process and control in lean 4.0 with artificial intelligence. In *2020 Third International Conference on Artificial Intelligence for Industries (AI4I)* (pp. 6–9). IEEE, Irvine, CA.

16

A Novel Deep Learning Structure for Detecting Human Activity and Clothing Insulation

Omar Mata, Diego López, Adan Medina, Pedro Ponce, Arturo Molina, and Ricardo Ramirez

Tecnológico de Monterrey

CONTENTS

DOI: 10.1201/b23345-19

16.1 Classification Algorithms for Computer Vision

The main objective of computer vision is to process images and analyze and comprehend the data contained in them. Several areas have taken advantage of this research field, such as healthcare, manufacturing, security, transportation, etc (Heidari, Navimipour, & Unal, 2022). Most of these computer vision applications can be summarized as image segmentation, detection, and classification. These processes seem simple enough when you take the human approach. However, a computer works with information differently. In other words, a computer cannot immediately look at an image and detect the edges, shapes, colors, forms, etc., with a simple look. So, what the computer sees is a matrix full of numbers corresponding to the intensity of a pixel, the location of that pixel, etc. To illustrate this, Figure 16.1 shows an image as a human sees it, and Table 16.1 illustrates the same image, but with the numeric values, the computer is to retrieve and analyze. Therefore, simple tasks for a human, such as object recognition and classification, become complicated tasks for a computer.

FIGURE 16.1
Example of a small image as a human sees it.

TABLE 16.1

Example of an Image as a Computer Sees It

255	255	255	0	0	255	255	255
255	255	0	255	255	0	255	255
255	0	255	0	0	255	0	255
0	255	0	255	255	0	255	0
0	255	0	255	255	0	255	0
255	0	255	0	0	255	0	255
255	255	0	255	255	0	255	255
255	255	255	0	0	255	255	255

This is a grayscale image, so only one value is obtained per pixel; in the case of a color image, three values are obtained for each pixel considering the RGB image format.

Two of the most commonly used algorithms to allow computers to analyze this type of data are convolutional neural networks (CNN) and recurrent neural networks (RNN) (Bhatt et al., 2021). However, it is important to remember that the architecture of CNN and RNN changes depending on the application of the algorithms. The following sections describe the general and more specialized architectures of CNN and RNN.

16.1.1 Convolutional Neural Networks

CNNs are a unique type of deep learning algorithm that are usually used for image segmentation, classification, and recognition. This algorithm takes an image as input; then it assigns some weights and biases to pixels/objects, which are tuned through a learning process (Indolia, Goswami, Mishra, & Asopa, 2018). There are other processes involved in between, such as pooling, flattening, and fully connected layers, which will be described in more detail in the following sections. All of this can be visualized in Figure 16.2.

16.1.1.1 Convolutional Layer

Convolution is a mathematical operation in which a filter or kernel is passed through the pixels of an image to extract its most important features of it. The filter consists of a matrix of weights in which the height and width are hyper-parameters while the dimension (depth) is defined by the input. However, the height and width of the kernel are not the only hyper-parameters involved in this stage. The other two layers are very important for the convolution operation and its result: padding and stride (Dumoulin & Visin, 2016). Before discussing these two parameters, let us give a simple example of how a simple convolution operation works.

Let's take Figure 16.3 as an example. Here, you can see a 3×3 image and a 2×2 kernel. In order to obtain the first value of the output (top left corner), the convolution operation is $0*0+1*1+3*2+4*3=19$, where the first

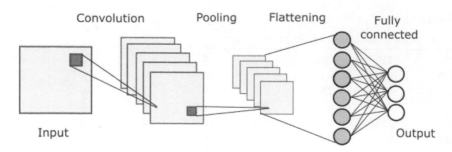

FIGURE 16.2
CNN basic architecture.

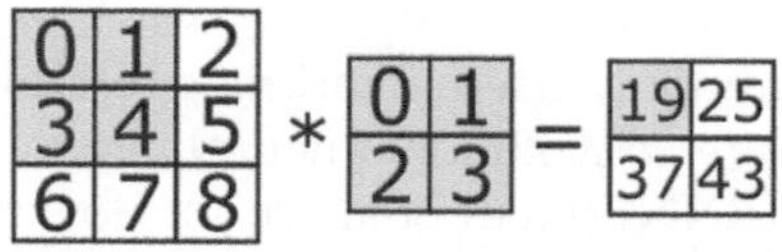

$$\begin{array}{|c|c|c|} \hline 0 & 1 & 2 \\ \hline 3 & 4 & 5 \\ \hline 6 & 7 & 8 \\ \hline \end{array} * \begin{array}{|c|c|} \hline 0 & 1 \\ \hline 2 & 3 \\ \hline \end{array} = \begin{array}{|c|c|} \hline 19 & 25 \\ \hline 37 & 43 \\ \hline \end{array}$$

FIGURE 16.3
Convolution operation.

numbers of each operation are the values of image pixels while the second numbers are the weights of the filter. Then, the filter slides over the image, and the convolution operation is done again until the whole image is processed by the filter. As you can see, applying a convolution operation on a 3×3 image using a 2×2 kernel results in an output representation of size 2×2. Having said that, it can be assumed that, in general, if you have an image of size $H \times W \times D$ (where H is the height, W is the width, and D is the dimension or depth of the input image) and a filter of size $F_h \times F_w \times F_d$ (where F_h is the height of the filter, F_w is the width of the filter, and F_d is the depth of the filter), then the output matrix would be of size $H - F_h + 1 \times W - F_w + 1 \times 1$. However, this only applies when there is no padding, and the stride is equal to 1.

As per previous example, there is some pixel loss during the convolution. In other words, the output size is reduced compared to the one input. This may not seem like a serious problem, but having several convolutional layers in a single network can be troublesome because of the loss of information. One solution to this problem is to add zero-value pixels around the input image. This is known as zero padding. By doing this, it can increase the effective size of the input image and, therefore, increase the size of the output. Generally, a total of P_h rows and P_w columns are added to the image. Typically, P_h (vertical padding) and P_w (horizontal padding) are even numbers so that can be added an equal amount of columns and rows in the left/right and top/bottom of the image. Applying padding, the output after the convolution results in a matrix of size $H - F_h + P_h + 1 \times W - F_w + P_w + 1 \times 1$. This is represented in Figure 16.4.

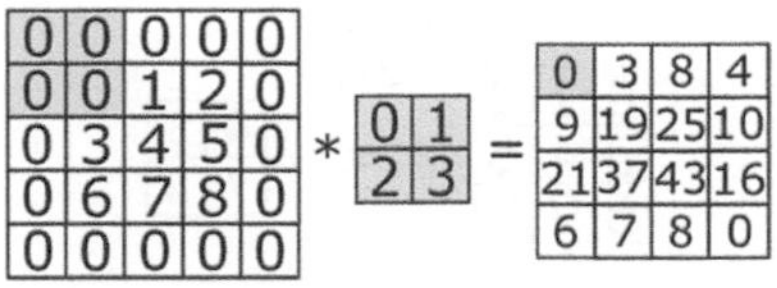

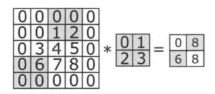

FIGURE 16.4
Convolution operation with padding.

$$\begin{array}{|c|c|c|c|c|} \hline 0&0&0&0&0\\ \hline 0&0&1&2&0\\ \hline 0&3&4&5&0\\ \hline 0&6&7&8&0\\ \hline 0&0&0&0&0\\ \hline \end{array} * \begin{array}{|c|c|} \hline 0&1\\ \hline 2&3\\ \hline \end{array} = \begin{array}{|c|c|} \hline 0&8\\ \hline 6&8\\ \hline \end{array}$$

FIGURE 16.5
Convolution operation with padding and different strides. The horizontal stride of 2 and vertical stride of 3.

Now, when applying the convolution operation, let's start at the top left of the input and then slide it to the right until reaching the end of the tensor. Then, it slides down and begins the process all over again. In the previous two examples, the filter was slid in one position at a time. This is a default stride of 1. Nevertheless, the horizontal stride (S_w) and vertical stride (S_h) can be changed to save computational time, avoid overlapping, or down-sample the image. When applying padding and changing the stride of the convolution, it arrived at a general expression for the size of the output which is $\dfrac{H - F_h + P_h + S_h}{S_h} \times \dfrac{W - F_w + P_w + S_w}{S_w} \times 1$. This can be seen in Figure 16.5.

16.1.1.2 Pooling and Flattening Layer

After the convolutional layer, it is usual to have a pooling layer. This layer reduces the number of parameters (down-sampling) resulting from the previous layer while retaining the important information. Therefore, pooling layers allow less computational power to process the data while extracting the main features, which may be invariant to position and rotation (Zafar et al., 2022). There are several types of pooling, such as max pooling, min pooling, average pooling, sum pooling, and global pooling.

- **Max pooling**: It takes the maximum value inside the feature map.
- Min pooling: It takes the minimum value inside the feature map.
- Average pooling: It takes the average value inside the feature map.
- **Sum pooling**: It takes the sum of all the values inside the feature map.

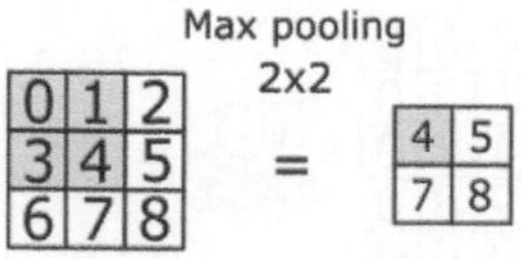

Max pooling
2x2

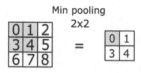

FIGURE 16.6
Max pooling with a 2×2 filter and stride of 1.

Min pooling
2x2

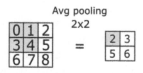

FIGURE 16.7
Min pooling with a 2×2 filter and stride of 1.

Avg pooling
2x2

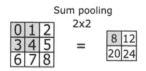

FIGURE 16.8
Average pooling with a 2×2 filter and stride of 1.

Sum pooling
2x2

FIGURE 16.9
Sum pooling with a 2×2 filter and stride of 1.

- **Global pooling:** It down-samples the whole feature map to a single
  value by applying any of the previous pooling strategies.

The following images are shown as examples of different types of pooling
(Figures 16.6–16.9).

After the previous steps, the flattening process takes place. This step will
flatten the pooled feature map into a single column. This procedure aims to
obtain a single vector of data that you can pass to the fully connected layer
that will be in charge of classifying the input image.

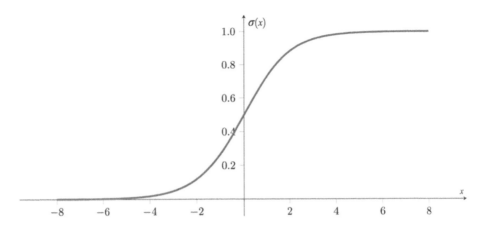

FIGURE 16.10
Sigmoid activation function.

16.1.1.3 Activation Functions

As its name implies, this non-linear function determines which neurons will fire depending on the function's input. These activation layers are used after the convolutional layers, and to allow an optimum training process through back-propagation, they must be easily differentiable. The most common activation functions are:

Sigmoid: it is an S-shaped function that maps a real number input to a number between zero and one (Figure 16.10).

$$\sigma(x) = \frac{1}{1+e^{-x}}$$

ReLu: It is also known as a ramp function and maps all negative numbers to zero while the output will remain the same as the input if the number is positive (Figure 16.11).

$$f(x) = \max(0, x)$$

Tanh: Similar to the Sigmoid function, it maps the output to a value between −1 and 1 (Figure 16.12).

$$f(x) = \frac{e^x - e^{-x}}{e^x + e^{-x}}$$

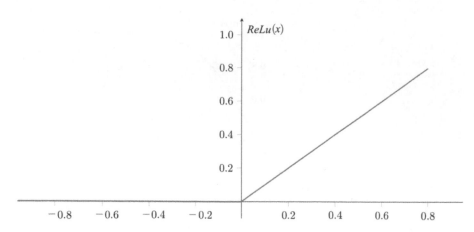

FIGURE 16.11
ReLu activation function.

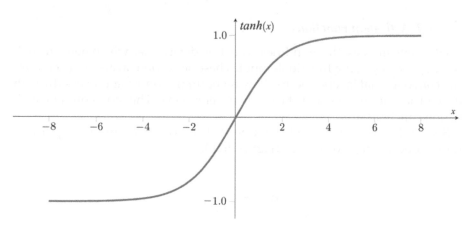

FIGURE 16.12
Tanh activation function.

16.1.1.4 Fully Connected Layer and Loss Functions

The final step of a general CNN is a fully connected layer. The input of this layer is the flattened pooled feature map, which will be processed by the fully connected layer, just as in a basic artificial neural network. There can be multiple fully connected layers; this depends on the depth of the classification model. At the very end, the output of this layer is sent to an activation function to obtain the probability distribution of the classification. Then, finally, a loss function is used to calculate the error during the training process. This error is the difference between the real value and the prediction.

16.1.2 CNN Regularization

Regularization is a process used in CNN training to avoid over-fitting. This last phenomenon happens when a given model performs well in training but performs poorly when used on testing data (Feng, Peng, & Gu, 2019). Some regularization methods that have been proposed are:

- **Data augmentation**: One of the easiest ways to avoid over-fitting is to have a large amount of different data. However, sometimes, it is difficult to get a large variety of images for CNN training. Therefore, data augmentation techniques are used to artificially increase the quantity of training data. Some of these techniques include:
  - Scaling: Resize the image, e.g., duplicate the width or height of the image. Cropping: An important portion of the image is selected while the other part of the image is discarded.
  - Flipping: Flip the image vertically or horizontally. Translation: Move the image around the x-axis and y-axis.
  - Color augmentation: Change the brightness, saturation, contrast, hue, grayscale, etc.
- **Dropout**: During each training step, randomly selected neurons are dropped or turned off to force the model to learn the features through different configurations. Nevertheless, during testing, the network is used without any dropout.
- **Weight dropout**: Similar to dropout, however, instead of turning off neurons, randomly chosen connections between neurons are dropped from the network.

16.1.3 Object Recognizers

Object detection or object recognition combines two very important computer vision techniques: object classification and object localization. As discussed in this chapter, object classification can be done through a CNN. However, object localization consists of detecting an object in an image and producing a bounding box containing the located object. Then, that object must be classified. To do this, a new type of CNN must be used: region-based CNN (R-CNN).

16.1.3.1 R-CNN

Since the first famous network architecture in 1998, CNNs have evolved to handle more difficult tasks. Various changes have been made to the original proposal. For example, new and novel network blocks and different optimization and regularization techniques have been implemented to improve

CNN performance. One of the mainly used networks for object recognition tasks is the R-CNN (Bharati & Pramanik, 2020).

The process can be thought of during object recognition tasks as three separate stages: localization, feature extraction, and classification. Therefore, the three main modules that can be found in an R-CNN are:

- **Region proposal**: The very first step of an R-CNN is to find regions in which the image contains the objects of interest. However, some of the proposed regions may not contain objects, so through the learning process, the void regions can be eliminated and only keep the ones that contain them. These regions are often represented as bounding boxes, as shown in Figure 16.13. The region proposal process is done using an algorithm called selective search, which proposes initial regions and then finds similarities between a given region and its neighbors to group them together. Then, new similarities are calculated, and the process is repeated until the whole image becomes a single region. Finally, the algorithm goes backward, searching for the initial regions and the merging probability. If an object's score is high, the regions involved will merge to get that object; if not, they remain as separate regions (Uijlings, Van De Sande, Gevers, & Smeulders, 2013).

- **Feature extraction**: This step is done through a CNN. The proposed regions are taken from the original image, resized, wrapped, and then introduced into the CNN to obtain the most representative features.

- **Classifier**: The features are finally classified using a classification algorithm. The R-CNN generally uses a support vector machine (SVM) as a classifier. However, this one can be modified if it is desired.

Despite being designed for object recognition, R-CNN suffers from some major drawbacks. One of them is that it has a low computational performance because of all the proposed regions that are passed to the CNN. Also,

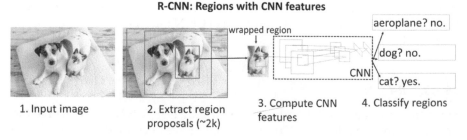

R-CNN: Regions with CNN features

1. Input image 　　2. Extract region proposals (~2k) 　　3. Compute CNN features 　　4. Classify regions

wrapped region
aeroplane? no.
dog? no.
CNN
cat? yes.

FIGURE 16.13
RCNN four main stages.

the three different modules of the network must be trained for each of their tasks, which consumes a lot of computational resources and time. Finally, because of these two issues, R-CNN cannot be used for real-time object detection tasks, causing it to only be applicable for asynchronous classification (Bharati & Pramanik, 2020). Therefore, other R-CNN architectures have been proposed to tackle this problem: fast R-CNN, faster R-CNN, and mask R-CNN.

16.1.3.2 Fast R-CNN

As told previously, basic R-CNN training is slow and demands a lot of computational resources. To tackle this problem, fast R-CNN feeds into the CNN the entire image to generate a feature map instead of feeding the region proposals. In other words, the image is first processed through diverse convolutional and max pooling layers to produce a convolutional feature map. Then, the fast R-CNN generates the proposals from this map using the selective search algorithm described above and then wraps them into a fixed-sized feature map using a new layer called a region of interest (ROI) pooling layer that often uses a max pooling strategy. Finally, each of these feature maps is fed into a series of fully connected layers that generate two outputs: one for the bounding box positions, which is obtained through a linear regression model, and the other for the probability of the object classification, which is obtained using a softmax activation function (Girshick, 2015) (Figure 16.14).

16.1.3.3 Faster R-CNN

The main drawback that fast R-CNN shares with R-CNN is the selective search algorithm. It is a computationally expensive algorithm that decreases the performance of both methods. To diminish this problem, faster R-CNN applies a separate network to propose the regions. Then, these regions are passed to the ROI pooling layer; the following process is the same as for the fast R-CNN (Ren, He, Girshick, & Sun, 2015) (Figure 16.15).

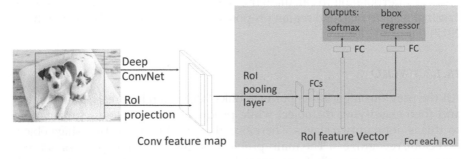

FIGURE 16.14
Fast R-CNN architecture.

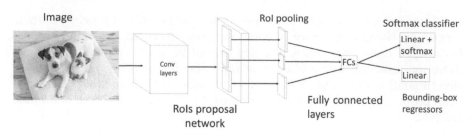

FIGURE 16.15
Faster R-CNN architecture.

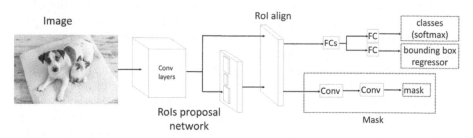

FIGURE 16.16
Mask R-CNN architecture.

16.1.3.4 Mask R-CNN

Mask R-CNN is an extension of faster R-CNN. While faster R-CNN has two outputs, just as fast R-CNN, mask R-CNN has three outputs: one for the bounding box, one for the object classification, and the last one for an object mask. This new branch allows the algorithm to improve its efficiency for object detection tasks. Also, another contribution of mask R-CNN variation is the ROI align layer. Instead of using an ROI pooling layer, which may force the boundaries of the regions and feature maps to be aligned, the ROI align layer allows the boundaries to not be forced to align and instead uses bilinear interpolation to calculate the new feature map values based on the original feature map and region proposals (He, Gkioxari, Doll'ar, & Girshick, 2017) (Figure 16.16).

16.1.3.5 YOLO

All the previous algorithms perform object detection by presenting a region and then classifying the object while producing finer image segmentation, so this is done in two stages. Therefore, they are known as two-stage object detectors/recognizers. The main problem is that the whole process is time-consuming, which leads to computational speed. Therefore, to accomplish faster object detection, one-stage classifiers were proposed.

Also, due to the reduced size of parameters, the one-stage classifiers have been implemented in different embedded systems, such as the work presented by Jin et al. (Jin, Wen, & Liang, 2020) where they implement it in NVIDIA Jetson TX2, or the work presented by Shafiee et al. (Shafiee, Chywl, Li, & Wong, 2017) where they use NVIDIA Jetson TX1. Due to the advancements in hardware capabilities, cheaper embedded systems have been tried such as in work presented by Medina et al. (Medina, Méndez, Ponce, Peffer, & Molina, 2022) where they use both NVIDIA Jetson Nano and Raspberry Pi 4, obtaining similar results from both systems.

Redmon et al. proposed an algorithm called You Only Look Once, commonly known as YOLO (Redmon, Divvala, Girshick, & Farhadi, 2016) where the main contribution is to divide the entire image in a grid and produce both the class probability map and bounding box predictions in a single pass, which reduces the time needed to produce object detection. The main problem with this is that each grid cell produces a single object prediction, so even though there could be two objects sharing the grid cell, only one will be detected, so problems with smaller objects can arise. However, this algorithm was very fast and worked better than other alternatives at that time with background noise.

16.1.3.6 YOLOv2

Following the research from the first YOLO version, the second iteration of YOLO called YOLOv2 was proposed by Redmon and Farhadi, (2017). This algorithm presented improvements by adding batch normalization as well as replacement of pooling layers and fully connected layers with convolutional layers to avoid missing any information. Another important aspect that changed is that instead of presenting arbitrary bounding boxes in each grid cell, five anchor boxes per cell are used with offsets to achieve a better training process for the bounding boxes. It is important to notice that this iteration established the backbone feature extraction algorithm that most YOLO versions use (in different and upgraded versions), which is darknet.

16.1.3.7 YOLOv3

While YOLOv2 used Darknet-19, YOLOv3 used a deeper feature extractor (Darknet-53). This iteration was presented by Redmon et al. (Redmon & Farhadi, 2018). This new iteration's main problem tackled is the object scale problem by producing object classification at three different scales, minimizing the errors in small objects. Also, this version got rid of two anchor boxes, working with only 3; however, since the prediction occurs at three different scales, this actually turns into 9 anchor boxes. Although this algorithm is slower than the previous iterations, its accuracy is much, much higher; therefore, it is considered a better algorithm.

16.1.3.8 YOLOv4

Although the original author of the YOLO algorithm stopped doing research on computer vision, new research teams continued to work on the algorithm, and Bochkovsiy proposed a new iteration (Bochkovskiy, Wang, & Liao, 2020). This new iteration of YOLO presented a modification of the base feature extractor by adding spatial pyramid pooling to increase the convolutional layer's perceptive field and extract more features. This also allowed increasing the resolution of the input image and therefore working with more pixels and information. The authors also tried out different activation functions and skip connections to find the best to increase accuracy while maintaining a high detection speed.

16.1.3.9 YOLOv5

This algorithm was developed by Jocher as part of Ultralytics LLC. and was developed with the idea to decrease the hardware requirements a person needs to train a custom model and the time it requires; they also use the PyTorch framework instead of darknet one as all previous iterations had; this makes it easier to train a model locally, but the accuracy and speed suffer a bit. This version managed to maintain great accuracy and speed while achieving its main goal of reducing training time and hardware requirements; however, it is still considered that YOLOv4 is more accurate and faster. It is important to note that this iteration was released without a paper and just uploaded to the GitHub repository of Ultralytics, and this results in the version being omitted from published comparisons.

16.1.3.10 PP-YOLO

This alternate YOLO version was released by Long et al. (Long et al., 2020); the authors claim that this version of YOLO does not look to propose a state-of-the-art object detector but to explore different techniques to achieve a higher speed and accuracy by replacing the framework to Pytorch and the backbone to ResNet. The authors also worked to improve the bounding box predictions and feature extraction capabilities.

16.1.3.11 YOLOX

At this point, all versions of YOLO worked with YOLOv3 as a starting point and improved upon it but Zhang et al. (2020) state that most object detection advances focus on anchor-free detection so instead of using anchors the author propose a new version (Ge, Liu, Wang, Li, & Sun, 2021) where they switch the head of the backbone network by dividing it into three different parts: the bounding box calculations, instead of having a single stage to produce the predictions of class probability, the objectness score (how likely it isthere is an object inside the bounding box), and the bounding box location it splits this to have a feature pyramid network for each prediction, improving all scores while getting rid of the anchor boxes and therefore eliminating a lot of parameters and operations that took time and delayed the classification.

16.1.3.12 YOLOv6

Nowadays, new YOLO versions have been published, and although the YOLOv7 version was published first by a month, YOLOv6 is discussed first. This new version of YOLO, proposed by Li et al. (C. Li et al., 2022), takes the same approach as YOLOX by changing the anchor box system by the same division of feature pyramid networks to obtain each probability separated. The authors also change the CSP blocks to one of two options depending on the size of the model; for the small version of YOLOv6, they use RepBlock, and for the larger version of YOLOv6, they use CSPStackRep Block, and they don't scale the RepBlock as older models do because the number of parameters and operations would increase exponentially. Finally, the authors address the bounding box problems of labeling and different losses to improve the bounding box accuracy and reduce the inference time.

16.1.3.13 YOLOv7

Finally, YOLOV7 was proposed by Wang et al. (C. Y. Wang, Bochkovskiy, & Liao, 2022); this new iteration offers a new block added to the backbone architecture called extended efficient layer aggregation network. Also, this iteration uses scaling to fit various hardware devices. Just like YOLOv4 tried different things in order to optimize the network to achieve the best performance. The authors also perform a re-parametrization in order to achieve the best inference time.

These last two models are considered the best ones, with the main difference considering that YOLOv7 takes less time to train.

Another thing that is worth noting is that starting with YOLOv3 almost every iteration has a reduced version of the algorithm; some are just scaled versions of the full architecture, often called tiny. These small versions aim to be implemented in hardware-constrained machines or embedded systems.

Also, it is important to note that there are other one-stage object classifiers outside the YOLO family, but most real-time implementations use a YOLO version or a modified version of one; also it has a growing community that keeps adding to the iterations and makes the algorithm evolve quickly, producing a lot of version on only 6 years.

16.1.4 Recurrent Neural Networks (RNNs)

RNNs are used to analyze a series of data in which the present state depends on a past state and have been successfully used for computer vision (Vinyals,

Toshev, Bengio, & Erhan, 2015). In general, RNN are thought as networks that have "memory" because the previous outputs can be used as inputs for the following steps. So, contrary to the traditional neural networks in which the inputs and outputs are considered independent of each other, the outputs of RNN are dependent on prior elements. Also, another difference that exists between traditional networks and RNNs is that the traditional ones usually map one input to one output while RNNs can have inputs and outputs that vary in dimension. Therefore, depending on the number of inputs and outputs, RNN can be classified into four different types:

- **One-to-one**: It is the basic RNN, which takes only one input and maps it to one output.
- **One-to-many**: This architecture is applied when having multiple outputs that come from a single input.
- **Many-to-one**: Here, multiple inputs are sent to the network while giving a single output as result.
- **Many-to-many**: This architecture maps many inputs to many outputs. However, there is a case in which the number of inputs and outputs is the same or a case in which it is not.

It does not matter which type of RNN is used; the main components of these networks can be summarized into the input, the hidden state, and the tanh activation function. The input consists of vectors that will enter the network. Remember that it can be one single input or several. The hidden state acts as the "memory" of the network. In other words, it contains information about the previous processes. Finally, the input of the current time step is concatenated with the previous hidden state and the resulting vector is passed to a tanh activation function that will squish the values of the vector to values between –1 and 1. The main objective of this step is to regulate the behavior of the network. Finally, the output of the tanh becomes the new hidden state that will enter the next step of the network and the process is repeated.

Figure 16.17 represents a many-to-many RNN where the number of inputs is equal to the number of outputs. Here, the letters W, V, and U represent the

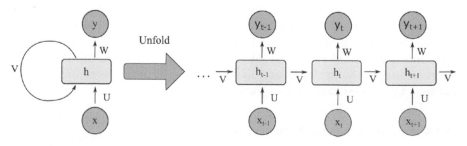

FIGURE 16.17
Recurrent neural network.

weights associated with the output, hidden, and input units respectively. Also, the rectangle cell that contains the hidden state h is where all the processes of tanh activation function are carried out. Despite RNN being a *good* option to handle sequences of data, it may encounter exploding and vanishing gradient problems (Takeuchi, Yatabe, Koizumi, Oikawa, & Harada, 2020). This means that during the process of learning, the gradient may grow exponentially and become too large or it can become too small so that the algorithm stops learning. Therefore, long-short-term memory (LSTM) networks are introduced to tackle this problem.

16.1.4.1 LSTM Networks

LSTM is used in order to allow RNNs to memorize long-term dependencies and forget unnecessary information while overcoming vanishing and exploding gradient problems (Yu, Si, Hu, & Zhang, 2019). This is done through new components that are added to traditional RNN architecture. These components and how they work can be explained as follows:

- **Forget gate**: This is the first step of LSTM and its main objective is to decide which information will be kept and which will be discarded. This gate takes the previous hidden state and the input of the current step. Then, it concatenates them and passes the resulting vector through a sigmoid function that will map the values to new values between 0 and 1. For this gate, it uses the sigmoid function because mapping a value to 0 means that it is going to forget it while mapping it to 1 means it is going to stay the same and, therefore, it is kept.
- **Input gate**: This gate passes the previous hidden state and current input through a sigmoid function. Then, these two vectors are also passed through a tanh function. Tanh function is able to regulate the network, while the sigmoid is in charge of keeping the important data during the current step.
- **Cell state**: The cell state is the "memory" of this network. It carries information through all the network sequences. The main difference between this state and the hidden state of a traditional RNN is that the cell state can carry information from the very first steps to the final time steps, thus reducing the short-term memory problem. In order to update the cell state, one has to do two operations:
  - First, the cell state is point-wise multiplied by the output of the forget gate. This allows the cell state to forget the non-important information and retain the relevant data.
  - Secondly, the output of the input gate is added to the cell state to update it to the new relevant data of the network.

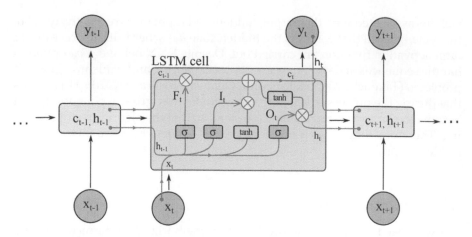

FIGURE 16.18
Long-Short-Term Memory network.

- **Output gate**: Finally, the output gate is obtained. For this one, first, it passes the previous hidden state and the current input through a sigmoid function to select relevant data. Then, the new cell state is passed through a tanh function to regularize it. At last, the output of the tanh function and the sigmoid function are multiplied to obtain the new hidden state that will be passed to the next time step.

So, in summary, the LSTM gates allow controlling of how information flows through the network and, therefore, controlling the gradient problems along with the short-term memory issue of a traditional RNN. All the previously described steps for the LSTM network can be seen in Figure 16.18.

16.2 Clothing Classification

Clothing classification refers to the computer being able to state which clothing garment a person is wearing. This is already implemented in the fashion industry by analyzing the clothing garments a person uses to offer recommendations suited to their tastes; however, the possibilities do not end there. This could also be used in security to verify construction or factory workers who are using the required garments to ensure their safety. Another use is missing person's descriptions that always include the clothing garments the person was wearing when last seen so this could help track that person. Finally, another use found for this implementation is energy saving by using this to determine the clothing insulation values of a clothing ensemble and determining if the user is inside the thermal comfort range; this concept will be discussed later in the case study presented.

16.2.1 Methods Used

Some of the solutions proposed for the clothing classification involve edge detection by using algorithms such as histogram of oriented gradients (HOG) and then some simple machine learning algorithms to classify the features obtained such as SVM (Silva, Cruz, Gutiérrez, and Avendanõ, 2020).

HOG is an image processing algorithm that allows the computer to obtain image features in order to detect objects. This algorithm focuses on the shape of the object in order to detect objects. The way this algorithm finds the shape of the object is by using the magnitude and the orientation of the gradients to obtain the features. Two gradients are calculated: the vertical and the horizontal gradients of each pixel in the image using the equations shown in Eqs. (16.1) and (16.2), where I is the pixel intensity, r is the row, and c is the column.

$$G_x(r,c) = I(r,c+1) - I(r,c-1) \tag{16.1}$$

$$G_y(r,c) = I(r-1,c - I(r+1,c)) \tag{16.2}$$

After obtaining each gradient, the magnitude and angle of each pixel are calculated using Eqs. (16.3) and (16.4) to obtain a gradient matrix and then the whole image is divided into an 8×8 grid; from each block, a 9-point histogram is calculated with 9 bin histograms dividing each bin by 20 degrees (angle values range from 0 to 180). After assigning the values to the bins, normalization is done to reduce the effect of changes in the contrast between the images of the same object, obtaining a final map as shown in Figure 16.19.

$$\text{Magnitude}(\mu) = \sqrt{(G_x)^2 + (G_y)^2} \tag{16.3}$$

$$\text{Angle}(\theta) = \left| \tan^{-1}(G_y/G_x) \right| \tag{16.4}$$

After the HOG is obtained, the features are passed on to SVM which is a simple neural network that is used to propose classifications by using a simple line to divide the features just like Figure 16.20 illustrates. The line can be a straight line or a polynomial function line that has a lot of curves this depends on the function used to classify also known as a kernel.

(a)   (b)

FIGURE 16.19
Example of HOG algorithm to extract features. (a) Original image, (b) HOG of the image.

Linear   Polynomial

FIGURE 16.20
SVM classification.

This type of clothing classification can be seen in Feifei, Pinghua, and Xuemei (2019) and is one of the most used classification methods for its simplicity and accuracy in any type of classification so that is why it is explored in a detailed manner in this chapter.

Other methods that have tried clothing classification have used segmentation in different forms, for example, Borras et al. (Borras, Tous, Lladós, & Vanrell, 2003) where they use clothing garment segmentation by using texture and edges to separate clothing garments. Another method that works with a similar process but with different techniques is the work presented by Zhang et al. (W. Zhang, Matsumoto, Liu, Chu, & Begole, 2008). Another project that worked with different kinds of segmentation was the research presented by Hu et al. (Hu, Yan, & Lin, 2008) and the one presented by Kalantidis et al. (Kalantidis, Kennedy, & Li, 2013).

Later on, Yang et al. presented a method to recognize clothing garments in Yang and Yu (2011) by detecting the edges, filtering information to propose regions, and finally removing the background and then using an SVM to produce the classifications. This achieved a good speed (16–20 FPS) with a recall of 80%, which is acceptable.

However, this process can take time and run into some problems when there is a lot of noise contained in the image and sometimes it may be a slow process in order to calculate the histogram of an image containing a lot of objects to be implemented as an object classifier in large noisy images or video analysis. Also, segmentation takes a lot of time and may not be ideal when the features used to separate the garments are very similar.

With this in mind, machine learning implementations have been tried, just like the work presented by Zhang et al. (Y. Zhang, Zhang, Yuan, & Wang, 2020) and the one presented by Wang et al. (W. Wang, Xu, Shen, & Zhu, 2018), where the authors propose the texture and shape recognition with a custom built network or attribute classification for clothing garments; however, they also state that the results are very slow to be implemented in video analysis. Other machine learning approaches to classify clothing garments can be seen in the work presented by Yamaguchi et al. (Yamaguchi, Okatani, Sudo, Murasaki, & Taniguchi, 2015) or the work presented by Hidayati et al. (Hidayati, You, Cheng, & Hua, 2017) where they extract features with machine learning but classify using SVM.

All of these later methods use machine learning but are fashion-oriented and leave a lot of potential uses for a clothing classifier outside their scope, and no object detector was found in the literature review, considering the flexibility these algorithms offer an implementation was done.

16.2.2 Clothing Detection Implementation

To start the object detector training, the first step was to decide which algorithm to use; for this implementation, the Tiny YOLOv4 algorithm was used due to its effectiveness in embedded systems while still achieving high accuracy scores.

To begin with the training, a search for a dataset was performed even though there were some datasets available on platforms like Kaggle or even well-established datasets like Fashion-MNIST had the problem of just providing single clothing garments with no other information such as a person wearing it, background information, and more things that could affect the performance of the algorithm if trained with these datasets and then implemented with a surveillance camera for example. So, a custom dataset was built, downloading images from the internet of different people wearing different clothing garments in different environments.

The dataset started with 2,000 labeled images and YOLO has its own labeling format where the bounding box center coordinates, width, and height are required as well as the class the object in it belongs to since the number of classes affects the number of iterations the algorithm needs to be trained for just 8 classes are proposed: highly insulating jacket, highly insulating shoes, jacket, shirt, trousers, skirt, hat, and shoes. However, 2,000 images can be considered a small dataset for deep learning training, so data augmentation techniques were used to increase the size of the training dataset.

Data augmentation is a technique used not only to increase the dataset but also to avoid overfitting of the algorithm—a serious problem where the algorithm gets almost every recognition on the training images perfect but with any other image it struggles to get right. Also, this helps address some possible problems with the camera or lightning conditions, like adding blur to some images to address a lower resolution, reducing image contrast to address poor lightning conditions, flipping the image to address different orientations of the camera, and avoiding problems if every image has a person looking right and struggles with people looking left for example. The image transformations performed were blurring, horizontal flip, vertical flip, both flips, contrast reduction, gamma correction, and hue changes. After applying these transformations to every image in the original dataset, the resulting dataset contains 15,000 images which is considered by many authors a reasonable size to train a deep neural network.

After obtaining the dataset to train the algorithm, the hyper-parameters need to be decided, especially the number of epochs the algorithm is going to be trained for, authors recommend 2,000 epochs per class, and since the dataset has 8 classes, 16,000 epochs were the number chosen for the training, as for starting the learning rate the recommended value used is 0.001, and for the optimizer, the most used is the Adam optimizer.

The method to train a YOLOv4 model using the darknet framework is by using a machine that has Linux operating system, and in the case of not having one, Google Colab can be used with Linux commands and can make the environments train the model with access to a remote GPU following the instructions of the author in its GitHub repository found at https://github.com/AlexeyAB/darknet; however, the access to the Google Colab GPU is restricted to four hours a day, so loading the previously generated weights to continue training is needed. The training time for the Tiny YOLOv4 model is around 9–12 hours for the 16,000 epochs.

The resulting network had a training accuracy of 93% and a test accuracy of 87% which shows no indication of overfitting and is an accuracy that can be considered good compared with the literature review.

For the real-time video implementation, OpenCV library was used with its tensor flow compatibility, as for the camera a Logitech HD Pro C920 was used which can provide 1080p images with 30 FPS.

An example of the obtained results can be seen in Figure 16.21.

16.3 Activity Recognition

Human pose estimation refers to the process of estimating the configuration of the body from a single observation, typically an image. Hence, detecting the posture of a person has gained importance because of the abundance of

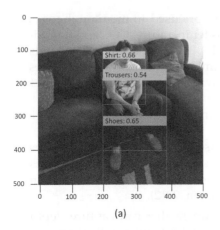 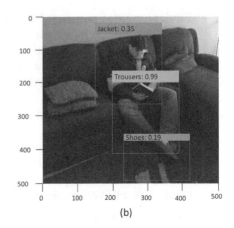

FIGURE 16.21
Results of the object detector trained for clothing garments. (a) Clothing set 1 classification results, (b) Clothing set 2 classification results.

applications that can benefit from it in different fields such as sports, security, entertainment, and health, among others. Furthermore, human pose estimation allows for higher-level processes to be identified as is the activity recognition.

Human activity recognition (HAR) is the process of interpreting human motion which can be activities, gestures, or behaviors from a series of observations. Hence, the data needed should be in the form of a time series of parameter values. Typically, the way this data is gathered can be divided into two approaches: vision-based and sensor-based.

The sensor-based solutions are generally divided into three categories (X. Li, Zhang, Marsic, Sarcevic, & Burd, 2016):

- **Wearable sensors**: The advances in embedded systems have let the development of sensors that can be incorporated into small devices attached to the user. The most common sensors used are an accelerometer, magnetometer, and gyroscope (Hassan, Uddin, Mohamed, & Almogren, 2018). Moreover, portable devices include smartphones, smartwatches, smart bands, custom devices, etc.

- **Object sensors**: These types of sensors are attached to a particular object with the objective to identify the activity related to the interaction with the said object (W. Wang, Liu, Shahzad, Ling, & Lu, 2017). Unlike wearable sensors that are attached to the user, the object sensor depends on the interaction of the user with the object to infer the activity.

- **Environmental sensors**: Refer to sensors used to monitor changes in environmental parameters such as temperature, CO_2, and humidity (Hao, Bouzouane, & Gaboury, 2018). The adoption of these kinds of sensors can complement the information of the previous sensors rather than inferring directly the activity.

On the other hand, vision-based sensors can be divided into two categories:

- **RGB sensor (Camera)**: A camera is used to obtain image sequences that are processed through the HAR system. This type of system is built with two main processes: feature extraction and classification (Preis, Kessel, Werner, & Linnhoff-Popien, 2012).
- **Depth sensor**: It uses an infrared camera that can capture depth information (Jalal, Uddin, Kim, & Kim, 2012). Unlike an RGB camera, infrared uses beams to calculate and measure the distance of any object in a 3D environment.

Both approaches of HAR have been studied significantly and reviewed in some previous works (Bux, Angelov, & Habib, 2017; Cheng, Wan, Saudagar, Namuduri, & Buckles, 2015; Cornacchia, Ozcan, Zheng, & Velipasalar, 2016; Kang & Wildes, 2016; Lun & Zhao, 2015; Vital et al., 2017), presenting several methodologies according to the source of the input data, the machine learning strategy, the feature extraction process, and the type of sensor used. The general agreement is that object sensors or contact-based systems are being abandoned due to difficult implementation. On the other hand, wearables and vision-based approaches are more popular. The wearable sensors are easy to implement, efficient, reduced in size, and most likely to be accepted by the user. Besides, the vision-based sensors do not require the user to wear several devices being more comfortable and less intrusive, although their acceptance is being on the work. Hence, this section shows the use of a vision-based system using LSTM networks for HAR.

16.3.1 Deep Learning Model for HAR

As previously stated, the HAR process is essentially a time series problem. Hence, the use of RNN which is a deep learning model typically used to solve problems with sequential data. Besides, this type of neural network can experience a loss in information referred to as the vanishing gradient problem caused by the multiplicative nature of the back-propagation algorithm. Therefore, the use of LSTM is widely accepted for the improvement of learning and eliminates the vanishing gradient problem.

The general process for HAR follows the steps shown in Figure 16.22. The next sections explain each step in detail.

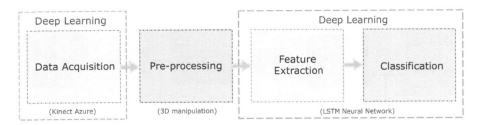

FIGURE 16.22
HAR general process.

16.3.1.1 Data Acquisition

Of the vision-based systems available for HAR, a depth sensor, specifically, an Azure Kinect camera was selected to obtain the 3D data of a skeleton model. The skeleton model was selected out of the four most common types of human body models:

- **Volume-based model.** The system estimates the 3D structures of an object based on various geometric shapes.
- **Contour-based model.** It gives a 2D representation of the body shape detecting the contour of the body, torso, head, and limbs.
- **Point cloud-based model.** The human model is represented in 3D by various points that form the shape.
- **Skeleton-based model.** This does not rely on object segmentation; instead, it detects the body parts and encodes the features to estimate the 3D skeletal joints of the model.

As the most significant challenges related to vision-based sensors are,

- Variability of human visual appearance
- Variability in lighting conditions
- Variability in the human physique (physiology)
- Partial occlusions

the infrared technology that uses depth sensors and the 3D information obtained is processed to be able to face these challenges and make a good classification. This data processing is explained in the next subsection.

Figure 16.23 shows the data acquisition process using the Azure Kinect SDK that implements a CNN with the IR data to estimate the skeletal joint of a person to later add the depth information to fit the joints into a 3D model. The result of the skeleton joint consisted of 32 joint data with 3-axis information is shown in Figure 16.24.

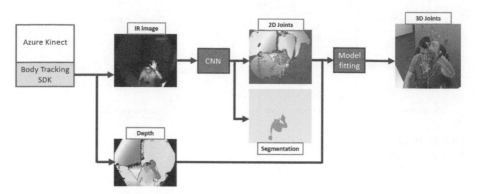

FIGURE 16.23
Skeletal joint architecture (Liu, 2020).

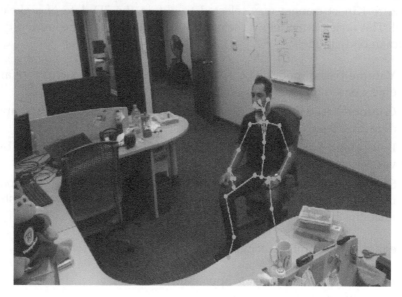

FIGURE 16.24
Skeleton joint estimation.

16.3.1.2 Pre-processing

The pre-processing of the data consists of an algorithm of 4 steps that transforms the 3D data to make it invariant with different orientations of the depth sensor, the variability of human visual appearance, and even the relative position of the body.

The first step is to center the pelvis joint into the origin (0,0,0) as the joint position and orientation are estimated relative to the depth sensor frame

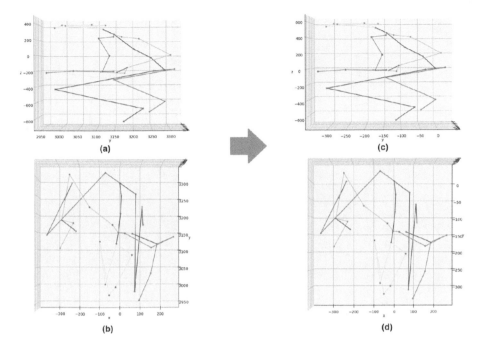

FIGURE 16.25
(a) Lateral view of original data, (b) Top view of original data, (c) Lateral view of centered data,
(d) Top view of centered data.

of reference. Figure 16.25 shows the resulting transformation of translating
each joint centered on the pelvis.

The second step is to correct the roll-and-pitch angle of the skeleton due
to the positioning of the depth sensor in the room. This can be noted in
Figure 16.26 on the left side as the line corresponding to the spine should be
completely vertical as shown in the model in Figure 16.24.

This angle correction is made using the incorporated inertial motion unit's
accelerometer of the device. Equation (16.5) represents the pitch angle using
the normalized accelerometer readings while Equation (16.6) represents the
roll angle with the same accelerometer reading.

$$\theta = \tan^{-1}\left(\frac{A_x}{\sqrt{A_y^2 + A_z^2}}\right) \tag{16.5}$$

$$\varphi = \tan^{-1}\left(\frac{A_y}{\sqrt{A_x^2 + A_z^2}}\right) \tag{16.6}$$

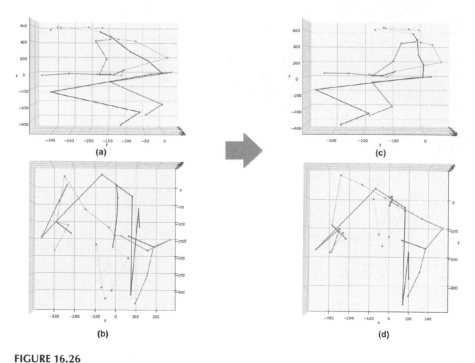

FIGURE 16.26
(a) Lateral view of previous data, (b) Top view of previous data, (c) Lateral view of roll-and-pitch rotation, (d) Top view of roll-and-pitch rotation.

Figure 16.26 shows the result of the pitch-and-roll correction, where in the right side can be noted that the line corresponding to the spine now is vertically aligned as was expected. The rotation was applied to the whole skeleton with Euler angle transformation.

The third step consists of a yaw rotation of the skeleton (the Z-axis) with the intention to make the model always face in the same direction (negative Y-axis). This process takes the relative position of the right and left joints corresponding to the clavicle to form a vector that should be parallel to the X-axis while the corresponding nose–head joint vector should be pointing into the negative Y-axis. Figure 16.27 shows the resulting skeleton rotated by the Z-axis.

Finally, the fourth step's purpose is to eliminate the variability of the human physique like height by normalizing the model. Figure 16.28 depicts the result of normalizing all joint data from –1 to 1 using the Eq. (16.7)

$$x' = 2\frac{x - \min(x)}{\max(x) - \min(x)} \qquad (16.7)$$

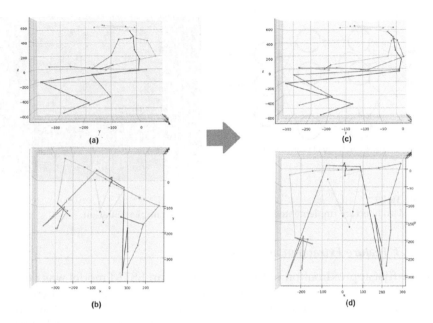

FIGURE 16.27
(a) Lateral view of previous data, (b) Top view of previous data, (c) Lateral view of yaw rotated data, (d) Top view of yaw rotated data.

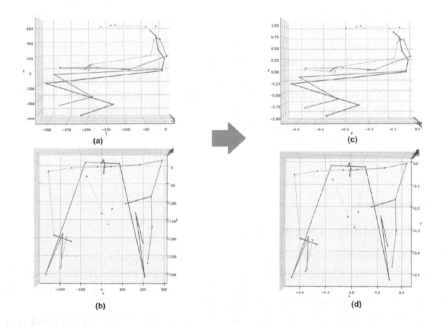

FIGURE 16.28
(a) Lateral view of previous data, (b) Top view of previous data, (c) Lateral view of normalized data, (d) Top view of normalized data.

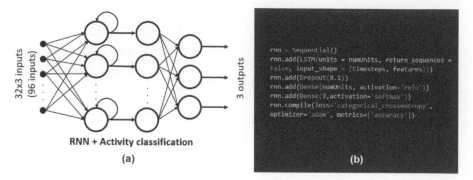

FIGURE 16.29
(a) LSTM architecture implemented, (b) Code in python for implementation.

16.3.1.3 Feature Extraction and Classification

For the first approximation of the classification, an LSTM model was defined with a single hidden layer followed by a dropout layer of 10% intended to reduce the overfitting of the model to the training data. Then, a dense fully connected layer was used to interpret the features extracted by LSTM and a final output layer with a softmax function to classify three different activities for this test—sitting, walking, raising hands—while the inputs are 96 data consisting of 3 axis values for each of the 32 joints of the skeleton model. Figure 16.29 shows the architecture previously described and its implementation in code.

The model was trained with 201,600 data corresponding to 40 repetitions of each activity to classify: sitting, walking, and raising arms. Then, the model was tested with 16 repetitions of each activity obtaining an overall accuracy of 94.79% which represents just 1 erroneous classification as shown in Figure 16.30. More testing data are needed to validate a higher accuracy but the pre-processing of the data permitted to train the network with just a few amount of data compared to other methods.

16.4 Case Study: Thermal Comfort

Thermal comfort describes the human satisfactory perception of the thermal environment locally and is therefore considered a subjective assessment (ISO, 1984). There have been two main approaches to thermal comfort: the steady-state model and the adaptive model.

The most widely applied index for the evaluation of indoor thermal conditions in moderate environments for steady-state models is the predicted mean vote (PMV). The main physical parameter on which the PMV index

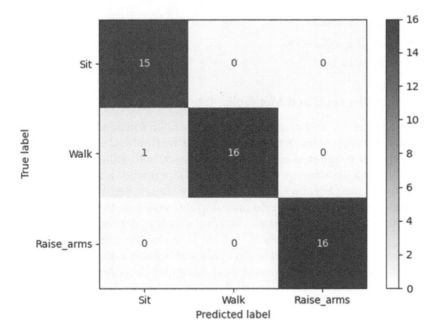

FIGURE 16.30
Confusion matrix of classified activities.

evaluation relies is the thermal load of the human body, and it is based on the deviation between heat loss and metabolic rate. Moreover, the predicted percentage of dissatisfied (PPD) is an index that assesses a quantitative prediction of the percentage of subjects who may declare themselves thermally dissatisfied with the indoor environment conditions. The PPD is therefore a function of PMV. On the other hand, the adaptive model is said to be a human-centered approach and is mainly based on the theory of the human body's adaptability to outdoor and indoor climates (Méndez, Peffer, Ponce, Meier, & Molina, 2022).

Thermal comfort is mainly influenced by six variable factors independently of the approach. The most common indicator is air temperature for its ease of sensing but this factor alone is not sufficient to accurately indicate a thermal comfort value. Therefore, the six factors, both environmental and personal needed for a valid indicator, are (Taleghani, Tenpierik, Kurvers, & Van Den Dobbelsteen, 2013):

- Environmental factors
  - Air temperature
  - Air velocity
  - Radiant temperature
  - Humidity

- Personal factors
  - Clothing insulation
  - Metabolic heat

16.4.1 Clothing Level and Metabolic Rate Estimations

Clothing insulation and metabolic rate are variables needed to obtain a thermal comfort range calculation in both used methods: the human-centered and the building-centered approaches. However, both values used in most of the model's calculations are fixed values obtained from a table and not estimated in real time because of the lack of easy-to-use sensors. Therefore, the use of deep learning algorithms on computer vision is to obtain the values in real time to get more accurate thermal comfort decisions. As standards such as ASHRAE 55 or ISO 9920 include tables of the most common clothing garment values expressed in clo units and human activities expressed in met units, the previously proposed methodologies described using computer vision are paired with the values of those tables to use them as a framework to calculate the thermal comfort in real time.

The clothing classification is trained to detect 8 classes of garments, as shown in Table 16.2. Each class detected is then paired to a clo value to finally sum the values and obtain an estimated total clothing value. This process is depicted in Figure 16.31 where three different garments are detected estimating a total of 0.32 clo value.

As for metabolic rate estimation, the proposed framework classifies the activity detected on a sequence of images. Then, the activity is paired with a metabolic value in Table 16.3. The resulting value is depicted in Figure 16.32 where the accuracy of the activity is shown with the metabolic value in met units.

TABLE 16.2

Clothing Insulation Values Considered for the Classes

Label	Garment	(clo)[a]
0	Highly insulating jacket, multicomponents	0.40
1	Highly insulating shoes, boots	0.10
2	Jacket, no buttons	0.26
3	T-shirt	0.09
4	Trousers (straight, fitted)	0.19
5	Shoes	0.04
6	Warm winter cap	0.03
7	A-Line, knee-length	0.15

[a] 1 clo = 0.155 m2 C/W.

FIGURE 16.31
Clothing insulation recognition.

TABLE 16.3

Metabolic Rate Values

Label	Activity	MET[a]
0	Sitting	1.3
1	Walking	2.3
2	Raising hands	2.1

[a] 1 met = 1 kcal/kg/hr.

16.5 Discussion

This chapter proposes a framework that focuses on the two most difficult factors to be obtained: clothing insulation and metabolic heat. It is done by detecting clothing level values and estimating metabolic rate values using deep learning algorithms applied to computer vision. The classification of different garments is achieved using an embedded system for processing in addition to the classification of three human activities. More activities and garments can be added to the framework as it is only necessary to train the algorithm with new classes. According to Nicol (1993), there are three main reasons for studying thermal comfort: to suggest and set standards, to provide instruments to satisfy conditions for people, and to manage energy consumption. Therefore, this framework supposes a great advantage for the calculation of real-time thermal comfort models as an instrument for

FIGURE 16.32
Metabolic rate estimation.

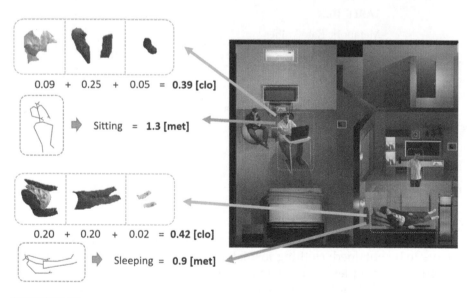

FIGURE 16.33
Computer vision for thermal comfort detection on smart houses.

new methodologies to save energy (Medina et al., 2022; Méndez et al., 2022). Figure 16.33 shows the possible application of this framework inside a smart home where a thermal comfort analysis can be obtained in real time to modify the set point of an HVAC system with the purpose of saving energy.

References

Bharati, P., & Pramanik, A. (2020). Deep learning techniques—R-CNN to mask R-CNN: A survey. *Computational Intelligence in Pattern Recognition, 999*, 657–668).

Bhatt, D., Patel, C., Talsania, H., Patel, J., Vaghela, R., Pandya, S., ... Ghayvat, H. (2021). CNN variants for computer vision: History, architecture, application, challenges and future scope. *Electronics, 10* (20), 2470.

Bochkovskiy, A., Wang, C.-Y., & Liao, H.-Y. M. (2020). YOLOv4: Optimal speed and accuracy of object detection. *arXiv preprint arXiv:2004.10934*.

Borras, A., Tous, F., Lladós, J., & Vanrell, M. (2003). High-level clothes description based on colour-texture and structural features. In *Iberian Conference on Pattern Recognition and Image Analysis,* Mallorca (pp. 108–116).

Bux, A., Angelov, P., & Habib, Z. (2017). Vision based human activity recognition: A review. *Advances in Computational Intelligence Systems*, 341–371.

Cheng, G., Wan, Y., Saudagar, A. N., Namuduri, K., & Buckles, B. P. (2015). Advances in human action recognition: A survey. *arXiv preprint arXiv:1501.05964*.

Cornacchia, M., Ozcan, K., Zheng, Y., & Velipasalar, S. (2016). A survey on activity detection and classification using wearable sensors. *IEEE Sensors Journal, 17* (2), 386–403.

Dumoulin, V., & Visin, F. (2016). A guide to convolution arithmetic for deep learning. *arXiv preprint arXiv:1603.07285*.

Feifei, S., Pinghua, X., & Xuemei, D. (2019). Multi-core SVM optimized visual word package model for garment style classification. *Cluster Computing, 22* (2), 4141–4147.

Feng, Q., Peng, D., & Gu, Y. (2019). Research of regularization techniques for SAR target recognition using deep CNN models. In *Tenth International Conference on Graphics and Image Processing (ICGIP 2018),* Chengdu. (Vol. 11069, pp. 1066–1073).

Ge, Z., Liu, S., Wang, F., Li, Z., & Sun, J. (2021). YOLOx: Exceeding YOLO series in 2021. *arXiv preprint arXiv:2107.08430*.

Girshick, R. (2015). Fast R-CNN. In *Proceedings of the IEEE International Conference on Computer Vision,* Santiago (pp. 1440–1448).

Hao, J., Bouzouane, A., & Gaboury, S. (2018). Recognizing multi-resident activities in non-intrusive sensor-based smart homes by formal concept analysis. *Neurocomputing, 318*, 75–89.

Hassan, M. M., Uddin, M. Z., Mohamed, A., & Almogren, A. (2018). A robust human activity recognition system using smartphone sensors and deep learning. *Future Generation Computer Systems, 81*, 307–313.

He, K., Gkioxari, G., Dollár, P., & Girshick, R. (2017). Mask R-CNN. In *Proceedings of the IEEE International Conference on Computer Vision,* Venice (pp. 2961–2969).

Heidari, A., Navimipour, N. J., & Unal, M. (2022). Applications of ml/dl in the management of smart cities and societies based on new trends in information technologies: A systematic literature review. *Sustainable Cities and Society 85*, 104089.

Hidayati, S. C., You, C.-W., Cheng, W.-H., & Hua, K.-L. (2017). Learning and recognition of clothing genres from full-body images. *IEEE Transactions on Cybernetics, 48* (5), 1647–1659.

Hu, Z., Yan, H., & Lin, X. (2008). Clothing segmentation using foreground and background estimation based on the constrained Delaunay triangulation. *Pattern Recognition, 41* (5), 1581–1592.

Indolia, S., Goswami, A. K., Mishra, S. P., & Asopa, P. (2018). Conceptual understanding of convolutional neural network-a deep learning approach. *Procedia Computer Science, 132,* 679–688.

ISO (1984). International standard 7730. *Ergonomics of the Thermal Environment— Analytical Determination and Interpretation of Thermal Comfort Using Calculation of the PMV and PPD Indices and Local Thermal Comfort Criteria.*

Jalal, A., Uddin, M. Z., Kim, J. T., & Kim, T.-S. (2012). Recognition of human home activities via depth silhouettes and transformation for smart homes. *Indoor and Built Environment, 21* (1), 184–190.

Jin, Y., Wen, Y., & Liang, J. (2020). Embedded real-time pedestrian detection system using YOLO optimized by LNN. In *2020 International Conference on Electrical, Communication, and Computer Engineering,* Istanbul (ICECCE) (pp. 1–5).

Kalantidis, Y., Kennedy, L., & Li, L.-J. (2013). Getting the look: Clothing recognition and segmentation for automatic product suggestions in everyday photos. In *Proceedings of the 3rd ACM Conference on International Conference on Multimedia Retrieval,* Dallas, TX (pp. 105–112).

Kang, S. M., & Wildes, R. P. (2016). Review of action recognition and detection methods. *arXiv preprint arXiv:1610.06906.*

Li, C., Li, L., Jiang, H., Weng, K., Geng, Y., Li, L., ... Others (2022). YOLOv6: A single-stage object detection framework for industrial applications. *arXiv preprint arXiv:2209.02976.*

Li, X., Zhang, Y., Marsic, I., Sarcevic, A., & Burd, R. S. (2016). Deep learning for RFID-based activity recognition. In *Proceedings of the 14th ACM Conference on Embedded Network Sensor Systems,* CaliforniaCD-ROM (pp. 164–175).

Liu, Z. (2020). *3D skeletal tracking on azure Kinect.* Retrieved from https://www.microsoft.com/en-us/research/uploads/prod/2020/01/AKBTSDK.pdf.

Long, X., Deng, K., Wang, G., Zhang, Y., Dang, Q., Gao, Y., ... Others (2020). PP-YOLO: An effective and efficient implementation of object detector. *arXiv preprint arXiv:2007.12099.*

Lun, R., & Zhao, W. (2015). A survey of applications and human motion recognition with Microsoft Kinect. *International Journal of Pattern Recognition and Artificial Intelligence, 29* (05), 1555008.

Medina, A., Méndez, J. I., Ponce, P., Peffer, T., Meier, A., & Molina, A. (2022). Using deep learning in real-time for clothing classification with connected thermostats. *Energies, 15* (5), 1811.

Medina, A., Méndez, J. I., Ponce, P., Peffer, T., & Molina, A. (2022). Embedded real-time clothing classifier using one-stage methods for saving energy in thermostats. *Energies, 15* (17), 6117.

Méndez, J. I., Peffer, T., Ponce, P., Meier, A., & Molina, A. (2022). Empowering saving energy at home through serious games on thermostat interfaces. *Energy and Buildings, 263,* 112026.

Nicol, F. (1993). *Thermal comfort: A handbook for field studies toward an adaptive model.* University ofEast London.

Preis, J., Kessel, M., Werner, M., & Linnhoff-Popien, C. (2012). Gait recognition with Kinect. In *1st International Workshop on Kinect in Pervasive Computing* (pp. 1–4).

Redmon, J., Divvala, S., Girshick, R., & Farhadi, A. (2016). You only look once: Unified, real-time object detection. In *Proceedings of the IEEE Conference on Computer Vision and Pattern Recognition* (pp. 779–788).

Redmon, J., & Farhadi, A. (2017). YOLO9000: Better, faster, stronger. In *Proceedings of the IEEE Conference on Computer Vision and Pattern Recognition* (pp. 7263–7271).

Redmon, J., & Farhadi, A. (2018). YOLOv3: An incremental improvement. *arXiv preprint arXiv:1804.02767.*

Ren, S., He, K., Girshick, R., & Sun, J. (2015). Faster R-CNN: Towards real-time object detection with region proposal networks. *Advances in Neural Information Processing Systems, 28,* 91–99.

Shafiee, M. J., Chywl, B., Li, F., & Wong, A. (2017). Fast YOLO: A fast you only look once system for real-time embedded object detection in video. *arXiv preprint arXiv:1709.05943.*

Silva, D. C. B., Cruz, P. P., Gutiérrez, A. M., & Avendanõ, L. A. S. (2020). *Applications of human-computer interaction and robotics based on artificial intelligence.* Editorial Digital del Tecnológico de Monterrey. Monterrey.

Takeuchi, D., Yatabe, K., Koizumi, Y., Oikawa, Y., & Harada, N. (2020). Real-time speech enhancement using equilibriated RNN. In *ICASSP 2020-2020 IEEE International Conference on Acoustics, Speech and Signal Processing (ICASSP)* (pp. 851–855).

Taleghani, M., Tenpierik, M., Kurvers, S., & Van Den Dobbelsteen, A. (2013). A review into thermal comfort in buildings. *Renewable and Sustainable Energy Reviews, 26,* 201–215.

Uijlings, J. R., Van De Sande, K. E., Gevers, T., & Smeulders, A. W. (2013). Selective search for object recognition. *International Journal of Computer Vision, 104* (2), 154–171.

Vinyals, O., Toshev, A., Bengio, S., & Erhan, D. (2015). Show and tell: A neural image caption generator. In *Proceedings of the IEEE Conference on Computer Vision and Pattern Recognition* (pp. 3156–3164).

Vital, J. P., Faria, D. R., Dias, G., Couceiro, M. S., Coutinho, F., & Ferreira, N. M. (2017). Combining discriminative spatiotemporal features for daily life activity recognition using wearable motion sensing suit. *Pattern Analysis and Applications, 20* (4), 1179–1194.

Wang, C.-Y., Bochkovskiy, A., & Liao, H.-Y. M. (2022). YOLOv7: Trainable bag-of-freebies sets new state-of-the-art for real-time object detectors. *arXiv preprint arXiv:2207.02696.*

Wang, W., Liu, A. X., Shahzad, M., Ling, K., & Lu, S. (2017). Device-free human activity recognition using commercial WIFI devices. *IEEE Journal on Selected Areas in Communications, 35* (5), 1118–1131.

Wang, W., Xu, Y., Shen, J., & Zhu, S.-C. (2018). Attentive fashion grammar network for fashion landmark detection and clothing category classification. In *Proceedings of the IEEE Conference on Computer Vision and Pattern Recognition* (pp. 4271–4280).

Yamaguchi, K., Okatani, T., Sudo, K., Murasaki, K., & Taniguchi, Y. (2015). Mix and match: Joint model for clothing and attribute recognition. In Xianghua Xie, Mark W. Jones, and Gary K. L. Tam, editors, *Proceedings of the British Machine Vision Conference (BMVC)* (pp. 51.1–51.12).

Yang, M., & Yu, K. (2011). Real-time clothing recognition in surveillance videos. In *2011 18th IEEE International Conference on Image Processing* (pp. 2937–2940).

Yu, Y., Si, X., Hu, C., & Zhang, J. (2019). A review of recurrent neural networks: LSTM cells and network architectures. *Neural Computation, 31* (7), 1235–1270.

Zafar, A., Aamir, M., Mohd Nawi, N., Arshad, A., Riaz, S., Alruban, A., ... Almotairi, S. (2022). A comparison of pooling methods for convolutional neural networks. *Applied Sciences*, 12 (17), 8643.

Zhang, W., Matsumoto, T., Liu, J., Chu, M., & Begole, B. (2008). An intelligent fitting room using multi-camera perception. In *Proceedings of the 13th International Conference on Intelligent User Interfaces* (pp. 60–69).

Zhang, Y., Zhang, P., Yuan, C., & Wang, Z. (2020). Texture and shape biased two-stream networks for clothing classification and attribute recognition. In *Proceedings of the IEEE/CVF Conference on Computer Vision and Pattern Recognition* (pp. 13538–13547).

17

Building Predictive Models to Efficiently Generate New Nanomaterials with Antimicrobial Activity

Gildardo Sanchez-Ante, Edgar R. López-Mena, and Diego E. Navarro-López
Tecnológico de Monterrey

CONTENTS

DOI: 10.1201/b23345-20

17.1 Introduction

Without a doubt, the discovery in 1928 of penicillin as an agent capable of inhibiting the growth of certain pathogenic organisms forever changed the quality of life of humans in a very positive way. Until then, it was known that about 90% of children who were infected with bacterial meningitis died. And among those who survived the disease, many continued to have severe and long-lasting disabilities, ranging from deafness to mental retardation. Similarly, something as simple as throat infections often turned deadly, and some ear infections sometimes spread to the brain, causing serious problems. Not to mention infections such as tuberculosis, pneumonia, and whooping cough are caused by highly aggressive bacteria with accelerated reproduction capacity and therefore cause very serious conditions in people and very often death. The success of penicillin made pharmaceutical companies keen to manufacture it, and research over the years has made it possible for there to be dozens of antibiotics today that can be used by doctors to treat a variety of diseases.

However, bacteria, like other living things, undergo changes over time in response to environmental changes. At present, it is considered that because of the widespread use of antibiotics and often without strict medical control, bacteria have been constantly exposed to these drugs, and although many of them die after being exposed to antibiotics, some develop resistance to its pharmacological effects (Ventola, 2015; Murray et al., 2022). A case of this circumstance occurs with the bacteria *Staphylococcus aureus* (*S. aureus*), which frequently causes skin infections but can also cause other conditions such as bloodstream infections, endocarditis, and osteomyelitis, to name a few. Until about fifty years ago, this bacterium was very sensitive to penicillin, so treatment with it was very effective. However, over time, some strains of this bacterium developed an enzyme capable of breaking down penicillin, rendering the drug ineffective. When a person carrying the bacteria takes antibiotics, they kill the strains that are not resistant, with the resistant strains surviving. If the bacteria of that strain manage to proliferate and cause an infection, it will be much more difficult to treat. This is a serious problem. Today, 70%–80% of *S. aureus* isolates are resistant to penicillin (Smith et al., 1999). As a result, some researchers developed a new form of penicillin that the enzyme could not break down, but within a few years, the bacteria adapted and became resistant even to this modified penicillin. This phenomenon has also occurred with other bacteria.

The resistance that some pathogenic organisms are developing against antibiotics is a matter of great concern to institutions such as the World Health Organization (Murray et al., 2022). This resistance has increased alarmingly in the last decade, which is seriously threatening the effectiveness of various treatments for both the prevention and cure of many diseases.

Given this situation, various strategies are being developed to alleviate this problem, including the design of new antibiotics that ideally have fewer side effects and that also do not allow the generation of resistance by pathogens. A promising alternative is based on the use of various nanomaterials such as silver nanoparticles (Díez-Pascual, 2018), graphene and its derivatives (Seifi, 2021), titanium oxide (TiO_2), ferrous–ferric oxide (Fe_3O_4), some biopolymers, and zinc oxide (ZnO) doped with rare earth elements (Shinde, 2015; Kaushik et al., 2019). Many of these materials have gained popularity due to their great chemical stability, catalytic activity, and high conductivity, which in the end means that they have greater antibacterial activity compared to other compounds.

17.1.1 The Process to Generate a New Material with Certain Desired Properties

Generating a new material that has certain properties is a complex, costly, and time-consuming task. An example of this is Teflon, which took nearly 20 years from conception to commercialization (Correa-Baena et al., 2018). Along with that, there are other notorious cases of materials that have required very important investments over long periods of time. This circumstance has several implications. One of them is that in many cases only very large and consolidated companies can undertake such long-term projects. In universities, perhaps the approach is to be able to conclude only some of the stages that this process takes.

This chapter describes a specific case in which the aim is to develop nanomaterials that exhibit antimicrobial properties. It is important to point out that this chapter summarizes the experience of the nanobiotechnology group at the Tecnológico de Monterrey, in which experts from areas such as nanotechnology, biotechnology, and computer science participate with their knowledge. Detailed descriptions of the process, experiments, and results can be found in Refs. (Navarro-López et al., 2021; Sánchez-López et al., 2022).

The canonical process to generate a new nanomaterial is described in Figure 17.1, in which three main areas can be observed:

1. Preparation of the nanomaterial through a synthesis route.
2. Physicochemical and biological characterization of the material.
3. Data analysis.

17.1.1.1 Synthesis Route

During this process, there are several points where decisions must be made, for example, what would be the chemical composition? Is the material completely new? Is it based on previously reported material by modification? The periodic table of chemical elements includes around 100 options to work with them. However, many of them can be toxic, expensive, or difficult to manipulate in normal conditions. Then, exhaustive research is necessary about

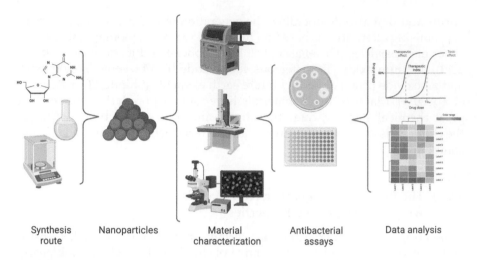

FIGURE 17.1
Schematic diagram to illustrate the workflow from synthesis to antibacterial data analysis of nanoparticles.

which materials can be combined due to chemical and physical interaction to achieve the goal. In the specific case of antibacterial applications, it turns out that elements such as zinc, titanium, iron, copper, and magnesium and their combinations have shown excellent properties.

Another important decision has to do with the synthesis route. In many ways, this could be considered the most complicated to be chosen. The synthesis route affects variables such as the crystallite's size, microstructure, porosity, energy consumption, and chemical waste, among others. Even more, nowadays, one of the goals of nanotechnology is to avoid chemical waste and carbon dioxide emissions during the synthesis process. For example, a simple route to obtain ZnO nanoparticles is using a high acidic solution, where different chemical compounds are used. Instead, an acidic solution can be obtained using plant or fruit extracts, resulting in a more eco-friendly process.

17.1.1.2 Nanomaterial Characterization

There exist several material techniques to get information about crystal structure, morphology, particle size, optical parameters, chemical composition, porosity, etc. For this reason, it is crucial to perform the necessary analyses to get information related to antibacterial activity. The basic characterization of nanomaterials to be tested as potential antibacterial agents is done by using X-ray diffraction (XRD), Fourier-transform infrared spectroscopy (FT-IR), scanning electron microscopy (SEM), X-ray photoelectron spectroscopy (XPS) and UV-visible spectrophotometry (UV-vis), and zeta potential

(Z-potential). In what follows, a brief description of the information provided by each characterization is given:

- **XRD**: This technique allows identification of the crystal structure of the nanoparticles and calculation of the lattice parameters, bond length between atoms, stress, strain, dislocation density, theoretical surface area, and preferential growth direction, among other structural parameters. A typical XRD pattern for ZnO is shown in the next graph, where each peak corresponds to any plane (3D-atom position) for the crystal structure.

- **FT-IR**: In the synthesis process, it is common that organic compounds are still on the surface of the nanoparticles. This can be identified with FT-IR. Knowing these chemical compounds is important because the surface charge can vary and consequently the antibacterial activity.

- **SEM**: With this analysis, it is possible to obtain details of the microstructure such as grain size distribution, particle agglomeration, and porosity (visual). A small particle size can promote more reactive oxygen species (ROS), which is related to antibacterial activity.

- **XPS**: This powerful technique allows calculation of structural defects such as oxygen and metal vacancies, bonding energy, and chemical composition, among others. The vacancies play a key role in antibacterial activity.

- **UV-vis**: A lot of nanoparticles can improve their antibacterial activity when exposed to UV light and are called photosensible nanoparticles. With this study, it is possible to calculate the optical band gap, conduction, and balance band energies. These energies can allow the formation of ROS and it has a direct effect on antibacterial activity.

- **Z-potential**: This analysis shows the surface charge of nanoparticles, which can be positive or negative. Remembering that the bacteria have a specific surface charge, the z-potential can give information about a good or bad interaction between the nanoparticle and the bacteria.

Figure 17.2 illustrates the results of each technique for CeO_2 nanoparticles. Together, these results help to understand the antibacterial activity of nanoparticles.

17.1.1.3 Antibacterial Assays

Different experiments must be carried out to verify the effectiveness of the nanoparticles against bacteria. Figure 17.3 illustrates the assays.

- **Bacteria types**: Usually, two types of bacteria are used as references, gram negative or gram positive. The principal difference between these bacteria is the single (gram positive) and double layer

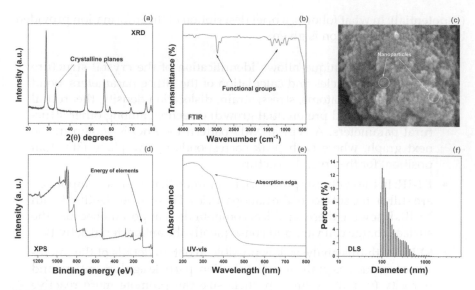

FIGURE 17.2
Representative results of (a) XRD, (b) FTIR, (c) SEM, (d) XPS, (e) UV-Vis, and (f) DLS characterization of CeO_2 nanoparticles.

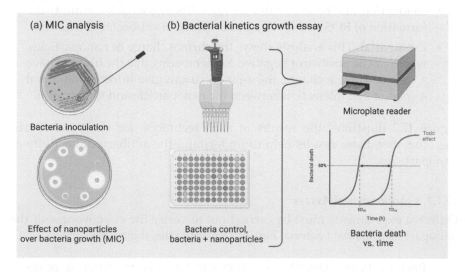

FIGURE 17.3
Schematic representations of antibacterial assays.

of lipopolysaccharide that surround the peptidoglycan wall (gram negative).

- **Nanoparticle concentration:** One of the goals of nanoparticles against bacteria is to reduce the amount of this to be used for this purpose. Then, a wide range of nanoparticles are tested.
- **Minimum inhibitory concentration:** It is the lowest concentration of nanoparticles to see antibacterial activity.
- **Zone the inhibition:** It is a qualitative method used to measure nanoparticles' effectiveness to inhibit bacterial growth.
- Bacterial growth kinetics is the relationship between the specific growth rate and the concentration of nanoparticles.

Finally, after all these processes have taken place, the research team is capable of collecting, ordering, and analyzing the information generated, and this is done with the help of artificial intelligence (AI) tools. In the following paragraphs, a short introduction to them will be given.

17.2 Artificial Intelligence: The Tool for Complex Problems

Today, the term "artificial intelligence" is used in so many and varied environments, with such broad interpretations that its essence and complexity seem to have been oversimplified. The term as such was first used in 1956 when John McCarthy, Marvin Minsky, and Claude Shannon, three leading scientists of the time, convened a month-long working session. It was during this seminar that the participants made their first effort to define the foundations as well as the lines of work that were considered a priority to achieve the maturation of this incipient discipline. An important aspect to point out is that both during that session and in the time following it, there was great enthusiasm and optimism, which led to an overestimation of the results of AI. As the years went by, it became clear that these goals were not being achieved, and therefore a disillusionment ensued, which was reflected in a period that has been called the "winter of artificial intelligence." Since that experience, AI scientists have been more cautious in forecasting its progress, although recently there has been a similar phenomenon with the development of deep learning, which has led on the one hand to the use of the term computational learning as a synonym for artificial intelligence and on the other hand it is the case that such techniques are being used in many problems that most likely would admit a simple solution and above all, more likely to be understood by humans. But this will be discussed briefly later in this chapter.

As the reader may imagine, the term "artificial intelligence" might be somewhat difficult to be defined. In fact, there are many definitions of AI (Legg and Hutter, 2007). For instance, Elaine Rich proposed a simple one in 1991: "It is the study of how to get computers to perform tasks that, at the moment, humans do better." John McCarthy, considered the father of AI, proposed the following definition in 1998: "It is the science and engineering of making intelligent machines, particularly programs."

Truth is that even today, after almost 70 years, there are several definitions (Kumar, 2013). In recent years, there has been some consensus in the scientific community that definitions of AI could fall into one of two groups:

Strong artificial intelligence: It is also known as general AI. In this case, scientists are referring to a form of AI whose aim is to create intelligent machines that are indistinguishable from the human mind. It means by consequence that the machines would perfectly mimic the processes of the human brain. This is a very serious assertion, given that it may imply that for instance, the machines would have a self-aware consciousness that could solve problems, learn, and plan for the future. An excellent example of this type of AI is what Professor Marvin Minsky stated in several of his publications. For instance, in "The Society of Mind" (Minsky, 1988), considered a "transcendent" exploration of brain structure and function, he refers that in order to achieve computers that surpass human intelligence, our brains must be deciphered to try to mimic the neural processes of that extraordinary "common sense." Despite having devoted much of his life to the contributions of science, in his last statements, he maintained that: "There is only one thing certain: anyone who says that there are basic differences between the minds of men and the machines of the future is wrong. If this has not already been achieved, it is only due to a lack of financial and human resources." (Minsky, 85) This type of vision, therefore, considers that cognitive and reasoning processes can be decoded and perfectly replicated on a computer. It is important to note that ultimately this has not yet been achieved and that we are probably a long way from reaching this goal if it were achievable.

Weak AI: Also known as narrow AI, it focuses on performing a specific task, such as answering questions based on user input or playing a game such as Go. Weak AI is an approach to AI research and development considering that AI is and always will be a simulation of human cognitive function and that computers can only appear to think but are not really conscious in any sense of the word. Weak AI simply acts and is subject to the rules imposed on it and cannot go beyond those rules. A good example of weak AI is computer game characters that act believably in the context of their game character but cannot do anything more than that.

Weak AI is a form of AI designed specifically to focus on a narrow task and appear very intelligent to you. Contrast this with strong AI, where an AI is capable of any and all cognitive functions that a human might have and in essence is no different from a real human mind. Weak AI is never taken as general intelligence but rather as a construct designed to be intelligent in the limited task to which it is assigned. At the end of the day, weak AI still relies on human intervention. Humans are needed, for example, to create the datasets used to train the learning algorithms, or to adjust the parameters on such models, and in many cases, to perform what is called "feature engineering" which is the process of defining what variables are considered relevant to build the machine learning (ML) models. Good examples of the applications of weak AI could be self-driving cars and virtual assistants.

In contrast to strong AI, which can learn to perform any task performed by a human, weak AI is limited to one or a few specific tasks. This is the type of AI we have today. In fact, deep learning, which is named after (and often compared to) the human brain, is very limited in its capabilities and is not even close to performing the kind of tasks that a human child's mind can perform. And that's not a bad thing. In fact, AI can focus on specific tasks and do them much better than humans, for example, feed a deep learning algorithm enough skin cancer images, and it will be better than experienced doctors at detecting skin cancer. This does not mean that deep learning will replace doctors because it needs intuition, abstract thinking, and many more skills to be able to decide what is best for a patient, but deep learning algorithms will surely help doctors do their job better and faster and help them see more patients in less time. It will also reduce the time it takes to educate and train professionals in the healthcare industry.

In a book that has become a must-read in AI, Russell and Norvig describe a model of intelligent systems or agents which considers three important elements: the agent must be able to *perceive* his/her environment to some extent, with the information received he/she performs a *reasoning* process that allows him/her to decide what *actions* to take to get closer to his/her goal. Actions change the state of the world and the cycle repeats. Given this conceptualization, they propose a classification of intelligent systems based on two dimensions. On the one hand, they consider reasoning capabilities and on the other hand, behavior. Thus, for them, there would be four quadrants under these assumptions, which would be (Russell and Norvig, 2010):

Systems that think like humans: The first type of systems refers to those that could be developed if strong AI could be achieved since this would include systems that are capable of perfectly emulating the cognitive, thought, and reasoning processes of humans. It is evident that to achieve this, it would be necessary to know precisely how these processes occur in humans, something that is not fully known at this time. This type of conception of AI requires the participation of other disciplines, such as psychology and cognitive sciences. In some sense, this approach would be the one that tries to reproduce human intelligence in a computer by fully copying it, with its pros and cons.

Systems that think rationally: To explain this type of intelligent systems, it is important to first establish the concept of *rationality*. For the purposes considered in this document, it will be understood that a system is rational if, given a certain perception, the system selects an action that maximizes a measure of performance, given the evidence provided by that perception and the system's own knowledge. That said, the systems that think rationally would be those that follow certain laws of thought, in the sense that Aristotle proposed. It is common for this type of system to use different types of logic to develop the reasoning processes. The approach presents a series of difficulties, from how to translate empirical knowledge and often considered common sense to a logical formalism, and then in the reasoning stage, logic usually has restrictions that prevent them from being always applied.

Systems that act like humans: It refers to systems in which the goal is for the system to act as a human under certain circumstances, without the condition that the thought and reasoning processes are faithful copies of how a human performs them. The well-known Turing test fits into this type of conception of AI. According to her, if a computer can trick a human into thinking that he is interacting with another human, then the system running on that computer can be said to be intelligent. It is clear that one thing is to faithfully emulate what happens in the human brain and another is to make the system act as if it were a human.

Systems that act rationally: In this case, we will have a system that decides its actions based on the perceptions it may have, its previous knowledge, and the consideration of which possible action would maximize a measure of performance, or in the case of uncertainty, maximize expectancy performance (in the probabilistic sense).

From a practical point of view, the latter type is the one most often used in solving real problems today. And it is within this framework that the techniques that will be referred to in this work are aligned. Develop systems that maximize a metric through the actions chosen.

Now, AI is actually a very broad and diverse area depending on whether your study is more interested in the strong version and disciplines such as cognitive science and its relationship with neuroscience, philosophy, and psychology. If the focus is more on the weak version, then math and computer science will be more relevant. At the end of the day, there are many ways to group the areas that AI studies. A rather simple one considers systems that can reason when they have complete information about the environment, for example, by resorting to propositional logic, or both blind and informed search strategies. In another category are the methods that allow reasoning under imprecision or uncertainty, in which case fuzzy logic and probabilistic reasoning appear. In other cases, it is a matter of seeking optimal solutions to problems that is where evolutionary computation fits in. Another great area, which has also received great attention in recent years, is the case in which the system learns from data, which is known as ML. As it is in this area that the application that concerns us in this chapter falls, some of its fundamentals will be described to facilitate the understanding of the application of these techniques to the design of nanomaterials.

ML has proven to be a powerful problem-solving tool in many disciplines. The most interesting feature of this area is that it allows computers to learn to perform tasks without being directly programmed to do so. ML algorithms receive data, process it according to certain algorithms, and develop models that represent patterns and, in many cases, complex mathematical relationships among the variables, here called *features*. This is often called predictive analytics. The term machine learning was used for the first time in 1959. However, since then, it has gained relevance due to several factors, including the increase in computing capacity and the increase in the amount of data available. Let us describe some basic concepts in this area of AI.

17.2.1 Machine Learning Algorithms

ML algorithms fall into three categories, with the first two being the most common:

- **Supervised learning**: It is a subset of ML that consists of deducing information from training data. These algorithms have prior learning based on a system of labels associated with data that allow them to make decisions or make predictions. The data are categorized into two sections: training data and testing data. Training data are used to train a model, and test data are used to determine the effectiveness of the model created. The goal of supervised learning is to create a program that can solve any input variable after being subjected to a training process. An example is a spam detector that labels an email as spam or not depending on the patterns it has learned from the email history. There are two subtypes of supervised learning: classification and regression. For classification, the algorithm tries to label the examples by choosing between 2 or more classes. It uses the information learned from the training data to choose the correct label. Regression, on the other hand, is the training of an algorithm to predict an outcome from a range of possible real (numeric) values.

- **Unsupervised learning**: These methods base their training process on a dataset without previously defined labels or classes. That is, no objective or class value is known a priori, be it categorical or numerical. They face the chaos of data to find patterns that allow them to be organized in some ways. Unsupervised learning is dedicated to grouping tasks, also called clustering or segmentation, where its goal is to find similar groups in the dataset. For example, in the field of marketing, they are used to extract massive data patterns from social networks and create highly segmented advertising campaigns.

- **Learning by reinforcement**: In reinforcement learning, there is not an "output label," so it is not of the supervised type, and although these algorithms learn by themselves, they are not of the unsupervised type either, where they try to classify groups considering some distance between samples. The objective of reinforcement learning is that an algorithm learns from its own experience. That is, to be able to make the best decision in different situations according to a trial-and-error process in which correct decisions are rewarded. This type of ML is applied in a variety of problems, including the navigation systems of drones or multi-robot systems, as well as in adversarial games.

17.2.2 Supervised Machine Learning and Nanomaterial Development

Computational learning is changing many areas of science, including physics and chemistry. Specifically, in materials science, a very important aspect is material characterization and the understanding of atomic structures since the properties of materials depend on this. For some time now, computational sciences have been applied to the generation of new materials. An example of this is density functional theory, although its implementation is usually limited to rather small models containing a few hundred atoms, due to the computational complexity involved in solving these models. This is where ML appears as an alternative that can offer reasonable accuracy for a limited computational cost. Previous examples of ML applications in this area include the analysis of XRD data, in which by using convolutional neural networks it has been possible to identify three-dimensional structures of certain compounds based on patterns analyzed in about 150,000 XRD results (Park et al., 2017). Similarly, models have been developed for other types of characterization techniques, such as RAMAN spectroscopy.

In the work reviewed in this chapter, the approach is simpler because it has been decomposed into three phases. In the first one, it is desired to test the modeling of one or some properties of the materials being generated, through the characterization data obtained in the laboratory with tests such as those mentioned above. In the second stage, already having a model that represents that property with reasonable accuracy, it is intended to optimize the laboratory work. It is common in problems involving the development of laboratory experiments to design them. This design can therefore determine the combinations of process parameter values that would be important to obtain statistical representativeness of the results. Depending on several factors, the recommended number of experiments can be high, and two or three repetitions are usually done for each experiment to reduce the effect of errors. This involves a great deal of time and cost. However, if the model is available, it can be used to determine which of these experiments would be most promising in terms of the information they would provide. With this, the experimental design could be optimized, reducing costs and, above all, time. Finally, in the third stage, what is desired is to incorporate data from different materials to see if a much more general model can be generated in which certain chemical elements can be substituted for others and to have a good idea of whether a material with that conformation would be promising. To be more specific, the group has been working on nanomaterials based on zinc oxide (ZnO), doped

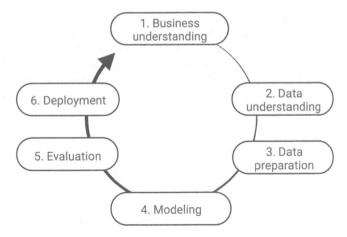

FIGURE 17.4
The phases in a data mining process.

with rare earth elements. So far, results have been obtained from the incorporation of erbium, ytterbium, cesium, and samarium. So, the next step would be, for example, to test whether using other chemical elements of the same family could provide antimicrobial activity. So far, progress has been made on the first objective, with the other two remaining.

In this section, we will describe the most common process for applying ML techniques to data modeling problems and make brief references to how this process has been applied to the data generated in the laboratory. This process comprises seven steps, illustrated in Figure 17.4, and described in the following paragraphs.

17.2.2.1 Phase 1: Understanding the Problem

It is very important to understand the problem to solve. It is common that the problem is not in the domain of the data scientist. The problem might be in areas such as medicine, engineering, chemistry, business, or any other discipline. It is therefore a stage where there must be a close interaction between the data science team and the owner of that information to understand what is being attempted and for what purpose. In the case reported in this paper, the data science team had to understand some basic aspects of nanomaterials and biological testing. Specifically, the problem is to be able to predict some properties of the nanomaterial from several parameters or attributes in order to optimize the experiments performed in the laboratory. The design of experiments normally happens under a methodology in which, from the definition of the variables and their ranges, it is calculated how many experiments to develop and with which specific values for those variables in order to guarantee a certain statistical representativeness of the results at the end.

However, it is very likely that some of these experiments will not work well for the variable considered as a response. If by using computational learning the function of that response variable is modeled, then it is possible to use the model to predict the probability of success of the test. This gives priority to the experiments that are most likely to produce a material with the desired properties and saves resources, especially time but also money since several of the analyses that are applied are costly.

17.2.2.2 Phase 2: Understanding the Data

Just as important as understanding the problem is understanding the data available. It is common to do an exploratory data analysis to familiarize ourselves with the data. In an exploratory analysis, graphs, correlations, and descriptive statistics are often done to better understand what story the data are telling. It also helps to estimate whether the data available are sufficient and relevant, to build a model.

17.2.2.3 Phase 3: Prepare the Data

Data preparation is one of the most challenging phases of ML. The main challenges faced in this phase are the following:

Incomplete data: In many cases, the information gathered is incomplete. In problems where the information is obtained directly from humans, for example, from an online survey, it may happen that there will be people who have not filled in all the fields. In other applications, the sensors may have failed to measure some values, or the network connection dropped, and data lost. Of course, having some data is better than having none, so the underlying question is how to make use of the data you have. One option is to delete those incomplete records. However, this option is almost never the best, especially if there are a lot of missing data, because the risk could be left with practically no meaningful information. Another option is to complete the missing information. This process is called imputation. It is clear that if the information is to be added or completed, it must have a reasonable basis. For example, sometimes the average of other values is used. If some of the missing data are thought to be the ages of persons, an assumption such as the one above might be plausible. In other cases, such an approach may not be the best. One more alternative is just to use a computational learning model. In that case, an ML model is trained to predict missing values. In addition to these options, it is also possible to look for an ML technique that is able to work with incomplete information.

17.2.3 Combine Data from Different Sources

It is extremely common in data science projects for information to come from different sources. Some might come from a database, some from spreadsheets, some from sensors, etc. Regardless of their origin, all these data must be combined and left in a format that can be used by ML algorithms.

Feature selection: One of the most relevant tasks is the selection of features, also called attributes. This process often implies that a subset of the original attributes is selected from the original attributes since this provides advantages such as improving the predictive performance of the model, building models more efficiently, and improving the understanding of the generated models. It is important to note that although using many attributes can bring a higher discriminative power, in practice, an excessive amount of attributes significantly slows down the learning process and can produce overfitting. Therefore, it can be said that the fundamental objective of this stage is to select the smallest subset of attributes such that the classification rate is not significantly affected and that the resulting class distribution is as close as possible to the original one. Put more formally, it means that you want to build a consistent, relevant, and nonredundant set of attributes.

Another common operation required is data normalization. Normalization is a technique often applied as part of data preparation for ML. The goal of normalization is to change the values of the numerical columns in the dataset to use a common scale, without distorting the differences in the value ranges or losing information. Normalization is also necessary for some algorithms to model the data correctly.

17.2.3.1 Phase 4: Building the Model

Depending on the problem, it is necessary to select the model(s) that could be the best candidate(s) to solve it. In the case described here, it is a regression problem, for which examples are available. Therefore, the models that can be applied are the supervised regression models. In particular, a multilayer perceptron (MLP), support vector machine (SVM), and linear regression have been tested. Of these, the one that has given the best results is the neural network, so it has continued to be used for various materials.

17.2.3.2 Phase 5: Error Analysis

The error analysis phase requires a relatively medium effort. Analyzing errors is important to understand what we need to do to improve ML results. Depending on the error, the available options will be to use a more complex model, use a simpler model, add more data, adjust the features selected, or even to question if the data science team has really understood the problem. The main purpose here is to identify if the trained model is able to generalize. Generalization is the ability of ML models to produce good results when using new data. It means that the underlying process behind the data has been correctly captured by the model. This means that sometimes it is needed to iterate over the previous phases several times. With each iteration, the understanding of the problem and the data will become better and better. This will enable the team to design better relevant features and reduce generalization errors. Greater understanding will also give the ability to choose the ML technique that best fits the problem. Almost always, having more data helps. In practice, more data and a simple model tend to work better than a complex model with few data.

17.2.4 Define Evaluation Criteria

Another important step is to define how the models will be evaluated. The evaluation criterion is usually a measure of error. Typically, the mean square error is used for regression problems and the cross-entropy for classification problems. For classification problems with 2 classes (which are very common), it is possible to use other measures such as precision and completeness.

Probably, the problem has already been solved in another way. Probably, the motivation to use ML to solve this problem is to obtain better results. Another common motivation is to obtain similar results automatically, possibly replacing boring manual work. If we measure the performance of the current solution (with the chosen evaluation criteria), we can compare it with the performance of the ML model. Then, we can know if it is worth using the ML model or if we stick with the current solution.

If there is no current solution, we can define a simple solution that is very easy to implement. For example, if we want to predict the price of a house with ML, we could compare it with a simple solution (e.g., median square meter value per neighborhood). Only then, when we have a finished ML model, we can say if it is good enough, if we need to improve it, or if it is not worth it.

If in the end it turns out that the current solution or a simple solution is similar to the solution that ML gives us, it is probably better to use the simple solution. It will almost always work better and be more robust.

17.2.4.1 Phase 6: Deployment

If the ML model is better enough than other solutions, then it will be ready to be integrated into a "production" system. A considerable part of the effort goes into building data interfaces. These interfaces are necessary so that the model can obtain data automatically and so that the system can use its prediction automatically. Although this is a considerable effort, it is essential. For ML and AI to be useful, in most cases, they must be integrated into a larger system.

17.3 Our Application

The multidisciplinary team involved in this work has allowed the generation of new nanomaterials, which are characterized through various analytical chemistry techniques. This has allowed the generation of data that can be the basis for the training of ML algorithms. The models have evolved as more experiments have been added and more results have been generated. It is important to note that so far all the experiments that have been performed are to analyze the behavior of the nanomaterial in question vs. *Staphylococcus aureus* and *Escherichia coli*.

In the first work, they generated a nanomaterial composed of ZnO doped with erbium particles. Variables such as treatment time, nanoparticle concentration, and type of bacteria were considered in the model. The output variable that was modeled was absorbance, which in the end is an indirect measure of the antimicrobial activity of the nanomaterial.

In this first attempt, models were trained for linear regression, SVM, and MLP. In each case, different combinations of their respective tuning parameters were tested, and it was found that the model that best represented the data (lower error) was the MLP, so that is the one reported in Sánchez-López et al (2022). In that case, the accuracy error of the model was around 2%, which is acceptable. In a second work (Navarro-López et al., 2021), doping ZnO with ytterbium was tested. The data collected had a similar structure to that of the previous experiment, only that the atomic concentration was added to the model. MLP was trained directly with these data. This time, a deeper analysis of the effect on model accuracy of perceptron parameters, such as the number of neurons in the hidden layer, and the activation function employed, was performed. The model error remained between 1% and 2%. Figure 17.5 shows the structure of a multilayer with its basic elements.

Subsequently, in a third work (Navarro-López et al., 2022), other rare earth elements were added: cerium, neodymium, and samarium. Again, with the data obtained, neural network models were generated in which the following parameters were optimized: the number of neurons in the hidden layer, activation function, solver, and alpha value. In this case, the error increased even 7%–10%, which is understandable since the model must be more general when incorporating more elements. But on the other hand, it also suggests that using only the five attributes that were selected is probably not enough. So, the next step is to incorporate other parameters of the chemical elements and morphological structure of the materials to try to generate a model that is applicable with a greater degree of generality to more materials.

17.4 The Future

In this chapter, the authors wanted to share an experience in which science, from different disciplines, cooperates to try to solve a complex problem. Materials science, especially nanotechnology, biotechnology, and AI complement each other very well in problems such as those addressed in this paper. Each of them contributes to part of the solution, and without the others, reaching the same results might not be possible. Multidisciplinary work is undoubtedly becoming increasingly necessary.

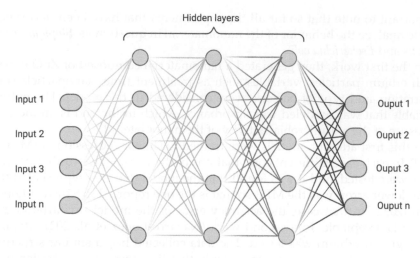

FIGURE 17.5
A multilayer perceptron illustrating the input layer, the hidden layers, and the outputs.

Although making predictions (beyond what computational learning models can model) is a complex task, what the authors see as the future is precisely this type of collaboration, where AI leaves a laboratory where the information is close to perfect, often with modest datasets and under very controlled conditions, to serve as support in more challenging problems and environments. It is time for AI to impact society by providing services and products that improve people's quality of life. This transition will also allow a better understanding of the limitations it faces and will surely raise important ethical questions, for example.

It is reasonable to think that with the increased availability of faster computers, with the appearance of a large number of software platforms, many of which offer at least part of their functionalities free of charge, and the ease of disseminating results not only through the traditional channels of the scientific community, such as specialized journals, but also through the Internet, it is reasonable to think that the development of new technologies will have an impact on the quality of life of people.

References

Correa-Baena, J. P., Hippalgaonkar, K., van Duren, J., Jaffer, S., Chandrasekhar, V. R., Stevanovic, V., … & Buonassisi, T. (2018). Accelerating materials development via automation, machine learning, and high-performance computing. *Joule*, 2(8), 1410–1420.

Díez-Pascual, A. M. (2018). Antibacterial activity of nanomaterials. *Nanomaterials*, 8(6), 359.

Kaushik, M., Niranjan, R., Thangam, R., Madhan, B., Pandiyarasan, V., Ramachandran, C., ... & Venkatasubbu, G. D. (2019). Investigations on the antimicrobial activity and wound healing potential of ZnO nanoparticles. *Applied Surface Science*, 479, 1169–1177.

Kumar, E. (2013). *Artificial intelligence*. IK International Pvt Ltd., New Delhi.

Legg, S., & Hutter, M. (2007). A collection of definitions of intelligence. *Frontiers in Artificial Intelligence and Applications*, 157, 17.

Minsky, M. (1988). *Society of mind*. Simon and Schuster., New York.

Murray, C. J., Ikuta, K. S., Sharara, F., Swetschinski, L., Aguilar, G. R., Gray, A., ... & Naghavi, M. (2022). Global burden of bacterial antimicrobial resistance in 2019: A systematic analysis. *The Lancet*, 399(10325), 629–655.

Navarro-López, D. E., Garcia-Varela, R., Ceballos-Sanchez, O., Sanchez-Martinez, A., Sanchez-Ante, G., Corona-Romero, K., ... & López-Mena, E. R. (2021). Effective antimicrobial activity of ZnO and Yb-doped ZnO nanoparticles against Staphylococcus aureus and Escherichia coli. *Materials Science and Engineering: C*, 123, 112004.

Navarro-López, D. E., Sánchez-Huerta, T. M., Flores-Jimenez, M. S., Tiwari, N., Sanchez-Martinez, A., Ceballos-Sanchez, O., ... & López-Mena, E. R. (2022). Nanocomposites based on doped ZnO nanoparticles for antibacterial applications. *Colloids and Surfaces A: Physicochemical and Engineering Aspects*, 652, 129871.

Park, W. B., Chung, J., Jung, J., Sohn, K., Singh, S. P., Pyo, M., Shin, N. and Sohn, K.-S. (2017). Classification of crystal structure using a convolutional neural network. *IUCrJ* 4, 486–494. https://doi.org/10.1107/S205225251700714X.

Russell, S. J. & Norvig, P. (2010). *Artificial intelligence a modern approach*. Prentice Hall series in artificial intelligence, Prentice Hall, New Jersey

Sánchez-López, A. L., Perfecto-Avalos, Y., Sanchez-Martinez, A., Ceballos-Sanchez, O., Sepulveda-Villegas, M., Rincón-Enríquez, G., ... & López-Mena, E. R. (2022). Influence of erbium doping on zinc oxide nanoparticles: structural, optical and antimicrobial activity. *Applied Surface Science*, 575, 151764.

Seifi, T., & Kamali, A. R. (2021). Anti-pathogenic activity of graphene nanomaterials: A review. *Colloids and Surfaces B: Biointerfaces*, 199, 111509.

Shinde, S. S. (2015). Antimicrobial activity of ZnO nanoparticles against pathogenic bacteria and fungi. *JSM Nanotechnology & Nanomedicine*, 3, 1033.

Smith, T. L., Pearson, M. L., Wilcox, K. R., Cruz, C., Lancaster, M. V., Robinson-Dunn, B., ... & Jarvis, W. R. (1999). Emergence of vancomycin resistance in Staphylococcus aureus. *New England Journal of Medicine*, 340(7), 493–501.

Ventola, C. L. (2015). The antibiotic resistance crisis: Part 1: causes and threats. *Pharmacy and Therapeutics*, 40(4), 277.

18

Neural Networks for an Automated Screening System to Detect Anomalies in Retina Images

Gildardo Sanchez-Ante and Luis E. Falcon-Morales
Tecnológico de Monterrey

Juan Humberto Sossa-Azuela
Instituto Politécnico Nacional

CONTENTS

18.1 Introduction

In Mexico, diabetes has a high prevalence, usually correlated with metabolic syndrome. According to figures reported by INEGI, in 2020, 151,019 people died from diabetes mellitus, equivalent to 14% of the total deaths (1,086,743) in the country. The mortality rate for diabetes for 2020 is 11.95 persons per 10,000, the highest figure in the last 10 years. There is an increase in the diagnosis of the disease as people get older; nationally, slightly more than a quarter of the population aged 60–69 years (25.8%) reported having a previous diagnosis of diabetes, representing 2.3 million people (INEGI, 2020).

These alarming figures present a series of challenges to the health system, especially the public health sector, which has the task of diagnosing and treating the disease. Likewise, through various public policies, there is the task of prevention. Regardless, it is essential to point out that one of the most common consequences of the lack of adequate follow-up of diabetes is the presence of a condition known as diabetic retinopathy.

This disease occurs because high blood sugar levels cause damage to the blood vessels in the retina. The retina is the layer of tissue at the back of the inside of the eye. It transforms light and images entering the eye into nerve signals sent to the brain.

When the patient presents retinopathy, the blood vessels in the retina can swell and leak fluid. They may also close and prevent blood from flowing. Sometimes, new abnormal blood vessels grow in the retina. All of these changes can cause the patient to lose vision. Therefore, it is crucial for people with diabetes to be monitored frequently to detect abnormalities in their retinas early. It has also been suggested that fundus examination should be done routinely in populations with a high prevalence of diabetes. The challenge is mainly due to the workload that implies for specialized ophthalmologists to analyze the images.

Fundus examination requires some training in the practice of direct ophthalmoscopy to take advantage of all the benefits that this procedure can provide, so it is essential the knowledge and proper handling of the direct ophthalmoscope, in addition to a piece of good equipment.

The ophthalmoscope is an instrument with several lenses and mirrors that illuminates the inside of the eye through the pupil and the crystalline lens, allowing the examination of the retina or fundus. There are several ways to perform the study, the most commonly used being the manual ophthalmoscope designed for direct magnified vision, with a light source projected through a mirror or prism to the back of the patient's eye, which is reflected on the retina and coincides with the observer's line of sight through the aperture.

The magnification obtained when using the direct ophthalmoscope is because the eye itself serves as a simple magnifying lens through the cornea and the crystalline lens. Its main application is the observation of the fundus. However, the rest of the ocular structures can also be examined, from the eyelids and the anterior segment of the eye to the intraocular media and the retina.

The direct ophthalmoscope consists of an illumination system comprising a spotlight located in the instrument itself, powered by batteries or electric current, whose light is focused by a convex condenser lens and goes to a prism that sends it to the eye under study. An observation system consisting of an orifice or transparent portion in the upper part of the prism, before which lenses of different concave or convex values are superimposed, inserted in a disc that can be moved at will by rotating the disc to focus precisely on the structures of the fundus. The light beam generated in an electric bulb is reflected by a 45° inclined mirror or a prism which performs the same function and is directed the pupil, passing through the lower half of the pupil. The light reflects off the back of the eye, exits again through the upper half of the pupil and passes flush over the mirror, and is received by the observer looking through an orifice immediately above the mirror. The center of the mirror is not silvered so that the illuminated retina can be seen through the mirror.

Good ophthalmoscopy depends on an adequate examination technique and constant practice to observe the fundus, even in less-than-ideal examination conditions. Its optical principle consists of projecting the light from the ophthalmoscope inside the eye so that through its reflection on the fundus, the observer can obtain an image of the internal structures. It provides a direct manifestation of the retinal structures magnified about 15 times. A satisfactory examination can usually be made through an undilated pupil as long as the media, cornea, aqueous, crystalline lens, and vitreous are transparent; however, with a dilated pupil, a greater extent of the peripheral posterior segment can be explored. If there is no medical contraindication, a fundus examination with dilated pupils is preferred.

Here is where artificial intelligence comes into the scene. In this chapter, the authors summarize an experience developing a prototype of an intelligent system capable of receiving retina images to classify them into one of two classes: with anomalies or without anomalies. The decision will be based on the structure of the vascular system in the retina, an approach that has been explored before (Vandarkuzhali & Ravichandran, 2005). The result of the system can be used as a screening criterion. Only those with abnormalities would require follow-up with the ophthalmologist. It is important to note that the system in no way is planned as a substitute for the specialist. It will only be an assistant to make their intervention more efficient. Figure 18.1 shows two retina images, the one on the left (Figure 18.1a) is for a healthy eye, while the one on the right (Figure 18.1b) shows various abnormalities. It was taken from the dataset (Benítez et al., 2021).

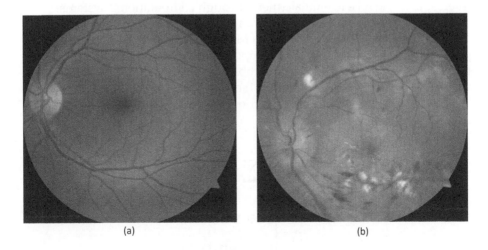

(a)   (b)

FIGURE 18.1
Retina images. Left image corresponds to a healthy eye. The image on the right corresponds to a sick eye.

In the following sections, a short introduction to artificial intelligence, and more particularly to a supervised learning model known as artificial neural networks, will be presented. A brief description of what a multilayer perceptron consists of will be given, which will detail the steps of the methodology followed to train and validate a model that a doctor can later use.

18.2 Artificial Intelligence and Machine Learning

Artificial intelligence has had a fascinating history. It is a relatively young area, compared to other disciplines such as mathematics or physics, since it was in 1956 when the term was used for the first time. In these nearly 60 years of existence, there have been significant developments but also moments in which it has been necessary to take a break to rethink the scope of some of its areas. There have been moments of euphoria in which it has been considered that various artificial intelligence tools could solve problems in a short time that were later discovered to be much more complex than they appeared to be.

One of the best-known cases is artificial neural networks (ANNs). As the name suggests, it is a model based on biological neuron networks, that is, those that exist in living beings such as humans. These networks have essential properties, such as learning from experience, fault tolerance, and handling high noise levels, among others. Therefore, there has been a latent interest in being able to emulate them through computational systems.

McCulloch and Pitts (1943) conceived an abstract and simple model of an artificial neuron (abbreviated here as MCP). This model is the fundamental processing element in an ANN. The inputs are given by vector $x_1, ..., x_n$, function Σ performs a summation over those inputs, and function ρ takes that value and fires or not (outputs value 1) depending on a step function, which is shown in Figure 18.2. The neuron outputs a 1 if the sum of the inputs exceeds a certain threshold θ.

Mathematically:

$$\Sigma = \sum_i x_i w_i$$

$$\rho = \begin{cases} 1 & \text{if } \Sigma \geq \theta \\ 0 & \text{otherwise} \end{cases}$$

The MCP was the first mathematical model of a neuron. At that time, the authors considered that neurons with a binary threshold activation function were analogous to first-order logic sentences. One of the difficulties with

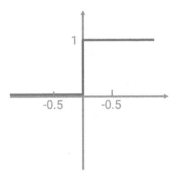

FIGURE 18.2
Step activation function illustrated. In this case, the neuron either fires (outputs 1) or do not fire (output 0), depending on the value of a certain threshold.

the McCulloch–Pitts neuron was its simplicity. It only allowed for binary inputs and outputs and used the threshold step activation function, shown in Figure 18.2. For those reasons, the model had many limitations.

Further research by Donald Hebb established that synaptic connections are strengthened when two or more neurons are activated contiguously in time and space. The results obtained by Hebb are an essential finding that, in terms of computer science, can be interpreted as a basis for learning. The adaptation of that strength can be achieved by modifying the weights of the connections in a network of neurons.

Therefore, these two results, the McCulloch–Pitts neuron model and Hebb's learning base, allowed Frank Rosenblatt to develop the perceptron (Rosenblatt, 1958). The perceptron is a new processing unit model that already includes the weights in the connections, even though the model still had a single layer of input functions and a single output. The weights could be changed manually to adjust the functioning of the neural network to a given task.

This process, to date, remains the foundation of more advanced learning models in neural networks. Figure 18.3a presents the perceptron, and Figure 18.3b shows several activation functions. In such figure, x_i represent inputs, w_i represent weights, Σ represents the summation of the inputs multiplied by their weights, and ρ is the activation function that computes the output of the neuron.

According to the perceptron Learning Rule, it is possible for an algorithm to automatically calculate the optimal weight coefficients, that is, those that minimize the error in the perceptron's predictions. To determine whether a neuron "fires" or not, the values of the inputs are multiplied by these weights. At the end of the day, the perceptron is a mathematical function. The input data (x) is multiplied by the weight coefficients (w). The result is a value. That value can be positive or negative. The artificial neuron is activated if the value is positive. It is only activated if the weighted weight of the

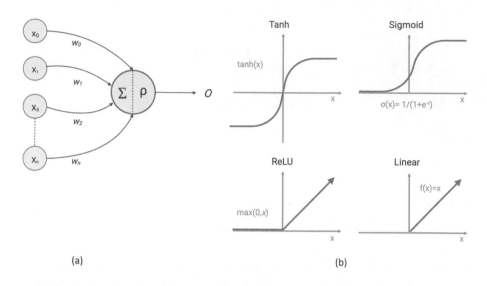

FIGURE 18.3
The image on the left shows a simple perceptron. The figure on the right shows some of the most common definitions of activation functions.

input data exceeds a certain threshold. The predicted result is compared with the known result. In case of a difference, the error is backpropagated to allow the weights to be adjusted. Initially, the perceptron was proposed considering a single layer. With this structure, it is possible to solve particular problems, such as emulating logical AND, OR, and NOT operations. With that in mind, some scientists considered that perceptron arrays could also solve any problem that Boolean equations or electrical circuits could solve. Imagination ran wild, and very optimistic goals were set for what could be done with perceptrons. Until 1969, a very relevant book was published by Marvin Minsky and Seymour Pappert (1969). In such book, the authors demonstrate that the perceptron is severely limited to solving only certain types of problems, that is, those that were linearly separable, that is, those in which a representation of the solutions in a specific type of space could be divided by a straight line. It turns out that a perceptron cannot be adjusted to represent a problem as simple as the XOR logic function. The XOR logic gate also called exclusive OR is a digital gate that implements true output (1) if one and only one of the inputs is true. If both inputs are false (0) or both are true, then the output is false. The XOR represents a Boolean function of inequality, that is, the output is 1 if the inputs are not equal; otherwise, the result is 0. These results discouraged the initial interest in ANNs, and for several years, there was a significant decline in research activity on the subject. It was later found that although a single-layer perceptron can teach only linear separable functions, a multilayer perceptron, also called a feed-forward neural network, overcomes this limit and offers superior computational power. In Figure 18.4, a multilayer

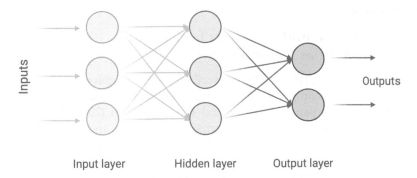

FIGURE 18.4
Multilayer perceptron with its elements: input, hidden, and output layers.

perceptron (MLP) is illustrated. As it is possible to note, in such a case, the network is composed of many neurons, which are distributed in layers. The layers that are between the input and output are called hidden layers. The learning process in an ANN is usually conducted by selecting some examples from the training set using any policy that guarantees that all training examples will eventually be chosen. If the output is correct, the weights are left as is. If the output unit answers incorrectly, the weights will be updated in proportion to the error. There are several different algorithms to perform such an update. One of the most well known is backpropagation (Haykin & Network, 2004). The MLP and the existence of several algorithms to automatically adjust the weights made a huge leap forward for this model to become a powerful tool for solving many problems that otherwise require highly experienced programmers and experts in the disciplines where the issues reside.

There are several types of neural networks, such as monolayer or single perceptron, multilayer perceptron (MLP), convolutional (CNN), recurrent neural networks (RNN), feedback or radial basis networks (RBF), Probabilistic Neural Networks (PNN), Self-Organized Maps (SOM), and Spiking Neural Networks (SNN), among others. From them, one interesting is the Morphological Neural Networks, also called Lattice Neural Networks with Dendritic Processing (LNNDP). The model was proposed by Gerhard X. Ritter and his collaborators (Ritter & Sussner, 1996). These neural networks use maxima or minima of sums, unlike previous neural networks, which base their operation on sums of products. With this new scheme and incorporating new ideas about dendritic neural networks, morphological neural networks have matched and sometimes surpassed the performance of backpropagation neural networks.

It is precisely through this powerful tool that a research project was proposed whose purpose was to develop an intelligent system based on neural networks to identify abnormalities in fundus images. The details are described in the following section.

18.3 Artificial Neural Networks for Retinal Abnormalities Detection

As mentioned above, the problem described in this chapter will be attacked using a supervised computational learning algorithm, which is ANNs.

Authors such as Vandarkuzhali and Ravichandran (2005) have argued that an efficient way to detect problems in the retina, and thus in the vascular system of individuals, is by analyzing the edges of the retina's blood vessels. Accurate segmentation of the vasculature is complicated for several reasons, including the presence of noise, the low contrast between the vessels and the image background, and the enormous variability in width, brightness, and shape in each individual. From a computational point of view, the task has usually been considered as one in which each image pixel must be classified as belonging or not to a vessel. In work reviewed here, the focus is to train a model that can categorize each pixel of the image, so that by segmenting the vasculature, it can be analyzed and determine whether there are abnormalities in the retina under study (Marín et al., 2010; Imran et al., 2019).

More specifically speaking, a MLP will be used. This kind of model requires data for which the correct classification is known. In other words, fundus images that have been labeled by a specialist are required. It should be noted that the project involved two stages. Retinal images from public databases were used in the first phase, and in the second phase, photos were provided by retinal specialists. For the illustrations in this chapter, only public images were used.

18.3.1 Data

Two public datasets were used in the first phase of the project. One of the datasets is called DRIVE, which stands for Digital Retinal Image for Vessel Extraction (Staal et al., 2004). It consists of 40 color images in JPEG format. The images were acquired using a Canon CR5 non-mydriatic 3CCD camera with a 45° field of view (FOV). The screening population consisted of 400 diabetic subjects between 25 and 90 years of age. The images were obtained from a diabetic retinopathy screening program in the Netherlands. This fact is important to be pointed out since, according to specialists, there are genetic differences that may affect the classification in the case of Latin patients. That is why it was necessary to add some images of local patients.

The second public dataset is STARE, an acronym for STructured Analysis of the Retina. It consists in about 400 raw images, which include a total of 44 different manifestations labeled by experts. It also contains 40 hand-labeled images for blood vessel segmentation. These are the ones that were used (Hoover et al., 2000).

A typical operation in all supervised learning models is the division of our dataset into two parts: One part will be the training set, which will correspond to the largest part of the dataset and will be used to train the model. The other part will be for testing, which is usually smaller in size and on which the trained model will be evaluated. This division can be done randomly or linearly, and the percentage is required for each subset is always chosen. Suppose the dataset does not have a temporal order, which will be more usual. In that case, it is convenient to perform the splitting randomly to ensure maximum data variability in the train/test sets. Typically, the dataset is split into 70% training and 30% test data, but the proportion can be varied case by case.

For classification tasks, a supervised learning algorithm examines the training dataset to determine, or learn, the optimal combinations of variables that will generate an excellent predictive model. Since the goal is to produce a trained (fitted) model that generalizes well to new and unknown data, the fitted model is evaluated using "new" examples to estimate the model's accuracy in classifying new data. It is therefore important to separate the two sets at the beginning.

The correct use of the two sets is important because by allowing for evaluating the model's accuracy, it also allows to understand if a more or less common problem in this type of learning technique, known as overfitting, has occurred.

Overfitting occurs when a machine learning system is trained too much or uses anomalous data, which causes the algorithm to "learn" patterns that are not general. It learns specific features but not the available patterns or the concept. Therefore, it is sometimes said that when a model is overfitted, it has actually "memorized" the training data rather than finding the general pattern that describes it. More complex models tend to overfit more than simpler models. In addition, for the same model, the smaller the data, the more likely it is that the model will overfit.

In some cases, the division into three sets is contemplated, the two already mentioned: train and test and a third one called "validation." This last set is used to adjust a classifier's hyperparameters (i.e., the architecture). An example of a hyperparameter for ANNs includes the number of hidden units in each layer. To avoid overfitting, when it is necessary to adjust some classification parameters, it is required to have a validation dataset in addition to the training and test datasets. For example, suppose the most suitable classifier for the problem is sought. In that case, the training dataset is used to train the different candidate classifiers, the validation dataset is used to compare their performances and decide which one to take, and finally, the test dataset is used to obtain the performance metrics, such as accuracy, sensitivity, and specificity, among others. The validation dataset works as a hybrid: It is training data used for testing, but not as part of the low-level training or as part of the final test. In this case, the original set was divided into two equally populated sets mainly due to the scarcity of available data (images). That is

20 images for the training set and 20 for the testing set. For the pictures of the training set, manual segmentation of the vasculature was performed by an ophthalmologist. In the case of the test cases, no annotations are available.

18.3.2 Preprocessing and Feature Extraction

It is widespread in computational learning applications to prepare the data to facilitate the rest of the process. In this case, it is even more critical because we have images, which eventually will have to be transformed into real numbers. Since it is not the purpose of this paper to go into all the technical details, it will only be briefly mentioned that to use these images, it was needed to perform some computer vision operations.

Computer vision encompasses an entire discipline of study within robotics and computer science, and artificial intelligence, whose advances have enabled the design of intelligent, dynamic, and versatile automated solutions that are leading to the resolution of increasingly ambitious objectives. Computer vision or computer vision refers to a group of technologies or tools that enable computers to capture images of the real world, process them, and generate information from them. In simple words, computer vision is a property of specific technologies that allow computerized equipment to "see." This has made it possible to design highly flexible computer systems capable of making intelligent decisions based on their environment at levels that are sometimes impossible for the human eye to match.

Concerning the problem at hand, various operations from computer vision were applied to prepare the images to be analyzed. The first is to decompose the image into three color channels: Red, Green, and Blue. Only the green channel was retained. After that, central light reflex removal, background homogenization, and vessel enhancement were applied in that order. Details of these operations can be found in Vega et al. (2015). Figure 18.5 shows three retina images, the one on the left is the original image, and the one in the middle is the green before the preprocessing. The one on the right shows the result of applying the three preprocessing operations.

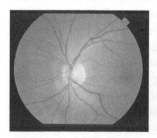 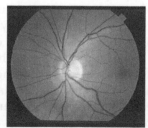 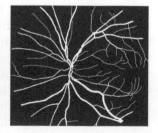

FIGURE 18.5
The image on the left shows one example of the initial image, the second image shows the green channel of the original image, and the image at the right shows the vascular structure after the preprocessing step is performed.

The goal of variable (attribute) selection is to create the subset with as few variables as possible that are necessary and sufficient to train the classifier. Within such variables, there may be some that are not relevant or are redundant for classification. Although common sense might indicate that the more variables added to the set of attributes, the better the results, this is not always the case. Sometimes adding irrelevant or redundant variables could negatively impact the classifier's performance. This is known as the curse of dimension. A relevant feature is neither irrelevant nor redundant; an irrelevant feature does not affect classifier training, and a redundant feature brings nothing new to training (Venkatesh & Anuradha, 2019). During this step, the variables that best represent the input patterns and explain the feedback variation must be selected. Considering that the retinal images give the original information, there must be a process to extract specific values from them to be used in the machine learning algorithm.

After analysis and some preliminary discussions, the research team decided to incorporate two types of features:

- **Gray-level features**: based on the gray-level intensity of the pixel of interest and a statistical analysis of its surroundings. There are five values calculated here.

- **Moment invariants-based features**: features based on moment invariants for small regions formed by a window centered at the pixel of interest. Two values are obtained in this case.

According to this, the final dataset will be composed of seven features calculated for each pixel in the 20 images. The next step in the process is to take that information and train a classifier.

18.3.3 Classification

Classification methods predict or explain the value of a class, that is, predict categorical values, for example, whether a person could have a disease. But classification methods are not limited to two possible values. They could help to evaluate, for instance, whether a given image contains a car or a truck, or neither. In classification, it is needed a master/supervisor. The data must be labeled with features so the machine can assign classes based on them. The algorithm learns from the given dataset and classifies the observations into various groups. The output variable of a classification algorithm is a category.

There are many classification algorithms, and in previous works, several of them were compared (Vega et al., 2015). Thus, in this report, the authors concentrate on only one standard classifier to illustrate the process and to show some results.

The dataset was generated as described in the former section for the application described in this chapter. Once the features were defined and computed, the next step was to take a reference image and extract 30,000 random pixels (half hand-labeled as blood vessels and half as non-blood vessels) to train several neural networks.

In the experiments performed by the authors, the better-saved neural network was a lattice neural network with dendritic processing (LNNDP), with an accuracy of 99.02% in the training phase. To achieve such results, 4,844 dendrites were in the neuron that classifies the pixels as blood vessels, and 4,786 dendrites in the neuron that classifies them as non-blood vessels. The weights obtained in the training phase were used to make the segmentation of the 20 images on the dataset. This LNNDP was compared with a MLP with 9 dendrites in the hidden layer (MLP-9).

Once an algorithm has been trained, it is always necessary to evaluate it to know if it has generalized correctly or if the famous overtraining has occurred.

There are several evaluation metrics, each of which has advantages over others in certain circumstances. However, when it comes to classification problems such as the one we are dealing with here, we usually start by calculating what is known as the confusion matrix. For this purpose, the data obtained by running the model with the test data are used, and four values are collected:

- **TP (True Positive)**: These are the values that the algorithm classifies as positive and are actually positive.
- **TN (True Negative)**: Values that the algorithm classifies as negative (0 in this case) and that are actually negative.
- **FP (False Positive)**: False positives, that is, values that the algorithm classifies as positive when they are actually negative.
- **FN (False Negative)**: False negatives, that is, values that the algorithm classifies as negative when they are actually positive.

From these values, several metrics can be calculated. For example, the precision, which is used to be able to know what percentage of values that have been classified as positive are really positive.

$$Precision = TP/(TP + FP)$$

Or the recall metric, also known as the ratio of true positives, is used to know how many positive values are correctly classified.

$$Recall = TP/(TP + FN)$$

Having these two values, it is possible to calculate another metric called F_1 Score. This metric is widely used in problems where the dataset to be

TABLE 18.1

Results Obtained with Two Models

	MLP-9	LNNDP
Average F_1 Score	0.5587	**0.6616**
Min F_1 Score	0.2579	0.4232
Max F_1 Score	0.7351	0.8123

A multilayer perceptron (MLP-9) and a Lattice neural network with dendritic processing (LNNDP).

analyzed is unbalanced. This metric combines precision and recall to obtain a much more objective value.

$$F_1 = 2 * \left(\left(\text{recall} * \text{precision} \right) / \left(\text{recall} + \text{precision} \right) \right)$$

Table 18.1 shows the values of the metrics obtained over the test set. It was decided to use metric F_1 Score because, in this case it gives a more realistic evaluation by calculating the relationship between true positives, false negatives, and false positives. It is important to notice that in the images, about 90%–95% of pixels are not blood vessels, so using metrics like accuracy may give the false impression of high performance even if it is not the case. Using F_1 Score, it will happen that if a model classifies all pixels as non-blood vessels, it will get a score of 0, while if it correctly classifies the pixels, it will get a score of 1.

These results say that the neural network trained in this experiment was able to get it right in almost 70% of the cases. This is not a bad result, as it probably exceeds that of a doctor without proper training, but it is also probably not a model that can be taken into medical practice yet. More experiments are needed in which more images labeled by experts are added so the system can generalize better. Also, perhaps it would be good to add other supervised learning models to the comparison, or better yet, to work with semi-supervised learning models so as not to rely so much on human-labeled images.

18.4 The Future

One of the applications of artificial intelligence that has generated the most interest among scientists and the public is medicine. Aspects range from early and efficient diagnosis of specific ailments, personalized treatment to minimize side effects, and the design of new drugs with better properties. Many of these tasks are now possible due to the incursion of very various artificial intelligence tools, which is not only automated learning, as is sometimes thought.

Thus, for example, in processing images to improve their properties, classification, and labeling, computational learning, whether traditional or the latest deep neural networks, has been of great support. Many projects have been developed over the years, helping in tasks such as identifying breast cancer, lung problems due to COVID-19, and brain problems, among others. Tools such as hidden Markov chains, natural language processing, metaheuristic planning, and scheduling algorithms have also found application in problems related to health sciences.

The interest in applying these and other artificial intelligence tools will continue to grow. The scenario is a population getting older but with a longer life expectancy. Therefore, medicine and its associated services will undoubtedly continue to have a preponderance. Physicians will have to continue to rely on computers, and these will have to expand and improve their intervention.

However, some aspects are very relevant, and their resolution becomes more pressing as these applications grow. Such is the case of explainability. Many models based on artificial intelligence behave like black boxes, in which information enters and decisions result. In some areas, this is not a concern, but the perspective changes when it comes to someone's health or life. Few people would be happy to know that it will recommend medical treatment after giving data to a computer. When asked for justification for that recommendation, the system will provide us with a matrix of real numbers. Surely, we would want a detailed explanation of what deductive processes the doctor went through to arrive at that decision. An issue that today, with specific IA tools, is not possible. On the other hand, some of the most straightforward tools for the common public to understand are, to some extent, simplistic codifications of processes that could be very complex and in which there may be many more variables to consider than those that are explicitly part of the model, such as a simple decision tree.

Thus, the authors of this chapter consider that AI will have more applications in medicine, but this will also trigger more research in the discipline itself to propose new models with more solid and traceable mathematical foundations. Undoubtedly, we expect significant advances in this direction in the coming years.

References

Benítez, V. E. C., Matto, I. C., Román, J. C. M., Noguera, J. L. V., García-Torres, M., Ayala, J., … & Grillo, S. A. (2021). Dataset from fundus images for the study of diabetic retinopathy. *Data in Brief, 36,* 107068.

Haykin, S., & Network, N. (2004). A comprehensive foundation. *Neural Networks,* 2(2004), 41.

Hoover, A. D., Kouznetsova, V., & Goldbaum, M. (2000). Locating blood vessels in retinal images by piecewise threshold probing of a matched filter response. *IEEE Transactions on Medical Imaging, 19*(3), 203–210.

Imran, A., Li, J., Pei, Y., Yang, J. J., & Wang, Q. (2019). Comparative analysis of vessel segmentation techniques in retinal images. *IEEE Access, 7,* 114862–114887.

Instituto Nacional de Estadística y Geografía. (2020). Características de las defunciones registradas en México durante 2020. https://www.inegi.org.mx/contenidos/saladeprensa/boletines/2021/EstSociodemo/DefuncionesRegistradas2020_Pre_07.pdf.

Marín, D., Aquino, A., Gegúndez-Arias, M. E., & Bravo, J. M. (2010). A new supervised method for blood vessel segmentation in retinal images by using gray-level and moment invariants-based features. *IEEE Transactions on Medical Imaging, 30*(1), 146–158.

McCulloch, J. L., & Pitts, W. (1943). A logical calculus of ideas immanent in nervous activity. *Bulletin of Mathematical Biophysics, 5,* 115–133.

Minsky, M., & Papert, S. (1969). An introduction to computational geometry. *Cambridge tiass., HIT, 479,* 480.

Ritter, G. X., & Sussner, P. (1996). An introduction to morphological neural networks. In *Proceedings of 13th International Conference on Pattern Recognition,* Vienna, Austria (Vol. 4, pp. 709–717). IEEE.

Rosenblatt, F. (1958). The perceptron: a probabilistic model for information storage and organization in the brain. *Psychological Review, 65*(6), 386

Staal, J., Abràmoff, M. D., Niemeijer, M., Viergever, M. A., & Van Ginneken, B. (2004). Ridge-based vessel segmentation in color images of the retina. *IEEE Transactions on Medical Imaging, 23*(4), 501–509.

Vandarkuzhali, D. C. S., & Ravichandran, T. (2005). Elm based detection of abnormality in retinal image of eye due to diabetic retinopathy. *Journal of Theoretical and Applied Information Technology, 6,* 423–428.

Vega, R., Sanchez-Ante, G., Falcon-Morales, L. E., Sossa, H., & Guevara, E. (2015). Retinal vessel extraction using lattice neural networks with dendritic processing. *Computers in Biology and Medicine, 58,* 20–30.

Venkatesh, B., & Anuradha, J. (2019). A review of feature selection and its methods. *Cybernetics and Information Technologies, 19*(1), 3–26.

19

Artificial Intelligence for Mental Health: A Review of AI Solutions and Their Future

Gerardo Castañeda-Garza, Héctor Gibrán Ceballos, and
Paola Gabriela Mejía-Almada
Tecnológico de Monterrey

CONTENTS

19.1 Introduction

While artificial intelligence (AI) may have reached a broader part of the population through the rules of robotics in the book of *I, Robot* (Asimov, 1950) or found a way to cover every emotional need in a relationship, as portrayed in the film *Her* (Jonze, 2013), the progress of AI in real life has not advanced as fast as expected. The term *"artificial intelligence"* is recognized to be used for the first time in 1955 by Prof. John McCarty, considered one of the founders of the field (McCarthy et al., 2006; Myers, 2011). Since its origin, one of the ultimate goals behind AI has been to *"make computer pro-grams that can solve problems and achieve goals in the world as well as humans"* (McCarthy, 2007). To pursuit this goal, the development of AI has needed to overcome multiple constraints, such as the cost of computing equipment, infrastructure (Anyoha, 2017), long-term funding, and energy consumption (Martin, 1995).

With the passage of time, many AI-like projects trying to resemble humans have appeared, such as Eliza, one of the first chatbots to which humans opened their hearts (The Tech, 2008); the speech capable of Japanese anthro-pomorphic robot, WABOT-1 (Waseda University, n.d.); even reaching to building of humanoid robots, such as Sophia (Hanson Robotics, 2022a), capa-ble of communicating, learning in different contexts, as well as joking about our future demise as humans (Fallon, 2017). However simple these actions may seem, achieving human-like communication is not an easy task, being that our communication process includes complex aspects, yet so familiar. For instance, just nonverbal communication can be broken down in dif-ferent broad categories, such as kinesics, haptics, vocalics, proxemics, and chronemics, each one of them filled with other characteristics (University of Minnesota, 2013); on the other hand, there is verbal communication, which inclusion requires another kind of specific approaches to help machine-learning models to learn on their own (Yao, 2017). Therefore, to resemble human communication in the best possible manner, different subsets of AI, such as machine learning (ML), natural language processing (NLP), and computer vision, have been developed along the years to process inputs (like voice, text, images, or videos) in interpretable data for computational models. Nevertheless, communication alone is not sufficient for a computer to com-pletely interact with the physical word, neither being emotionally accepted by humans in their daily-life interactions.

To fulfill the ambition of achieving AI, it is said that a machine should sur-pass the Turing test – being able to confuse a person to think that it is inter-acting with a human, instead of a machine (Oppy & Dowe, 2021). Thus, for AI in mental health, this would mean the dream of achieving the potential perception of receiving psychotherapy from AI with a similar quality as to the one provided by a human. Still, this dream would lead to other questions before occurring: *should AI or a robot be our therapist?*

19.2 Mental Health Overview

There are concerning reasons to evaluate the possibilities of introducing AI solutions for the treatment of mental conditions. For starters, among the lead-ing causes of global causes of burden and disability worldwide, mental dis-orders, such as depressive and anxiety disorders, figured as the main causes for the highest disability-adjusted life years (DALYs)[1] in 2019, accounting for 125.3 million DALYs (Ferrari et al., 2022). Across age groups, their rates affect people in working age, starting from 15 years old until late age, with a greater effect in females over men (Vos et al., 2020; Ferrari et al., 2022, p. 146). Around the world, it is estimated that 5% of the population (approximately 280 million people) are affected by depression (World Health Organization (WHO), 2021), with a majority of them being unaware of it for long periods of time, unable to receive treatment or effective care due to the lack of resources, such as available trained healthcare providers, financial barriers, or others such as social stigma. Unaware of the consequences, mental health disorders may continuously impact their quality of life in different areas of their lives. Now, with global events as the surge of the COVID-19 pandemic, the worldwide prevalence of only depression has increased among the population (Santomauro et al., 2021).

One of the reasons that impede to reach out for help is the accessibility to mental health professionals. The geographical distribution of licensed professionals can be uneven between rural and urban areas for different regions (Salinas et al., 2010; Chen et al., 2019, Morales et al., 2020; Zhang et al., 2021), limiting the information, education, and care that people could receive. In some cases, the availability of health professionals per capita may differ significatively across multiple countries. As an example, countries like Mexico reported 3.45 psychologists per 100,000 population, while others like Argentina and USA had 222.57 and 29.86, respectively (WHO, 2019). In such cases, remote therapy may offer an alternative to provide treatment to distant communities, but it still would require access to technology that covers minimum requirements, such as a smartphone, fast enough internet, private spaces (Watson et al., 2021), but mostly, licensed personnel available to provide it at different schedules. Scenarios like these have promoted the development of tools based on AI that may facilitate the work of mental health professionals through activities such as initial screenings (i.e., applying questionnaires for mental health), self-monitoring actions for wellness (i.e., guided activities based on cognitive-behavioral therapy (CBT)), organizational monitoring (services provided by companies), and in some cases, being able to provide guidance or referral to patients with acute symptoms, such as self-harming behaviors or suicidal thoughts.

By exploring these new possibilities, we may discover how AI, able to work continuously for longer periods compared to humans, accessible in any moment, could participate in our pursuit to deliver better healthcare services – or even more, able to perform them independently. To grasp the probabilities of these outcomes, in this chapter we discuss about the potential of AI for

---

[1] One DALY represents the loss of the equivalent of one year of full health. DALYs for a disease or health condition are the sum of the years of life lost due to premature mortality (YLLs) and the years lived with a disability (YLDs) due to prevalent cases of the disease or health condition in a population (WHO, 2020, p. 6).

mental health in different sections. For that purpose, this chapter has been divided into the following four sections:

- In the first section, we offer an introduction of CBT, an approach of psychotherapy that is commonly used for AI applications due to scientific approach and proved record in treating mental health disorders such as depression, anxiety, and many others.
- The second section describes examples how AI is providing mental healthcare, including information about the ML models, natural language algorithms, computer vision, and robotics to serve in the treatment of mental health disorders.
- For the third section, examples of available applications, products, and proof of concept of AI will be provided, such as chatbots, robot therapist, and multimodal perception virtual therapist.
- The fourth section includes a reflection concerned to the ethical aspects related to the use of AI for mental health, given that these technologies may provide an utmost value to people, but also carry the potential to severely harm if not properly tested and regulated.

At the end of this chapter, as a conclusion we offer a reflection related to the present state of AI for mental health and its future, in their possible utopic or dystopic outcomes, thinking about future scenarios and issues of AI, the human condition and dignity, and the following step if AI (as we may imagine it right now) is ever reached by any of our societies.

19.3 Part 1. Cognitive Behavioral Therapy (CBT)

The understanding of psychology as the scientific study of the mind and its processes is nearly as recent as AI, with its first experiments designed by Wilhelm Wundt nearly 100 years ago, in the last decade of the nineteenth century (The Editors of Encyclopaedia Britannica, 2022). Since then, multiple theories and therapeutical approaches have been elaborated with the purpose of explaining the underlying mechanisms of the mind, and furthermore, there have been approaches to treat cognitive, behavioral, or affective pathologies in a person. Therefore, while there is consensus in contemporaneous trends in psychology (such as psychoanalysis, cognitive behavioral psychology, and humanistic psychology, among others), one approach, cognitive behavioral psychology, stands out as it has been used consistently in recent years.

The appearance of CBT as a new approach in psychotherapy would first be recognized until 1977, after major clinical trials found that patients receiving CBT had a higher efficiency in the treatment of depression in comparison with a medical treatment (Rush et al., 1977; Blackburn et al., 1981). Ever since, CBT has gained recognition in the treatment of a range of problems, including depression, anxiety, drug use problems, and eating disorders. Nowadays, CBT is one of a few psychotherapeutic approaches recommended by the American Psychological Association for its scientific evidence, made both in research and in clinical practice (APA, 2022).

The goal in CBT is to help patients to develop coping skills that let them understand better and change their own affective, cognitive, and behavioral patterns (APA, 2017). To do so, a therapist could employ exercises to reevaluate reality, learning problem-solving skills, and other strategies. The purpose is to guide the patient through a process of (i) obtaining awareness of automatic thoughts, feelings, and core beliefs; (ii) performing an examination of these in the practice; and (iii) creating alternative thoughts that may allow the acquisition of new thought, behavior, or emotional patterns (Crane & Watters, 2019). To achieve this task, treatment with CBT commonly includes activities from which data generation is easily obtained and useful to provide feedback and accomplish goals. A few examples of these are monitoring logs for thoughts and behaviors, relaxation training, thought stopping, and activity monitoring. An extensive description of treatment strategies in CBT is presented in Table 19.1.

When using AI, programmers can take advantage of obtaining helpful data to understand the mental health state of the user. Therefore, from mental disorders screenings (i.e., clinical questionnaires) to complex algorithms (i.e., combinations of user smartphone behavioral activity with feedback), the use of ML models can promote the analysis of multiple factors with the purpose of attaining one or a few goals in mental health, since this continues to be a trend in therapy through AI. As an example of this, we may find

TABLE 19.1

Examples of Treatment Strategies in CBT

Behavioral Strategies	Cognitive Strategies
Activity monitoring and log	Recognizing mood shifts
Values identification	Combining thoughts and emotions
Mood log	Thought stopping
Pleasant activities	Downward arrow technique
Activity scheduling	Use of thought change records
Increasing pleasure and achievement	Identifying cognitive errors
Behavioral activation	Socratic questioning
Graded task assignments	Challenging questions
Relaxation training	Generating rational alternatives
Scheduled worry time	Life history
Problem solving	Modifying core beliefs
	Guilt vs Shame

Source: Crane & Watters (2019; p. 23).

smartphone applications focused on different activities and goals, such as journaling, meditation, managing anxiety, and conversation agents, among others (Wasil et al., 2022).

Once we understand some of the fundamentals behind CBT, and how this can be enhanced by AI, it is easier to understand how ML models could take advantage of proved psychological techniques for mass, accessible, and continuous replication.

19.4 Part 2. How Does Artificial Intelligence Provide Mental Health Support?

The mechanisms by which AI is able to provide support in mental health differ according to different attributes. Depending on the task, AI could take advantage of local processing through a costly hardware, or enable greater capabilities through a cost-effective solution, such as cloud computing. On another instance, AI characteristics are not the only to be considered, understanding that the human experience – by many of their cultural, age, and social differences – is important as well in the adoption or rejection of a new technology.

In this section of this chapter, it will be discussed how AI provides mental health – according to recent published papers on the matter. For this purpose, we have categorized AI in three areas, corresponding to the increasing level of complexity that each one of these requires:

1. **Natural Language Processing**: Focuses on communication processes in written or speech forms, through the use of different natural language algorithms. Considers solutions like conversational agents (i.e., chatbots).

2. **Virtual**: Compared to natural language algorithms, virtual AI adds another multimodal perception capabilities, such as visual recognition (i.e., machine vision), and allows AIs to communicate through virtual layers (i.e., virtual avatars).

3. **Robotics**: In contrast with virtual, AI in robotics adds a physical layer of action with humans, with enables further interaction, nonverbal communication, such as other patterns in robotic solutions (i.e., humanoid robots, commercial robots).

19.4.1 Mental Health Support through Natural Language Processing

With the rise of social media, the amount and variety of information available to NLP researchers has transformed (Hirschberg & Manning, 2015). Shaped by ML and NLP, le Glaz et al. (2021) affirm that NLP can be used with different data sources, going from static sources (such as electronic health records [EHRs], psychological evaluation reports, and transcribed interviews) to dynamic data (as an ongoing conversation with a person through a conversational agent). This information can be gathered and then analyzed with the support of medical dictionaries with the purpose of detecting specific terms, as those related to suicide (p. 3).

According to a literature review performed by Zhang et al. (2022), research about NLP for mental health has increased over the last 10 years with a major interest for the study of depression (p. 46). The extent of this research goes through multiple themes and disorders. In prevention, on the one hand, research has found that text analysis and NLP can be an effective tool to identify shifts to suicidal ideation (the act of thinking about taking your own life) (de Choudhury et al., 2016; Hassan et al., 2020), as well to discover valuable patterns, such as identifying people at-risk in social media, or underlying manifestations of symptoms expressed through language (Low et al., 2020). On the other hand, reaction settings, NLP sentiment analysis has proved useful to perform analysis in millions of messages posted in social media, making them useful to analyze great amounts of messages in real time, with the purpose of grasping the emotions due to traumatic events, as it occurred during the COVID-19 pandemic workplace and school reopening (Q. Chen et al., 2021). Properly used, these insights can offer potential information to authorities in order to design strategies for mental health programs, to future health policy.

Addressing more disorder-specific aspects of mental health, NLP has been used to predict episodes of violence by using EHRs, risk assessment scales, and NLP dictionaries (van Le et al., 2018), to build a suicide prevention system through text and voice analysis with a chatbot (Kulasinghe et al., 2019), early-depression detection systems with BERT transformers (El-Ramly et al., 2021), and late-life depression (LLD) by using speech analysis (DeSouza et al., 2021). Nonetheless, NLP applications have also been created to address other disorders, such as diagnosis of achluophobia and autism spectrum disorder (ASD) by using decisions trees (Mujeeb et al., 2017), insomnia treatment (Shaikh & Mhetre, 2022), and detection of anorexia in social media (Hacohen-Kerner et al., 2022) among many other disorders (Abd-alrazaq et al., 2019).

19.4.2 Mental Health Support through Computer Vision

Going beyond NLP algorithms, there are other channels from which machines can obtain multimodal data, such as visual or biometrical. In this sense, computer vision enables computers to derive meaningful information from visual inputs (such as images, videos, and others) and take action or make recommendations based on the analyzed data (IBM, 2022). By using this technology, computer vision permits conducting a variety of useful tasks for mental health, such as recognition of facial gestures, emotion recognition and prediction, eye tracking, and movement patterns, among others (Sapiro et al., 2019). Consequently, it is discussed that computer vision may enable the creation of low-cost (Hashemi et al., 2014), mobile health methods to assess disorders such as ASD (Sapiro et al., 2019; p. 15), early detection of depressions signs in students (Namboodiri & Venkataraman, 2019), as well as the detection of stress and anxiety from videos (Giannakakis et al., 2017).

Evidence of the value of reading facial features is continuously recognized in AI as a way to identify proof of potential emotional states in an individual. To perform these tasks, AI relies on the use of multiple algorithms, such as LBP, Viola-Jones algorithm, and support vector machine (SVM), for data classification in the research of Namboodiri and Venkataraman (2019). Additionally, the use of computer vision algorithms has been recognized as useful for the differential analysis of obsessive-compulsive disorder (OCD) symptoms, between subtypes of OCD such as compulsive cleaning and compulsive checking (Zor et al., 2011).

In another example of detection of mental health disorders, computer vision has demonstrated its benefits when identifying attention deficit hyperactivity disorder (ADHD) by using an extension of dynamic time warping (DT) to recognize behavioral patterns in children (Bautista et al., 2016). Similarly, dynamic deep learning and 3D analysis of behavior have been employed to diagnose ADHD and ASD, with classification rates higher than 90% for both disorders (Jaiswal et al., 2017). Likewise, Zhang et al. (2020) designed a system to perform functional test tasks through inputs of multimodal data, including facial expressions, eye and limb movements, language expressions, and reaction abilities in children. The data is then analyzed by using deep learning to detect specific behaviors, which can serve as a diagnosis complementary to the one of health professionals. Even more, whereas these cases have been addressed for human patients, it is suggested that these computational methods could also be used to detect cases of ADHD like behavior in dogs (Bleuer-Elsner et al., 2019), opening the umbrella for further approaches for well-being, not only in humans, but also in other species.

19.4.3 Mental Health Support through AI Robotics

As the advances in electronic technologies continue to occur, the boundaries between the digital and physical world continue to blur. Human beings used to a physical reality are now aiming to create new digital worlds in what we call "the metaverse, while we welcome computers to go further than just sensing our world – but also, moving and interacting with it." With the power of robotics, virtualized applications can go further than communicating with NLP, or identifying visual cues, and allowing them to participate in physical activities with people with the possibility of gaining an embodiment.

During the last few decades, advancements in AI with robotics grew rapidly in multiple areas, from household appliances to medicine and astronautical applications (Andreu-Perez et al., 2017). For the case of AI for mental well-being, researchers have published evidence of how robotics may give some advantages over the other methods. As a taxonomic description, Feil-Seifer and Mataric (2005) suggest that socially assistive robots can help users in multiple populations, such as (i) elderly, (ii) individuals with physical impairments, (iii) in convalescent care, (iv) with cognitive disorders, and (v) students in special education (p. 466). For instance, Fasola and Mataric (2013) found that robots can be used to promote physical exercise on elderly users who have a preference of physical embodied robot coach over virtual solutions; however, Kabacińska et al. (2021) and Okita (2013) observed that interventions using different types of robots with children were able to reduce anger and depression levels, distress, as well as pain in some cases. Similarly, other studies have discovered that robotic dogs can be able to reduce loneliness in a comparable fashion to a living dog that could do (Banks et al., 2008). As a fact, a decrease in feelings of loneliness would be beneficial in a variety of settings, being that it may affect greatly groups such as women, people living alone, or people without a partner. Moreover, its reduction could reduce risks associated with depression, anxiety, and suicidal ideation, among others (Beutel et al., 2017).

For researchers to continue understanding how these relationships between humans and robots may be favorable, Riek (2016) suggests that human-robot interaction (HRI) requires further work as an emerging area, as it is noted that the looks of robots (i.e., their morphology), their autonomy, and capabilities can differ greatly, between them. This is meaningful according to Šabanović et al. (2015), who suggest that design of robots might be challenging to use for older adults or people suffering from depression, even when there is willingness to participate or adopt these new technologies.

Now that some examples of how mental health if being supported by different technologies associated with AI has been presented, a description of some current applications of AI (such as mental health coaches (chatbots), virtual therapists, and AI robots) will be presented in the following section as examples of the point in which these technologies have been developed.

19.5 Part 3. Current Applications of AI

Over the last decade, many AI applications have been developed, with some of them focused on providing support for mental health issues. Virtual therapists, chatbots, and robots have slowly gained attention as alternatives for the treatment of specific mental conditions. In this part of this chapter, we will describe some examples of the applications developed over the last 10 years, how these are providing support for the care, and moreover, how this assistance could change to reach and ensure availability to more people over the next decade.

19.5.1 Chatbots

Considered as one of the technological solutions to mitigate the lack of mental health workforce are chatbots (Abd-alrazaq et al., 2019). Designed to communicate with humans, chatbots have surged during the last decade as an option to provide free, accessible, and immediate mental health support. To access them, several options are offered as smartphone applications, making them one of the most accessible options compared to other equipment (i.e., VR headsets, computers, robots). Supplying a therapist-like support, available 24/7 – or as long as there is internet – these intelligent products allow their users to express by using a chat interface. Through multiple interactions, chatbots may perform data analysis to provide a better sense of comprehension to their users, learning and invoking information from previous conversations (Woebot Health, 2022), with some cases demonstrating potential evidence of capabilities to establish a therapeutic bond with their users (Darcy et al., 2021).

Before continuing the discussion around chatbots, it is important to remark that due to their lower difficulty to be built, there is an increasing quantity of digital well-being apps that have being produced for major smartphone app stores over the last decade. In this sense, Martínez-Pérez et al. (2013) found that more than 1,500 commercial apps are targeted as "useful" for the treatment or reduction of symptoms of depression, with most of them lacking any notice of evidence in clinical trials to prove their effectiveness. Even more, many apps may not consider anonymity and privacy in their design, becoming a potential risk in safeguarding the data of their users. For this reason, this section will provide information in mental health chatbots that have undergone a process of academic research experiment (such as clinical trials) to prove their efficacy. To illustrate this task, we chose one chatbot, which has been previously studied by researchers over the last decade: Wysa (Tewari et al., 2021; Kulkarni, 2022).

19.5.1.1 Wysa – Mental Health Chatbot

Represented by a bluish penguin, Wysa is an AI-enabled mental health app that leverages evidence-based CBT techniques through its conversational agent (Malik et al., 2022). Requiring a smartphone and an internet connection, the app offers their users free features to interact with their chatbot, to reflect in their emotions, thoughts, and experiences. Consequently, these interactions feed a journal with conversations, which help the chatbot to improve, as well as their users to reflect on their experiences in life. The app includes premium features, such as guided self-care activities for different topics (i.e., overcoming grief, trauma, confidence, self-esteem) with the possibility of receiving remote therapy sessions through the app.

In comparison with human psychological interventions, chatbots require to surpass and maintain a certain level of quality to facilitate the effectiveness of therapeutic exercises and other activities. Kim et al. (2019) in a literature review identify three common themes in research paper about chatbots: (i) therapeutic alliance, which refers to the collaboration between the patient and therapist to achieve treatment goals; (ii) trust; and (iii) human intervention. As for this section, Beatty et al., (2022) found that Wysa, as a chatbot, was able to establish a therapeutic alliance with their users in a variety of treatment scenarios, including chronic pain management (Meheli et al., 2022; Sinha et al., 2022) and depression (Inkster et al., 2018). Nonetheless, it is important to note that the representativeness of the data needs to be judged carefully, due to the involvement of representatives of Wysa involved in some popular studies about the application. Despite this, Wysa still remains as one of the few mental health applications receiving multiple organization acknowledgements, such as one of the best apps for managing anxiety during COVID-19 (ORCHA, 2020), being compliant to Clinical Safety Standards in UK (Wysa, 2022b), among others.

As a supplemental option for mental healthcare, Wysa stands around as one of the accessible alternatives across chatbots and could continue being the first step for many people who have never experienced a therapeutical approach before. While chatbots are not able yet to listen carefully to us, and help us to reorganize or thoughts, or process our feelings, the data that they keep obtaining will probably help in their improvement. As long as a person has a smartphone – with an estimated of 6.64 billion users around the world (Turner, 2022) – chatbots could continue to be one of the most accessible and affordable forms to provide care for the preservation and recovery of human well-being.

19.5.2 Virtual AI Therapists

While a human therapist can be physically only in one place, virtual AI therapists (or virtual human therapists[2]) could be awake when we need them most, in any place with a screen, in anytime – with the added value of having a familiar look, resembling a human face through the screen. Swartout et al. (2013) define virtual humans as computer-generated characters designed to look and behave like real people and designed to be engaging. On these aspects, recent studies have found that trusting an AI may improve the user's commitment and usage of technology assistants, which can improve performance of intelligent assistants (Song et al., 2022). In a similar trend, virtual (AI) humans have been found to be a support in times of distress and increase the willingness of some patients to disclose (Lucas et al., 2014; Pauw et al., 2022).

---

[2] Although the term in the case of Ellie is "virtual human therapist," this construct may lead to confusion associated with its approach (e.g., a virtual human therapist could be a person providing remote therapy, and not an artificial intelligence). Due to this reason, the term "virtual AI therapist" may be used interchangeably in this chapter.

A popular example of virtual AI in the contemporary era is associated with "Ellie," a virtual human therapist presented by the University of Southern California in 2013, with the objective of giving support for people suffering mental disorders as post-traumatic stress disorder (PTSD) or depression (Brigida, 2013). Being able to have some perception of the physical world and react accordingly, Ellie performs its functions powered by two technological systems: (i) Multisense, a program that facilitates real-time tracking and analysis of facial expressions, body posture and movement, sound characteristics, linguistic patterns, and high-level behavioral descriptors (e.g., attention, agitation); and (ii) SimSensei, a virtual human model platform able to detect audiovisual signals in real time captured by Multisense (USCICT, 2014). Working together, both technologies have shown how the virtual AI may provide feedback and encouragement to a patient through comments, gesture reactions, and to be able to perform strategies that motivate participants to continue the conversation (Rizzo & Morency, 2013). As a result, Ellie aims to establish a similar level of human empathy, confidence, or *rapport*, as a therapist would do.

Though the possibility of having a therapeutic conversation with a virtual AI may seem convenient, neither the affordability nor accessibility issues were found widely discussed about the resources needed to provide mental health care through these intelligent applications. Consequently, it is not possible to affirm or predict if virtual AI therapists may represent a convenient and effective option in the future to receive mental care.

Meanwhile, in the near future, it is expected that with new advancements in algorithms, infrastructure, and equipment, virtual AI therapists could be better at tracking and analyzing data, with the possibility of including other channels for data gathering, for example, breath, heart frequency, or data from other sources (Lazarus, 2006). Also, it is expected that the aesthetics may change in the short term (Epic Games Inc, 2022; Estela, 2022), opening of a new research area to understand what effect different digital avatars may have in their users.

19.5.3 Robot AI Therapists

Probably, robots are still one of the most expected technologies that people have wished for a long time. Imagining them has accompanied us with different stories, from robots that turn companions through life, as in the *Bicentennial Man* (IMDB, 2022), to others as *HAL* (SHOCHIKU, 2013) created with the specific purpose of helping people to overcome grief and provide closure to traumatic events. Examples like these are just a few of many that explore how relationships between humans and robots could work in a future we have not reached yet. However, the expectation might continue for a little longer, due to personal robots and emotion AI being expected to take at least 10 years to offer practical benefits (Klappich & Jump, 2022). Nevertheless, recent research in the field of AI in robotics allows us now to see how robots could be used to improve, maintain, and recover mental health.

19.5.3.1 Commercial Therapy Robots

In this section, we define "commercial therapy robots" as those that are more affordable to consumers and are more related to the ideal of "personal robots," compared to others that may be acquired only by organizations such as research institutes, universities, companies, or governments. Along with chatbots, commercial therapy robots continue to be tested by people around the world, in a broad market (personal service robotics) expected to rapidly grow at a compound annual growth rate of 38.5% by 2030 (Globe Newswire, 2022). Therefore, it is relevant to understand how these robots are integrated in people's lives and comprehend which needs they help to solve through AI – from being the new best friend of the century, to human size assistants.

19.5.3.1.1 Japanese Robotic Companions

In an era in which many people fear the potential consequences of AI and autonomous robots in society (Liang & Lee, 2017), Japan embraces and adopts these technologies more as partners than work and life competitors. In fact, there is an expected shortage of elderly care givers, estimated in 370,000 nurses and other care professionals by 2025, that, along with other factors, continues to promote the development of "carebots" (Plackett, 2022). Moreover, Japan tops in the world rankings for the biggest proportion of people over 65 years, with 29.1% representing 36.27 million people (Kyodo News, 2022). Because of these differences, the case of Japanese robotic companions deserves a bit of attention, due to their positive assessment of robot participation in certain tasks for elderly care (Coco et al., 2018) while observing ethical limitations in naturalistic contexts (A. Gallagher et al., 2016).

Mentioned this, it should be easier to realize the motivations behind the advances and the role of robotic companions in Japan, and how these may provide insights for researchers around the world.

19.5.3.1.2 Aibo: A Robotic Companion

While the ownership or coexistence with a pet (as a dog) has its benefits for mental health, such as promoting physical activity, enhancing mood and psychological health (Knight & Edwards, 2008), it is highly possible that in the near future, we may see people interacting with a new breed of robotic pets, such as Aibo. Created by Sony Corporation, Aibo (**a**rtificial **i**ntelligence ro**bo**t) is robotic puppy made in Japan, able to interact with people, developing its own identity over time. Originally appearing in 1999, the robot reappeared in 2018, selling 11,111 puppies in the first three months after its launch (Kyodo, 2018) announcing a potential interest from the country population.

Since its beginning, the robot pet has attracted the interest, not only from children and a new kind of pet owners, but also from researchers who found that the robotic puppy could help in the improvement or treatment of mental health disorders. In one example of this, Stanton et al. (2008) found that children with ASD that used Aibo instead of other mechanical toy dogs exhibited behaviors commonly found in children without autism more frequently, while reducing autistic behaviors as well. Other studies explored how Aibo could reduce anxiety and pain in hospitalized children (Tanaka et al., 2022); as in other cases, its purpose has been focused on reducing loneliness and increasing social interaction in patients with dementia (Banks et al., 2008; Kramer et al., 2009).

Deciding whether a robotic dog or a live Australian Shepherd is a best option is a matter of debate and of research. Compared to a living dog, a robotic pet has less care requirements (such as no need for food, water, cleaning, etc.), attributes that could be convenient for some people (such as children, or people suffering memory loss) to whom taking care of a living being would prove a difficult endeavor, or instead, for those who are stimulated and engaged by certain pet behaviors, such as playing with a ball (Ribi et al., 2008). Thus, the acceptance of a robotic pet as a similar (or equal) to a living dog has been found high in children (Melson et al., 2005), noting the possibility that Aibo is capable of being present in controlled environment settings (such as hospitals), assisting in the improvement of the quality of life of pediatric inpatients (Kimura et al., 2004; Tanaka et al., 2022). In spite of this information, it is possible that the robotic puppy is not affordable for everyone, since its offering price remains still around $3,000 USD (Craft, 2022).

19.5.3.1.3 Paro: A Robotic Companion

Presented in the form of a small seal with huge eyes, Paro is an advanced interactive robot that includes different kinds of sensors, such as tactile, light, audition, temperature, and posture (PARO Robots, 2014). In research, the robotic white seal has been used as therapeutic tool in different settings. In elderly populations, Inoue et al., (2021) explored how Paro could support the care of people with dementia in a home context, in group therapy settings (Chang et al., 2013). Similarly, Paro seal has been studied as an option for the reduction of symptoms of depression in elderly adults (Bennett et al., 2017), reduction of loneliness, and as a potential substitute for animal therapy in specific settings (Shibata & Wada, 2011) (Figure 19.1).

Even though there are numerous publications providing evidence of its potential, there are also concerns to its use, and underlying effects in how it affects or could damage the dignity of elderly people (Sharkey & Wood, 2014). Lastly, in comparison with other robots, such as Aibo, the price of the therapeutic robot seal has been previously offered around $6,000 USD (Tergesen & Inada, 2010), becoming a less affordable option compared to other robotic companions.

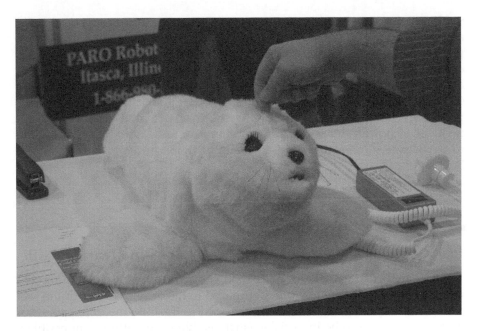

FIGURE 19.1
Paro therapy robot. (Trowbridge, 2010.)

19.5.3.2 Humanoid Therapy Robots

The ideal representation of a robot therapist is continuously depicted in media with a human form. While not limited to it, it is almost certain that a human face and shape will provide more assurance and familiarity to talk in a similar way as we would do with a therapist. In this aspect, there are examples available that exhibit the capabilities of a humanoid robot for therapy.

19.5.3.2.1 Sophia – First Robot Citizen

Created by Hanson Robotics, Sophia is a full-body human-like robot that combines advances both in robotics and in AI with the purpose of resembling the experience of interacting with a human. To achieve this task, the AI combines symbolic AI, neural networks, expert systems, machine perception, conversational NLP, and cognitive architecture, among others (Hanson Robotics, 2022b). Even though Sophia's main purpose has been to promote public discussion about the interaction between humans and robots – and being recognized as the UN Innovation Champion (UN, 2018) – some of her traits have been used to explore her role in the treatment of mental health disorders, such as those in the ASD. In this sense, Hanson et al. (2012) found that children with ASD were curious about robots, engaged in conversations with them and in most cases, did not show fear (p. 4,5). These results were positive despite cross-cultural environments (USA and Italy). Unfortunately, while

Sophia has made a spark in the debate of the ethics, governance, and other aspects of AI and robotics, other publications associated with her role as a potential aid in mental health treatments were not found. Moreover, news on the web expected the possibility of a mass production of the AI robot to offer assistance and care of the sick and elderly throughout the pandemic of COVID-19 (Hennessy, 2021) but neither news nor publications following this possibility were discovered, implying multiple potential difficulties (economical, technical, and social, among others) for humanoid robots as Sophia to participate in critical tasks as the care of living humans in middle of a healthcare crisis.

19.6 Concerns and Ethical Implications of AI in Mental Health

The potential of AI to generate both good and harm could be something that we have never witnessed before. Compared to the developments of the second and third industrial revolution, we need to imagine the implications of what was thought to be impossible (such as flying, something impossible a few centuries ago, or traveling around the world in less than 80 days, to using smartphones and other devices nearly as an extension of ourselves). Because of reasons like these, it stays relevant to reflect on what role we wish AI to have in something as personal our and others' mental health.

In this sense, governments of United States, United Kingdom, and the European Union have raised serious concerns on the applications of AI and robotics in multiple domains (Cath et al., 2018) with the participation of academic and corporate experts in the analysis. These reports provide recommendations of ethical issues that should be addressed in any AI project, such as fairness, accountability, and social justice, through transparency in the use of these techniques. In addition, the use of AI in robots envisions other challenges, with a combination of hard and soft laws to guard against possible risks, including the development of legal tools to assess liability issues associated with these technologies. Lastly, to prepare to future outcomes, these reports examine prospective problems and adverse consequences on the use of AI and robotics that may require prevention, mitigation, and governance.

Nevertheless, none of these reports seem to specifically analyze the concerns and ethical implications on the use of AI for therapeutic applications in mental health services. Even so, a first approach to this is offered by Fiske et al. (2019), who summarize these current and potential ethical issues, including (i) harm prevention and questions of data ethics, (ii) lack of guidance on the development of AI applications, and (iii) their clinical integration and training of health professionals, among others. Consequently, other challenges they address in this sense include risk assessment, referrals and supervision, loss of patient autonomy, long-term effects, and transparency in the use of algorithms. For instance, concerns about the potential harm that AI may provoke in a patient are due to a malfunction. A few examples of these critical functions are when a video recording fails to be preserved in confidentiality, affecting the privacy of a person; or when deciding how should AI support and refer a patient in crisis to a human or clinical institution. In this sense, as long as there exists a lack of understanding of which data privacy frameworks could safeguard our identities, a need to prevent future steps and its likely damage will remain of the utmost importance. As it is commented by some mental health applications, current AI maturity levels still remain to assist users and are not intended to replace mental health professional activities (Wysa, 2022a).

Where these consequences may occur, which other exist and how do we deal with them are questions open to future research. As Fiske et al. (2019) found, challenges for clinicians and patients will continue to emerge on the field of HRIs, with common topics (such as patient adherence to treatment, therapist bias, or attachment) appearing new settings, such as chatbot screens, through virtual platforms, or with robots and us in a safe room, listen closely to everything that we may have to express.

19.7 Conclusions

Should artificial intelligence be your therapist?

> *"Why do we still have a duty to treat the dead with dignity*
>
> *if they will not benefit from our respect?"*
>
> Michael Rosen in *"Dignity Its history and meaning (2018)"*

"How could artificial intelligence help everyone to have a better mental health around the world?" This was one of the main questions behind the development of this chapter, which purpose has been to provide an integrative foundation of what exists in AI for mental health, how does it work, and under what psychotherapy foundations. Designed to provide the basis for health and information technologies professionals alike, this chapter focused on establishing a common ground to understand what is behind the algorithms of AI, a few examples of their current applications in different levels, as well as ethical concerns and implications. However, despite providing the core themes of AI for mental health, this chapter would be incomplete without including a last comprehensive philosophical reflection of the utopic and dystopic futures that AI could bring us, depending in how careful we are. After all, AI therapy could have an impact – sooner or later – in ways that we have not thought or addressed before, including possible new ways of discrimination. As in the idea of Gallagher (2012), if we decide to use AI as way, not to increase the quality of our health services,

but more for convenience or cost reduction, we risk losing our own humanity while denying others dignity. In other words, only those with enough financial and economic resources to pay for human therapists will be served, reducing the perceived human dignity of those unable to afford for their care.

As this future could be seen as dystopic, it is not for many people around some geographies of the world – and even for developed countries. As this has been documented, the number of bankruptcies due to healthcare has reached in some regions up to 62%, while attention disparities continue to hinder efforts to provide the best healthcare for those who need it most (Shmerling, 2021). Moreover, concerns about potential biases in the development of chatbots remain in various areas, as governance boards in these applications may not be designed to be representative, or even unaware of hardships that their users could be living according to community and individual context factors, such as gender, development stage, health conditions, and others, changing their life conditions. Additionally, the question remains if their creators would (or do) use their own creations to fulfill their needs, as they intend to provide value to others. Therefore, while these services are continuously portrayed as useful for others, it remains to be seen if their creators have assimilated these technologies into their own lives.

Before we go further and beyond in constructing solutions through AI, reflection and critical thinking will be continuing to be more important than the entrepreneurial spirit conveyed towards its development. Not aware of what it means to have knowledge of the intimacies of someone, unregulated or unvalidated mental health applications could have a lasting impact in a society, due to a combination of high-power processing and the vulnerability factors in people. We don't need to go further than a few years ago, to remember when a trend of eating tide pods (a laundry detergent product) sparked and went viral on social media, affecting young people and leading to an estimate of 18 intentional poisoning cases in 2018 (Bever, 2018). What was the role of the underlying algorithms to prevent these incidents? In another example of the risks when algorithms, social media, and vulnerability are combined, Raman et al. (2020) evaluate how disinformation could be weaponized, leading to city-scale blackouts due to changes in energy consumption behaviors of users who receive false information about energy discounts. As a result of these scenarios, it is important to think about the role of present algorithms, and furthermore, the AI that is behind these systems. If the present capabilities of AI will seem elemental compared to what will be used in 5–10 years, risks will be higher, and those in a vulnerable situation due to a lack of education and digital literacy could suffer the most.

It will remain significant to imagine and appreciate the role of our technology frameworks for the future, and several important questions will require a careful thought process about the topic:

- Could AI discourage cases where humans are harmed, and decide between their creators' wishes and the benefits for humanity?

- What would be the ethical framework to assess the better outcome, from minimizing harm for the most or maintaining virtuous imperatives in favor of mankind?
- What would happen to our experiences with these AI, and what would happen to the data and (our) memories we generate with them?

Undoubtedly, the notion of what AI for mental health represents will continue to advance along the time, and potentially, we will grasp new possibilities in the development of these technologies that go in conjunction with others, as blockchains or internet 3.0 infrastructures are built and developed. Finally, the question remains also for the field of psychology and mental health professionals, since AI could open the opportunity to have, sooner or later, a psychologist who will learn across time, from multiple sources of data and without the limitations of a human lifetime span.

References

Abd-alrazaq, A. A., Alajlani, M., Alalwan, A. A., Bewick, B. M., Gardner, P., & Househ, M. (2019). An Overview of the Features of Chatbots in Mental Health: A Scoping Review. *International Journal of Medical Informatics, 132*, 103978. https://doi.org/10.1016/j.ijmedinf.2019.103978.

American Psychological Association (APA). (2017, July). *What is Cognitive Behavioral Therapy?* Clinical Practice Guideline for the Treatment of Posttraumatic Stress Disorder.

American Psychological Association (APA). (2022, April). *PTSD Treatment: Information for Patients and Families.* Clinical Practice Guideline for the Treatment of Posttraumatic Stress Disorder.

Andreu-Perez, J., Deligianni, F., Ravi, D., & Guang-Zhong Yang. (2017). Chapter 9: Robotics and AI. In *Artificial intelligence and robotics.* UK-RAS Whitepaper. https://www.researchgate.net/publication/318859042_Artificial_Intelligence_and_Robotics.

Anyoha, R. (2017, August 28). *The history of artificial intelligence.* Special Edition: Artificial Intelligence. https://sitn.hms.harvard.edu/flash/2017/history-artificial-intelligence/.

Asimov, I. (1950). *I, Robot* (First edition). Fawcett Publications. https://www.worldcat.org/title/i-robot/oclc/1384948.

Banks, M. R., Willoughby, L. M., & Banks, W. A. (2008). Animal-Assisted Therapy and Loneliness in Nursing Homes: Use of Robotic versus Living Dogs. *Journal of the American Medical Directors Association, 9*(3), 173–177. https://doi.org/10.1016/j.jamda.2007.11.007.

Bautista, M. A., Hernandez-Vela, A., Escalera, S., Igual, L., Pujol, O., Moya, J., Violant, V., & Anguera, M. T. (2016). A Gesture Recognition System for Detecting Behavioral Patterns of ADHD. *IEEE Transactions on Cybernetics, 46*(1), 136–147. https://doi.org/10.1109/TCYB.2015.2396635.

Beatty, C., Malik, T., Meheli, S., & Sinha, C. (2022). Evaluating the Therapeutic Alliance with a Free-Text CBT Conversational Agent (Wysa): A Mixed-Methods Study. *Frontiers in Digital Health*, 4. https://doi.org/10.3389/fdgth.2022.847991.

Bennett, C. C., Sabanovic, S., Piatt, J. A., Nagata, S., Eldridge, L., & Randall, N. (2017). A Robot a Day Keeps the Blues Away. *2017 IEEE International Conference on Healthcare Informatics (ICHI)*, pp. 536–540. https://doi.org/10.1109/ICHI.2017.43.

Beutel, M. E., Klein, E. M., Brähler, E., Reiner, I., Jünger, C., Michal, M., Wiltink, J., Wild, P. S., Münzel, T., Lackner, K. J., & Tibubos, A. N. (2017). Loneliness in the General Population: Prevalence, Determinants and Relations to Mental Health. *BMC Psychiatry*, 17(1), 97. https://doi.org/10.1186/s12888-017-1262-x.

Bever, L. (2018, January 17). *Teens are daring each other to eat Tide pods. We don't need to tell you that's a bad idea.* The Washington Post. https://www.washingtonpost.com/news/to-your-health/wp/2018/01/13/teens-are-daring-each-other-to-eat-tide-pods-we-dont-need-to-tell-you-thats-a-bad-idea.

Blackburn, I. M., Bishop, S., Glen, A. I. M., Whalley, L. J., & Christie, J. E. (1981). The Efficacy of Cognitive Therapy in Depression: A Treatment Trial Using Cognitive Therapy and Pharmacotherapy, Each Alone and in Combination. *British Journal of Psychiatry*, 139(3), 181–189. https://doi.org/10.1192/bjp.139.3.181.

Bleuer-Elsner, S., Zamansky, A., Fux, A., Kaplun, D., Romanov, S., Sinitca, A., Masson, S., & van der Linden, D. (2019). Computational Analysis of Movement Patterns of Dogs with ADHD-Like Behavior. *Animals*, 9(12), 1140. https://doi.org/10.3390/ani9121140.

Brigida, A.-C. (2013, October 18). *A Virtual Therapist.* USC Viterbi. https://viterbi.usc.edu/news/news/2013/a-virtual-therapist.htm.

Cath, C., Wachter, S., Mittelstadt, B., Taddeo, M., & Floridi, L. (2018). Artificial Intelligence and the 'Good Society': The US, EU, and UK Approach. *Science and Engineering Ethics*, 24(2), 505–528. https://doi.org/10.1007/s11948-017-9901-7.

Chang, W.-L., Sabanovic, S., & Huber, L. (2013). Use of Seal-Like Robot PARO in Sensory Group Therapy for Older Adults with Dementia. *2013 8th ACM/IEEE International Conference on Human-Robot Interaction (HRI)*, 101–102. https://doi.org/10.1109/HRI.2013.6483521.

Chen, Q., Leaman, R., Allot, A., Luo, L., Wei, C.-H., Yan, S., & Lu, Z. (2021). Artificial Intelligence in Action: Addressing the COVID-19 Pandemic with Natural Language Processing. *Annual Review of Biomedical Data Science*, 4(1), 313–339. https://doi.org/10.1146/annurev-biodatasci-021821-061045.

Chen, X., Orom, H., Hay, J. L., Waters, E. A., Schofield, E., Li, Y., & Kiviniemi, M. T. (2019). Differences in Rural and Urban Health Information Access and Use. *The Journal of Rural Health*, 35(3), 405–417. https://doi.org/10.1111/jrh.12335.

Coco, K., Kangasniemi, M., & Rantanen, T. (2018). Care Personnel's Attitudes and Fears toward Care Robots in Elderly Care: A Comparison of Data from the Care Personnel in Finland and Japan. *Journal of Nursing Scholarship*, 50(6), 634–644. https://doi.org/10.1111/jnu.12435.

Craft, L. (2022, January 3). *Robo-Dogs and Therapy Bots: Artificial Intelligence Goes Cuddly.* CBS News. https://www.cbsnews.com/news/robo-dogs-therapy-bots-artificial-intelligence/.

Crane, K. L., & Watters, K. M. (2019). *Cognitive Behavioral Therapy Strategies.* Mental Illness Research, Education and Clinical Center (MIRECC). https://www.mirecc.va.gov/visn16/clinicalEducationProducts_topic.asp.

Darcy, A., Daniels, J., Salinger, D., Wicks, P., & Robinson, A. (2021). Evidence of Human-Level Bonds Established With a Digital Conversational Agent: Cross-sectional, Retrospective Observational Study. *JMIR Formative Research*, 5(5), e27868. https://doi.org/10.2196/27868.

de Choudhury, M., Kiciman, E., Dredze, M., Coppersmith, G., & Kumar, M. (2016). Discovering Shifts to Suicidal Ideation from Mental Health Content in Social Media. *Proceedings of the 2016 CHI Conference on Human Factors in Computing Systems*, 2098–2110. https://doi.org/10.1145/2858036.2858207.

DeSouza, D. D., Robin, J., Gumus, M., & Yeung, A. (2021). Natural Language Processing as an Emerging Tool to Detect Late-Life Depression. *Frontiers in Psychiatry*, 12. https://doi.org/10.3389/fpsyt.2021.719125.

El-Ramly, M., Abu-Elyazid, H., Mo'men, Y., Alshaer, G., Adib, N., Eldeen, K. A., & El-Shazly, M. (2021). CairoDep: Detecting Depression in Arabic Posts Using BERT Transformers. *2021 Tenth International Conference on Intelligent Computing and Information Systems (ICICIS)*, 207–212. https://doi.org/10.1109/ICICIS52592.2021.9694178.

Epic Games Inc. (2022). *Metahuman: High-Fidelity Digital Humans Made Easy*. Unreal Engine. https://www.unrealengine.com/en-US/metahuman.

Estela, L. (2022, June 10). *Metahuman Designs for Therapy*. Lucy Estela. https://www.unrealengine.com/en-US/metahuman.

Fallon, J. (2017, April 25). *Tonight Showbotics: Jimmy Meets Sophia the Human-Like Robot*. YouTube.

Fasola, J., & Mataric, M. (2013). A Socially Assistive Robot Exercise Coach for the Elderly. *Journal of Human-Robot Interaction*, 2(2). https://doi.org/10.5898/JHRI.2.2.Fasola.

Feil-Seifer, D., & Mataric, M. J. (2005). Socially Assistive Robotics. *9th International Conference on Rehabilitation Robotics (ICORR)*, 465–468. https://doi.org/10.1109/ICORR.2005.1501143.

Ferrari, A. J., Santomauro, D. F., Mantilla Herrera, A. M., Shadid, J., Ashbaugh, C., Erskine, H. E., Charlson, F. J., Degenhardt, L., Scott, J. G., McGrath, J. J., Allebeck, P., Benjet, C., Breitborde, N. J. K., Brugha, T., Dai, X., Dandona, L., Dandona, R., Fischer, F., Haagsma, J. A., … Whiteford, H. A. (2022). Global, Regional, and National Burden of 12 Mental Disorders in 204 Countries and Territories, 1990–2019: A Systematic Analysis for the Global Burden of Disease Study 2019. *The Lancet Psychiatry*, 9(2), 137–150. https://doi.org/10.1016/S2215-0366(21)00395-3.

Fiske, A., Henningsen, P., & Buyx, A. (2019). Your Robot Therapist Will See You Now: Ethical Implications of Embodied Artificial Intelligence in Psychiatry, Psychology, and Psychotherapy. *Journal of Medical Internet Research*, 21(5), e13216. https://doi.org/10.2196/13216.

Gallagher, A., Nåden, D., & Karterud, D. (2016). Robots in Elder Care: Some Ethical Questions. *Nursing Ethics*, 23(4), 369–371. https://doi.org/10.1177/0969733016647297.

Gallagher, J. (2012, March 25). *Dignity: Its History and Meaning by Michael Rosen – Review*. The Guardian. https://www.theguardian.com/books/2012/mar/25/dignity-history-meaning-rosen-review.

Giannakakis, G., Pediaditis, M., Manousos, D., Kazantzaki, E., Chiarugi, F., Simos, P. G., Marias, K., & Tsiknakis, M. (2017). Stress and Anxiety Detection Using Facial Cues from Videos. *Biomedical Signal Processing and Control*, 31, 89–101. https://doi.org/10.1016/j.bspc.2016.06.020.

Globe Newswire. (2022, June 24). *Personal Service Robotics Market Anticipated to Hit USD 35.9 Billion at a Whopping 38.5% CAGR by 2030- Report by Market Research Future (MRFR)*. GLOBE NEWSWIRE. https://www.globenewswire.com/en/news-release/2022/06/24/2468841/0/en/Personal-Service-Robotics-Market-Anticipated-to-Hit-USD-35-9-Billion-at-a-Whopping-38-5-CAGR-by-2030-Report-by-Market-Research-Future-MRFR.html.

Hacohen-Kerner, Y., Manor, N., Goldmeier, M., & Bachar, E. (2022). Detection of Anorexic Girls-In Blog Posts Written in Hebrew Using a Combined Heuristic AI and NLP Method. *IEEE Access, 10*, 34800–34814. https://doi.org/10.1109/ACCESS.2022.3162685.

Hanson, D., Mazzei, D., Garver, C. R., Ahluwalia, A., de Rossi, D., Stevenson, M., & Reynolds, K. (2012, June). Realistic Humanlike Robots for Treatment of ASD, Social Training, and Research; Shown to Appeal to Youths with ASD, Cause Physiological Arousal, and Increase Human-to-Human Social Engagement. *5th International Conference on Pervasive Technologies Related to Assistive Environments (PETRA)*. https://www.researchgate.net/publication/233951262_Realistic_Humanlike_Robots_for_Treatment_of_ASD_Social_Training_and_Research_Shown_to_Appeal_to_Youths_with_ASD_Cause_Physiological_Arousal_and_Increase_Human-to-Human_Social_Engagement.

Hanson Robotics. (2022a). *Being Sophia*. Being Sophia.

Hanson Robotics. (2022b). *Sophia's Artificial Intelligence*. Hanson Robotics. https://www.hansonrobotics.com/sophia/.

Hashemi, J., Tepper, M., Vallin Spina, T., Esler, A., Morellas, V., Papanikolopoulos, N., Egger, H., Dawson, G., & Sapiro, G. (2014). Computer Vision Tools for Low-Cost and Noninvasive Measurement of Autism-Related Behaviors in Infants. *Autism Research and Treatment, 2014*, 1–12. https://doi.org/10.1155/2014/935686.

Hassan, S. B., Hassan, S. B., & Zakia, U. (2020). Recognizing Suicidal Intent in Depressed Population using NLP: A Pilot Study. *2020 11th IEEE Annual Information Technology, Electronics and Mobile Communication Conference (IEMCON)*, 0121–0128. https://doi.org/10.1109/IEMCON51383.2020.9284832.

Hennessy, M. (2021, January 24). *Makers of Sophia the Robot Plan Mass Rollout Amid Pandemic*. Reuters. https://www.reuters.com/article/us-hongkong-robot-idUSKBN29U03X.

Hirschberg, J., & Manning, C. D. (2015). Advances in Natural Language Processing. *Science, 349*(6245), 261–266. https://doi.org/10.1126/science.aaa8685.

IBM. (2022). *What Is Computer Vision?* IBM.

IMDB. (2022). *Bicentennial Man (1999)*. IMDB. https://www.imdb.com/video/vi783941913/.

Inkster, B., Sarda, S., & Subramanian, V. (2018). An Empathy-Driven, Conversational Artificial Intelligence Agent (Wysa) for Digital Mental Well-Being: Real-World Data Evaluation Mixed-Methods Study. *JMIR MHealth and UHealth, 6*(11), e12106. https://doi.org/10.2196/12106.

Inoue, K., Wada, K., & Shibata, T. (2021). Exploring the Applicability of the Robotic Seal PARO to Support Caring for Older Persons with Dementia within the Home Context. *Palliative Care and Social Practice, 15*, 263235242110302. https://doi.org/10.1177/26323524211030285.

Jaiswal, S., Valstar, M. F., Gillott, A., & Daley, D. (2017). Automatic Detection of ADHD and ASD from Expressive Behaviour in RGBD Data. *2017 12th IEEE International Conference on Automatic Face & Gesture Recognition (FG 2017)*, 762–769. https://doi.org/10.1109/FG.2017.95.

Jonze, S. (2013). *Her*. Warner Bros. Pictures. https://www.imdb.com/title/tt1798709/.

Kabacińska, K., Prescott, T. J., & Robillard, J. M. (2021). Socially Assistive Robots as Mental Health Interventions for Children: A Scoping Review. *International Journal of Social Robotics*, 13(5), 919–935. https://doi.org/10.1007/s12369-020-00679-0.

Kim, J., Park, S. Y., & Robert, L. (2019, November 9). Conversational Agents for Health and Wellbeing: Review and Future Agendas. *22th ACM Conference on Computer Supported Cooperative Work and Social Computing*. Austin, TX, https://deepblue.lib.umich.edu/handle/2027.42/151800

Kimura, R., Abe, N., Matsumura, N., Horiguchi, A., Sasaki, T., Negishi, T., Ohkubo, E., & Naganuma, M. (2004, August 4). Trial of Robot Assisted Activity Using Robotic Pets in Children Hospital. *SICE 2004 Annual Conference*. https://ieeexplore.ieee.org/abstract/document/1491448.

Klappich, D., & Jump, A. (2022). *Hype Cycle for Mobile Robots and Drones, 2022*. https://www.gartner.com/interactive/hc/4016694.

Knight, S., & Edwards, V. (2008). In the Company of Wolves: The Physical, Social, and Psychological Benefits of Dog Ownership. *Journal of Aging and Health*, 20(4), 437–455. https://doi.org/10.1177/0898264308315875.

Kramer, S. C., Friedmann, E., & Bernstein, P. L. (2009). Comparison of the Effect of Human Interaction, Animal-Assisted Therapy, and AIBO-Assisted Therapy on Long-Term Care Residents with Dementia. *Anthrozoös*, 22(1), 43–57. https://doi.org/10.2752/175303708X390464.

Kulasinghe, S. A. S. A., Jayasinghe, A., Rathnayaka, R. M. A., Karunarathne, P. B. M. M. D., Suranjini Silva, P. D., & Anuradha Jayakodi, J. A. D. C. (2019). AI Based Depression and Suicide Prevention System. *2019 International Conference on Advancements in Computing (ICAC)*, 73–78. https://doi.org/10.1109/ICAC49085.2019.9103411.

Kulkarni, S. (2022). Chatbots: A Futuristic Approach to Therapy. *International Research Journal of Modernization in Engineering, Technology, and Science*, 4(5). https://www.irjmets.com/uploadedfiles/paper/issue_5_may_2022/24692/final/fin_irjmets1656671957.pdf.

Kyodo. (2018, May 7). *Sales of Sony's New Aibo Robot Dog Off to Solid Start*. The Japan Times.

Kyodo News. (2022, September 19). *Over 75s Make Up over 15% of Japan's Population for First Time*. The Japan Times. https://www.japantimes.co.jp/news/2022/09/19/national/japans-graying-population/.

Lazarus, A. A. (2006). Multimodal Therapy: A Seven-Point Integration. In *A casebook of psychotherapy integration*. (pp. 17–28). American Psychological Association. https://doi.org/10.1037/11436-002.

le Glaz, A., Haralambous, Y., Kim-Dufor, D.-H., Lenca, P., Billot, R., Ryan, T. C., Marsh, J., DeVylder, J., Walter, M., Berrouiguet, S., & Lemey, C. (2021). Machine Learning and Natural Language Processing in Mental Health: Systematic Review. *Journal of Medical Internet Research*, 23(5), e15708. https://doi.org/10.2196/15708.

van Le, D., Montgomery, J., Kirkby, K. C., & Scanlan, J. (2018). Risk Prediction Using Natural Language Processing of Electronic Mental Health Records in an Inpatient Forensic Psychiatry Setting. *Journal of Biomedical Informatics*, 86, 49–58. https://doi.org/10.1016/j.jbi.2018.08.007.

Liang, Y., & Lee, S. A. (2017). Fear of Autonomous Robots and Artificial Intelligence: Evidence from National Representative Data with Probability Sampling. *International Journal of Social Robotics*, 9(3), 379–384. https://doi.org/10.1007/s12369-017-0401-3.

Low, D. M., Rumker, L., Talkar, T., Torous, J., Cecchi, G., & Ghosh, S. S. (2020). Natural Language Processing Reveals Vulnerable Mental Health Support Groups and Heightened Health Anxiety on Reddit During COVID-19: Observational Study. *Journal of Medical Internet Research*, 22(10), e22635. https://doi.org/10.2196/22635.

Lucas, G. M., Gratch, J., King, A., & Morency, L.-P. (2014). It's Only a Computer: Virtual Humans Increase Willingness to Disclose. *Computers in Human Behavior*, 37, 94–100. https://doi.org/10.1016/j.chb.2014.04.043.

Malik, T., Ambrose, A. J., & Sinha, C. (2022). Evaluating User Feedback for an Artificial Intelligence–Enabled, Cognitive Behavioral Therapy–Based Mental Health App (Wysa): Qualitative Thematic Analysis. *JMIR Human Factors*, 9(2), e35668. https://doi.org/10.2196/35668.

Martin, C. D. (1995). ENIAC: Press Conference that Shook the World. *IEEE Technology and Society Magazine*, 14(4), 3–10. https://doi.org/10.1109/44.476631.

Martínez-Pérez, B., de la Torre-Díez, I., & López-Coronado, M. (2013). Mobile Health Applications for the Most Prevalent Conditions by the World Health Organization: Review and Analysis. *Journal of Medical Internet Research*, 15(6), e120. https://doi.org/10.2196/jmir.2600

Morales, D. A., Barksdale, C. L., & Beckel-Mitchener, A. C. (2020). A call to action to address rural mental health disparities. *Journal of Clinical and Translational Science*, 4(5), 463–467. https://doi.org/10.1017/cts.2020.42McCarthy, J. (2007, November 12). *What Is Artificial Intelligence? - Basic Questions*. Stanford University.

McCarthy, J., Minsky, M. L., Rochester, N., & Shannon, C. E. (2006). A Proposal for the Dartmouth Summer Research Project on Artificial Intelligence, August 31, 1955. *AI Magazine*, 27(4). https://doi.org/10.1609/aimag.v27i4.1904.

Meheli, S., Sinha, C., & Kadaba, M. (2022). Understanding People With Chronic Pain Who Use a Cognitive Behavioral Therapy–Based Artificial Intelligence Mental Health App (Wysa): Mixed Methods Retrospective Observational Study. *JMIR Human Factors*, 9(2), e35671. https://doi.org/10.2196/35671.

Melson, G. F., Kahn, P. H., Beck, A. M., Friedman, B., Roberts, T., & Garrett, E. (2005). Robots as dogs?: Children's interactions with the robotic dog AIBO and a live Australian shepherd. *CHI '05 Extended Abstracts on Human Factors in Computing Systems*, 1649–1652. https://doi.org/10.1145/1056808.1056988.

Mujeeb, S., Hafeez, M., & Arshad, T. (2017). Aquabot: A Diagnostic Chatbot for Achluophobia and Autism. *International Journal of Advanced Computer Science and Applications*, 8(9). https://doi.org/10.14569/IJACSA.2017.080930.

Myers, A. (2011, November 25). *Stanford's John McCarthy, Seminal Figure of Artificial Intelligence, Dies at 84*. Stanford | News.

Namboodiri, S. P., & Venkataraman, D. (2019). A computer vision based image processing system for depression detection among students for counseling. *Indonesian Journal of Electrical Engineering and Computer Science*, 14(1), 503. https://doi.org/10.11591/ijeecs.v14.i1.pp503-512.

Okita, S. Y. (2013). Self–Other's Perspective Taking: The Use of Therapeutic Robot Companions as Social Agents for Reducing Pain and Anxiety in Pediatric Patients. *Cyberpsychology, Behavior, and Social Networking*, 16(6), 436–441. https://doi.org/10.1089/cyber.2012.0513.

Oppy, G., & Dowe, D. (2021, October 4). *The Turing Test*. Stanford Encyclopedia of Philosophy.

ORCHA. (2020, March 13). *Coronavirus: Apps to Help Self-Management*. ORCHA - News. https://orchahealth.com/coronavirus-apps-to-help-self-management/.

PARO Robots. (2014). *PARO Therapeutic Robot*. PARO Therapeutic Robot. http://www.parorobots.com/.

Pauw, L. S., Sauter, D. A., van Kleef, G. A., Lucas, G. M., Gratch, J., & Fischer, A. H. (2022). The Avatar Will See You Now: Support from a Virtual Human Provides Socio-Emotional Benefits. *Computers in Human Behavior, 136*, 107368. https://doi.org/10.1016/j.chb.2022.107368.

Plackett, B. (2022, January 19). *Tackling the Crisis of Care for Older People: Lessons from India and Japan*. Nature. https://www.nature.com/articles/d41586-022-00074-x.

Raman, G., AlShebli, B., Waniek, M., Rahwan, T., & Peng, J. C.-H. (2020). How Weaponizing Disinformation Can Bring Down a City's Power Grid. *PLoS One, 15*(8), e0236517. https://doi.org/10.1371/journal.pone.0236517.

Ribi, F. N., Yokoyama, A., & Turner, D. C. (2008). Comparison of Children's Behavior toward Sony's Robotic Dog AIBO and a Real Dog: A Pilot Study. *Anthrozoös, 21*(3), 245–256. https://doi.org/10.2752/175303708X332053.

Riek, L. D. (2016). Robotics Technology in Mental Health Care. In *Artificial intelligence in behavioral and mental health care* (pp. 185–203). Elsevier. https://doi.org/10.1016/B978-0-12-420248-1.00008-8.

Rizzo, A., & Morency, L.-P. (2013, February 7). *SimSensei & MultiSense: Virtual Human and Multimodal Perception for Healthcare Support*. USCICT (YouTube). https://www.youtube.com/watch?v=ejczMs6b1Q4.

Rosen, M. (2018). *Dignity Its History and Meaning*. Harvard University Press. https://www.hup.harvard.edu/catalog.php?isbn=9780674984059.

Rush, A. J., Beck, A. T., Kovacs, M., & Hollon, S. (1977). Comparative Efficacy of Cognitive Therapy and Pharmacotherapy in the Treatment of Depressed Outpatients. *Cognitive Therapy and Research, 1*(1), 17–37. https://doi.org/10.1007/BF01173502.

Šabanović, S., Chang, W.-L., Bennett, C. C., Piatt, J. A., & Hakken, D. (2015). *A Robot of My Own: Participatory Design of Socially Assistive Robots for Independently Living Older Adults Diagnosed with Depression* (pp. 104–114). https://doi.org/10.1007/978-3-319-20892-3_11.

Salinas, J. J., Al Snih, S., Markides, K., Ray, L. A., & Angel, R. J. (2010). The Rural-Urban Divide: Health Services Utilization Among Older Mexicans in Mexico. *The Journal of Rural Health, 26*(4), 333–341. https://doi.org/10.1111/j.1748-0361.2010.00297.x

Santomauro, D. F., Mantilla Herrera, A. M., Shadid, J., Zheng, P., Ashbaugh, C., Pigott, D. M., Cristiana, A., Adolph, C., Amlag, J. O., Aravkin, A. Y., Bang-Jensen, B. L., Bertolacci, G. J., Bloom, S. S., Castel, R., & Ferrari, A. J. (2021). Global Prevalence and Burden of Depressive and Anxiety Disorders in 204 Countries and Territories in 2020 due to the COVID-19 Pandemic. *Lancet (London, England), 398*(10312), 1700–1712. https://doi.org/10.1016/S0140-6736(21)02143-7.

Sapiro, G., Hashemi, J., & Dawson, G. (2019). Computer Vision and Behavioral Phenotyping: An Autism Case Study. *Current Opinion in Biomedical Engineering, 9*, 14–20. https://doi.org/10.1016/j.cobme.2018.12.002.

Shaikh, T. A. H., & Mhetre, M. (2022). Autonomous AI Chat Bot Therapy for Patient with Insomnia. *2022 IEEE 7th International Conference for Convergence in Technology (I2CT)*, 1–5. https://doi.org/10.1109/I2CT54291.2022.9825008.

Sharkey, A., & Wood, N. (2014). The Paro Seal Robot: Demeaning Or Enabling? *AISB 2014–50th Annual Convention of the AISB*. https://www.researchgate.net/publication/286522298_The_Paro_seal_robot_Demeaning_or_enabling.

Shibata, T., & Wada, K. (2011). Robot Therapy: A New Approach for Mental Healthcare of the Elderly – A Mini-Review. *Gerontology*, *57*(4), 378–386. https://doi.org/10.1159/000319015.

Shmerling, R. H. (2021, July 13). *Is Our Healthcare System Broken?* Harvard Health Publishing. https://www.health.harvard.edu/blog/is-our-healthcare-system-broken-202107132542.

SHOCHIKU. (2013, February 7). 『ハル』本予告. 松竹チャンネル. [HAL – Official trailer]. https://www.youtube.com/watch?v=tNXhoXiufGk.

Sinha, C., Cheng, A. L., & Kadaba, M. (2022). Adherence and Engagement with a Cognitive Behavioral Therapy–Based Conversational Agent (Wysa for Chronic Pain) Among Adults with Chronic Pain: Survival Analysis. *JMIR Formative Research*, *6*(5), e37302. https://doi.org/10.2196/37302.

Song, X., Xu, B., & Zhao, Z. (2022). Can People Experience Romantic Love for Artificial Intelligence? An Empirical Study of Intelligent Assistants. *Information & Management*, *59*(2), 103595. https://doi.org/10.1016/j.im.2022.103595.

Stanton, C. M., Kahn Jr., P. H., Severson, R. L., Ruckert, J. H., & Gill, B. T. (2008). Robotic Animals Might Aid in the Social Development of Children with Autism. *Proceedings of the 3rd International Conference on Human Robot Interaction - HRI '08*, 271. https://doi.org/10.1145/1349822.1349858.

Swartout, W., Artstein, R., Forbell, E., Foutz, S., Lane, H. C., Lange, B., Morie, J. F., Rizzo, A. S., & Traum, D. (2013). Virtual Humans for Learning. *AI Magazine*, *34*(4), 13–30. https://doi.org/10.1609/aimag.v34i4.2487.

Tanaka, K., Makino, H., Nakamura, K., Nakamura, A., Hayakawa, M., Uchida, H., Kasahara, M., Kato, H., & Igarashi, T. (2022). The Pilot Study of Group Robot Intervention on Pediatric Inpatients and Their Caregivers, Using 'New Aibo.' *European Journal of Pediatrics*, *181*(3), 1055–1061. https://doi.org/10.1007/s00431-021-04285-8.

Tergesen, A., & Inada, M. (2010, June 21). *It's Not a Stuffed Animal, It's a $6,000 Medical Device*. The Wallstreet Journal. https://www.wsj.com/articles/SB10001424052748704463504575301051844937276.

Tewari, A., Chhabria, A., Khalsa, A. S., Chaudhary, S., & Kanal, H. (2021). A Survey of Mental Health Chatbots using NLP. *SSRN Electronic Journal*. https://doi.org/10.2139/ssrn.3833914.

The Editors of Encyclopaedia Britannica. (2022, August 27). *Wilhelm Wundt*. Encyclopaedia Britannica. https://www.britannica.com/biography/Wilhelm-Wundt.

The Tech. (2008, March 14). *Joseph Weizenbaum*. The Tech. https://thetech.com/2008/03/14/weizenbaum-v128-n12.

Trowbridge, T. (2010, January 9). *Paro Therapy Robot*. Flickr. https://www.flickr.com/photos/therontrowbridge/4261112915/.

Turner, A. (2022, October). *How Many Smartphones Are in the World?* Bank My Cell. https://www.bankmycell.com/blog/how-many-phones-are-in-the-world.

United Nations (UN). (2018, October 10). *Robot Sophia, UN's First Innovation Champion, Visited Armenia*. United Nations Development Programme.

University of Minnesota. (2013). *Communication in the Real World: An Introduction to Communication Studies*. University of Minnesota Libraries Publishing. https://open.lib.umn.edu/communication/chapter/4-2-types-of-nonverbal-communication/.

USCICT. (2014). *SimSensei/MultiSense Overview 2014*. YouTube. https://www.youtube.com/watch?v=I2aBJ6LjzMw

Vos, T., Lim, S. S., Abbafati, C., Abbas, K. M., Abbasi, M., Abbasifard, M., Abbasi-Kangevari, M., Abbastabar, H., Abd-Allah, F., Abdelalim, A., Abdollahi, M., Abdollahpour, I., Abolhassani, H., Aboyans, V., Abrams, E. M., Abreu, L. G., Abrigo, M. R. M., Abu-Raddad, L. J., Abushouk, A. I., … Murray, C. J. L. (2020). Global Burden of 369 Diseases and Injuries in 204 Countries and Territories, 1990–2019: A Systematic Analysis for the Global Burden of Disease Study 2019. *The Lancet, 396*(10258), 1204–1222. https://doi.org/10.1016/S0140-6736(20)30925-9

Waseda University. (n.d.). *WABOT -WAseda roBOT-*. Waseda University Humanoid. Retrieved September 14, 2022, from http://www.humanoid.waseda.ac.jp/booklet/kato_2.html.

Wasil, A. R., Palermo, E. H., Lorenzo-Luaces, L., & DeRubeis, R. J. (2022). Is There an App for That? A Review of Popular Apps for Depression, Anxiety, and Well-Being. *Cognitive and Behavioral Practice, 29*(4), 883–901. https://doi.org/10.1016/j.cbpra.2021.07.001.

Watson, A., Mellotte, H., Hardy, A., Peters, E., Keen, N., & Kane, F. (2021). The Digital Divide: Factors Impacting on Uptake of Remote Therapy in a South London Psychological Therapy Service for People with Psychosis. *Journal of Mental Health*, 1–8. https://doi.org/10.1080/09638237.2021.1952955.

Woebot Health. (2022). *Life Changes. And So Do We, with the Help of AI*. Woebot Health. https://woebothealth.com/what-powers-woebot/.

World Health Organization (WHO). (2019, April). *Mental Health Workers - Data by Country*. World Health Organization.

World Health Organization (WHO). (2020, December). *WHO Methods and Data Sources for Global Burden of Disease Estimates 2000–2019*. https://cdn.who.int/media/docs/default-source/gho-documents/global-health-estimates/ghe2019_daly-methods.pdf?sfvrsn=31b25009_7.

World Health Organization (WHO). (2021, September 13). *Depression*. World Health Organization. https://www.who.int/news-room/fact-sheets/detail/depression.

Wysa. (2022a). *FAQs*. Wysa. https://www.wysa.io/faq.

Wysa. (2022b). *First AI Mental Health App to Meet Clinical Safety Standards*. Wysa. https://wysa.io/clinical-validation.

Yao, M. (2017, March 21). *4 Approaches to Natural Language Processing & Understanding*. FreeCodeCamp.

Zhang, T., Schoene, A. M., Ji, S., & Ananiadou, S. (2022). Natural Language Processing Applied to Mental Illness Detection: A Narrative Review. *Npj Digital Medicine, 5*(1), 46. https://doi.org/10.1038/s41746-022-00589-7.

Zhang, J., Zhu, L., Li, S., Huang, J., Ye, Z., Wei, Q., & Du, C. (2021). Rural–urban disparities in knowledge, behaviors, and mental health during COVID-19 pandemic. *Medicine, 100*(13), e25207. https://doi.org/10.1097/MD.0000000000025207

Zhang, Y., Kong, M., Zhao, T., Hong, W., Zhu, Q., & Wu, F. (2020). ADHD Intelligent Auxiliary Diagnosis System Based on Multimodal Information Fusion. *Proceedings of the 28th ACM International Conference on Multimedia*, 4494–4496. https://doi.org/10.1145/3394171.3414359.

Zor, R., Fineberg, N., Eilam, D., & Hermesh, H. (2011). Video Telemetry and Behavioral Analysis Discriminate between Compulsive Cleaning and Compulsive Checking in Obsessive-Compulsive Disorder. *European Neuropsychopharmacology, 21*(11), 814–824. https://doi.org/10.1016/j.euroneuro.2011.03.006.

20

What AI Can Do for Neuroscience: Understanding How the Brain Represents Word Meanings

Nora Aguirre-Celis and Risto Miikkulainen

University of Texas

CONTENTS

20.1 Introduction

People intuitively know that the meaning of the word *large* means something different when used to label a pencil, a house, or a mountain. For humans, the properties associated with *large* vary in context-dependent ways: they know what the words means, its use in context, and how the words combine to construct its word meaning. Alexa (digital voice assistant), on the other hand, responds that a house is larger ("Alexa, what is larger a house or a mountain?" Alexa's response: "Houses not only have bigger square footage, but they include more rooms such as a home office, a playroom, a den or a family room over a garage, as well as more bedrooms"). Alexa's response clearly demonstrates that it does not understand language, and specifically, it does not understand the central properties and combinations of words such as *large house* and *large mountain*. In fact, this is not a unique problem to Alexa; today's digital voice assistants are not programmed with the intrinsic properties of the structure and organization of the semantic knowledge as humans do.

Although many of the brain's operating principles are still opaque (i.e., how are concepts acquired, organized, and stored?), Artificial Intelligence (AI) can be used to model such phenomena providing neuroscientists the tools to make more discoveries and consequently, build more human-like AI applications of natural language understanding.

The conceptualization of AI emerged 72 years ago. The founding figures such as John McCarthy, Marvin Minsky, Warren McCulloch, and Donald Hebb, among others, were motivated mainly to understand how the brain works. Back in the 1950s, the brain was hypothesized as a powerful computer, and intelligence was thought to be a symbol processing program located somewhere in our brain. While this classical view of intelligence was probably suitable in formal domains like chess playing or logical reasoning and static knowledge bases, in the past 35 years, it has become evident that conceptualizing intelligence as a simple computational process does not explain the constantly changing human intelligence (i.e., systems acting in the physical world). As a result, the paradigm shifted from a computational to an embodied perspective, changing the focus of study from higher-level (i.e., symbolic approaches) to lower-level mechanisms (i.e., neural networks). The representations and algorithms used by the latter approach were somehow inspired by the brain.

The concept of embodiment in AI promotes the study of an intelligent mechanism that is aware of its surroundings, adapts in real time, and acts and communicates accordingly to new or changing situations. In other words, current challenges focus on how robots (and embodied agents in general) can give meaning to symbols (*dogs, mountain, birthday cake*) and internally construct meaningful language representations, among other tasks.

To show what AI can do for neuroscience, this chapter presents a case study from a collaboration between neuroscience research, its tools (functional magnetic resonance imaging – fMRI) and theories (Concept Attribute Representation model – CAR model), and an AI approach (context-dependent meaning representation in the brain – CEREBRA model). Such interaction produced new opportunities to allow researchers to gain insights and validate some hypotheses about the functioning of the brain (within the language domain) and delivered a unique class of dynamic word representations (based on the way word meanings are represented in the brain) that may improve current natural language processing (NLP) systems such as Siri, Google, and Alexa, by dynamically adapting their representations to fit context.

In the following sections, a general overview of the key ideas is presented first. Then, two different models of semantic representations addressing the questions on: "How concepts are represented in the brain?" and "How word meanings change in the context of a sentence?" are described. The first model, applied to today's digital voice assistants, uses word representations based on the Distributional Semantic Model (DSM). The second model, based on the Embodied approach, involves word representation based on a brain-based semantic model (CAR) proposed by Binder (2016) and Binder et al. (2016). Afterward, a case study produced by the collaboration between neurosciences and AI called CEREBRA is briefly described. The last part of this chapter talks about a proposed idea on semantic modeling to advance the way computers represent word meaning and ends with some speculations about the future of AI by trying to replicate more brain-like functions.

20.2 Overview of the Key Ideas

The purpose of this chapter is to present how neuroscience and AI build on each other's findings to produce new tools or techniques to advance both fields, particularly in the semantic representation of words. How individuals represent knowledge of concepts is one of the most important questions in the semantic memory research. Semantic knowledge consists of facts about the world, concepts, and symbols. Also, such knowledge is sensitive to context, task demands, and perceptual and sensorimotor information from the environment.

Many experimental studies suggest that there are two types of semantic knowledge: linguistic and experiential (Meteyard et al., 2012; Vigliocco et al., 2007, 2009). Humans acquire linguistic knowledge through a lifetime of linguistic exposure, and experiential knowledge is acquired through their perception and interaction with the physical world. Linguistic knowledge includes (spoken and/or written) words defined by their relations to other words in the sentence and in the context in which they are expressed. This knowledge provides individuals with the capacity to communicate about history, scientific terms, ideas, plans, emotions, objects, and everything. For example, the word *carrot* is defined as a root vegetable, usually orange, Dutch invented the orange carrots, it contains high alpha- and beta- carotene, human body turns carotene into vitamin A, K and B6, etc. (Wikipedia). Experiential knowledge, on the other hand, denotes the visual, motor, somatosensory, auditory, spatial, cognitive, emotional, and many more attributes of the experienced objects (the referents of words). For example, the word *carrot* refers to an object whose attributes describe it as orange, conical/cylindrical, juicy, crispy, sweet, and so on.

Likewise, for each type of semantic knowledge, word meanings arise differently. For linguistic-based knowledge, meaning comes from what people know about the world. For example, for the sentence *She was at the court*, different meanings associated with a single word such as *court* could refer to a building for legal proceedings, or a quadrangular area surrounded by buildings, or used for ball games such as tennis. On the other hand, for experiential-based knowledge, meaning comes from the word itself. In this case, there are context-dependent interpretations arising from the same underlying meaning. For example, the meaning of the word *food* for the sentences *The food is hot*, and *The food is delicious*, one sentence refers to the Temperature and the other to Taste and Smell in regard to the diner's perception and interaction with food. This is the mental representation of how people perceive and interact with objects, and this is the type of word meaning addressed by the brain-inspired model described in Section 20.4.

Do computers represent word meanings like the human brain does? Significant research on semantic representations has focused on issues related to meaning representation and has given rise to different classes of models. The following sections describe two models relevant to this chapter: the DSM, used by today's digital voice assistants, and the Embodied or brain-based semantic features model (CARs), supported by substantial evidence on how humans acquire and learn concepts through sensory-motor, affective, social, and cognitive interactions with the world (Binder, 2016; Binder & Desai, 2011; Binder et al., 2009, 2016).

20.3 Models of Semantic Representation

Humans have a remarkable ability to form new meanings. For example, the same concept can be combined to produce different interpretations: *red apple* means a fruit having a certain color, *apple basket* means a basket used to hold apples, and *apple pie* means a pie made from apples (Wisniewski, 1997). Since *basket* is an inanimate object, *apple* in this context is likely to have salient properties related to the inanimate category (Small, Shape, Pattern, Texture, Near). In contrast, *apple pie* is palatable, which suggests having salient properties related to animate objects and the food category (i.e., Smell, Taste, Color, Benefit, Pleasant, Happy). Thus, how do language users determine the category membership structure of such combinations of concepts, and how do they deduce their interpretation? As this example illustrates, there are no simple rules, e.g., for how *apple* combines with other concepts. Consider, for example, digital voice assistants such as Siri, Google, and Alexa. These applications are built to recognize speech and to give a response posed by humans in natural language. However, whereas humans process language at many levels, machines process linguistic data with no inherent meaning (i.e., not connection to the physical world). They have no idea what the color *red* actually is. Their linguistic interactions with users are therefore limited to simple responses. Given the ambiguity and flexibility of human language, modeling human semantic representations is essential in building AI systems that interact more effectively with humans.

There are many models of semantic representation in the literature, but most of them fall into two general classes: theories based on relations among words, i.e., those in which a word's meaning is represented by way of its relation to other words (Bruni et al., 2014; Mitchell et al., 2008; Mitchell & Lapata,

2010), and feature-based, i.e., those in which a word's meaning is represented with a set of features as basic components of meaning, which together make the meaning of the word. Feature-based models further differ in the way the features are defined, i.e., whether they are abstract or embodied (Binder et al., 2009). The following models are briefly reviewed with emphasis on relevant aspects to the case study presented in this chapter.

20.3.1 Distributional Semantic Models (DSMs)

Semantic representation is the foundation of many AI tasks such as information retrieval, question answering, machine translation, and document summarization. Former approaches were limited to algorithms that required the use of hand-annotated knowledge bases like WordNet (e.g., Resnik 1995, 1999) and part-of-speech treebanks (Marcus et al., 1993). Current techniques in language modeling (i.e., artificial neural networks) produce word representations (i.e., embeddings) based on the idea that lexical semantic information is reflected by word distribution (Harris 1954, Firth, 1957, Miller 1986). Specifically, DSMs work by extracting co-occurrence patterns and inferring associations between concepts (words) from a large text-corpus without explicitly specifying any knowledge.

Digital voice assistants' semantic representation is based on these models. This approach builds representations by directly extracting statistical regularities from a large natural language corpus (e.g., books, newspapers, encyclopedias, and online documents). The assumption is that large text corpora are a good approximation for the language humans are exposed during their lifetime. Essentially, DSMs built semantic representations by counting the co-occurrences of words within a sliding window. These co-occurrences produce a word-by-word or word-by-document co-occurrence matrix that serves as a spatial representation of meaning. Thus, semantically related words are closer in their representations (i.e., ear, eye, and nose all acquire very similar representations). Figure 20.1 shows the word representations based on the DSMs. (a) The matrix encodes the meaning of the words by directly extracting word co-occurrences from the sentence *the dog ran in the park* with a window size of 1 (left and right of the word under analysis). (b) The matrix encodes the meaning of the words by directly extracting word co-occurrences from a sample of documents (context). (c) Illustration of vector space of word representations. Semantically related words appear closer in the semantic space.

A recurrent neural network (RNN) model, word2vec that falls within the same DSM principles, has gained substantial popularity in the last few years due to its remarkable performance on a variety of semantic tasks. It uses novel techniques to deal with common problems faced by other models such as removing frequently used words, considering context by using a window to predict the intended word or predicting the surrounding context words given an input word, and refining representations by using

	the	dog	ran	in	park
the	0	1	0	1	1
dog	1	0	1	0	0
ran	0	1	0	1	0
in	1	0	1	0	0
park	1	0	0	0	0

(a) Word by word co-occurrence matrix

	C1	C2	C3	C4	C5
W1	1	0	0	2	0
W2	0	4	1	0	0
W3	2	0	0	1	0

(b) Word by context (document) co-occurrence matrix

(c) Illustration of vector space word representations

FIGURE 20.1
Word representations based on the Distributional Semantic Models (DSMs). (a) The matrix encodes the meaning of the words by directly extracting word co-occurrences from the sentence *the dog ran in the park* with a window size of 1 (left and right of the word under analysis). (b) The matrix encodes the meaning of the words by directly extracting word co-occurrences from a sample of documents (context). (c) Illustration of vector space of word representations. Semantically related words appear closer in the semantic space.

negative sampling so unrelated words are suppressed from learning. These meaningful representations are able to solve higher-level semantic relationships such as analogies: *man is to king* as *woman is to*?? – a characteristic that was not present in previous models. However, the cognitive plausibility of this model and most of the DSM has been addressed by some researchers pointing out that (i) negative sampling, (ii) training on vast amounts of data, (iii) static semantic representations, and (iv) lack of sensorimotor experience do not adequately account for how humans generate rich semantic representations.

Humans have lesser linguistic input compared to the corpora that modern semantic models are trained on (Lake et al., 2016). The Embodied perspective (Binder & Desai, 2011) states that humans have access to an abundance of non-linguistic sensory and environmental input, which is likely responsible for their rich semantic representations (Lake et al., 2016). The following section presents how semantic representations are formed through such a model.

20.3.2 Embodied Semantic Model

All DSMs rely on large text corpora to construct semantic representations. A consistent criticism of these models comes from the grounded (embodied) cognition view (Barsalou, 2008), which rejects the idea that meaning can be represented by means of deriving semantic representations from only linguistic texts that are not grounded in perception and action but only in other words. Rather, grounded (embodied) cognition research suggests that sensorimotor modalities, the environment, and the body play a central role for the construction of meaning.

One alternative proposed by Binder and Desai (2011) is to model concept representations based on known brain systems. The CAR theory (a.k.a. The Experiential Attribute Representation Model) is supported by substantial evidence on how humans acquire and represent concepts through visual, auditory, sensory-motor, affective, social, and cognitive interactions with the world (Binder, 2016; Binder & Desai, 2011; Binder et al., 2009, 2016). The central axiom of this theory is that concept knowledge is built from experience; as a result, knowledge representation in the brain is not static. This process starts from birth: babies learn about objects through sensory input (e.g., *parrot* is *loud*). As they develop, the dimensions to such concepts expand through more modalities, including visual, somatosensory, and auditory (e.g., *parrot* is *green*, has *feathers*, and is *musical*). Later in life, humans connect previous concepts to new ones (e.g., *parrots* are similar to *penguins*), while also learning how to relate and differentiate between concepts.

This theory suggests that conceptual knowledge can be decomposed into a set of features that are mapped to individual brain systems (i.e., sensory, motor, visual, spatial, temporal, and affective). It is based on these assumptions: (i) recalling a concept stimulates the features that were active when the concept was first experienced; (ii) concepts with similar features produce similar neural patterns; and (iii) context modifies the baseline meaning of a concept. In this brain-based semantic model, the features are weighted according to statistical regularities. The semantic content of a given concept is estimated from ratings provided by human participants. For example,

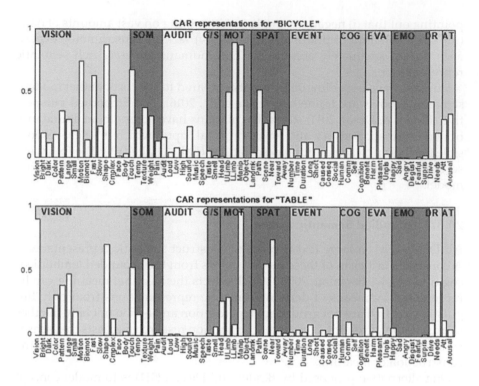

FIGURE 20.2

Bar plot of the 66 semantic features for the words *bicycle* and *table* (Binder & Desai, 2011; Binder et al., 2009, 2016). Given that both concepts are objects, they have low weightings on animate attributes such as Face, Body, Speech, Human, and emotions including Sad and Fear and high weighting on attributes like Vision, Shape, Touch, and Manipulation. However, they also differ in expected ways, including stronger weightings in Motion, Fast, Lower Limb, and Path for *bicycle* and stronger weightings in Smell, Scene, Near, and Needs for *table*. Weighted features for the words *bicycle* and *table*.

concepts referring to things that make sounds (e.g., *explosion, thunder*) receive high ratings on a feature representing auditory experience (Noise, Loud), relative to things that do not make a sound (e.g., *milk, flower*).

CAR theory models each word as a collection of 66 features (shown in Figure 20.2) that captures the strength of association between each neural attribute and word meaning. Specifically, the weight of activation of each feature associated with the concept can be modified depending on the linguistic context, or combination of words in which the concept occurs. Thus, people weigh concept features differently to construct a representation specific to the combination of concepts in the sentence. For example, *plastic bottle* is a bottle made out of plastic (inanimate-related features are strongly activated, such as Size, Texture, Small), but *baby bottle* is not a bottle made out of babies (animate-related features are clearly activated such as Biological Motion,

Affective, Social). There are some general principles that govern such combinations as part of people's semantic knowledge.

Figure 20.2 shows the weighted CARs for the concepts *bicycle* and *table*. The weight values represent the average human ratings for each feature. Given that both concepts are objects, they get low weighting on animate attributes such as Face, Body, Speech, Human, Communication, and emotions such as Sad, Angry, Disgust and Fear, and high weighting on attributes like Vision, Shape, Touch, and Manipulation. However, they also differ in expected ways, including stronger weightings for *bicycle* on Motion, Biomotion, Fast Motion, Lower Limb and Path, and stronger weightings for *table* on Large, Smell, Head, Scene, Near, and Needs.

The terms concept, word, and word meaning have specific instantiation in CAR theory, and this instantiation is used throughout this chapter. The relation of thought to language is seen as the relation of concepts to meanings. Concepts are seen as a collection of individual features encoded in different neural systems according to the way they are experienced. Words are the symbolic names of concepts, and word meanings are generated when a word is recognized in the interaction with its context (Ogden & Richards, 1923). CAR theory thus integrates concepts and word meanings in the same semantic representation. The weights given to the different features of a concept collectively convey the meaning of a word (Binder, 2016; Binder & Desai, 2011; Binder et al., 2009, 2016; Yee & Thompson-Schill, 2016).

Building on this theory of grounded word representation, the neuroscientists aimed to understand how word meaning changes depending on context. The next section reviews the main contributions of the collaboration between AI and neuroscience and concludes with a general view of the opportunities brought by this combined effort.

20.4 Case Study: The CEREBRA Model

A great deal of research in neuroscience has focused on how the brain creates semantic memories, and what brain regions are responsible for the storage and retrieval of the semantic knowledge. Early studies focused on behavior of individuals with brain damage and with various types of semantic disorders, but recent studies employ neuroimaging techniques to study the brain mechanisms involved.

This study follows the theory proposed by Binder & Desai (2011) to develop a computational framework to address the following question: How do word meanings change in the context of a sentence? If indeed they do, it should be possible to see such changes by directly observing brain activity (fMRI patterns) during word and sentence comprehension.

The study consisted of three parts: first, Binder and his team collected brain imaging data from several subjects reading everyday sentences by recording visual, sensory, motor, affective, and other brain systems. Those recordings were limited to the regions considered as part of the semantic network (Binder & Desai, 2011). Eleven participants took part in this experiment. The fMRI images consisted of 240 sentences per participant (e.g., *The flood was dangerous, The mouse ran into the forest, The dangerous criminal stole the television*). To obtain the neural patterns, subjects viewed each sentence on a computer screen while in the fMRI scanner. The sentences were presented word-by-word, and participants were instructed to read the sentences and think about their overall meaning.

Second, the CEREBRA computational model was trained with the collected sentence fMRI patterns and the CAR semantic feature patterns to characterize how word meanings are modulated within the context of a sentence. With CARs of words as input, a neural network is trained to generate first approximations of fMRI patterns of subjects reading sentences. Then, the same neural network with back-propagation extended to inputs was used to modify the CARs to predict the fMRI patterns more accurately (Aguirre-Celis, 2021). These changes represent the effect of context; it is thus possible to track the brain dynamic meanings of words by tracking how the CARs change across contexts.

Third, CEREBRA was evaluated through three computational experiments. They were designed to demonstrate that words in different contexts have different representations, that the changes observed in the concept attributes encode unique conceptual combinations, and that the new representations changed towards their respective context.

An individual example visualizing how CEREBRA characterizes the effect of similar context for the words *boat and car* is shown in Figure 20.3. The salient attributes of these concepts are compared under the semantic category of transportation vehicles as expressed in 57: *The boat crossed the small lake* and 142: *The green car crossed the bridge*. Context draws attention to a subset of attributes, which are then heightened, forming the basis for object categories (Binder, 2016). The figure shows the results averaged across 11 subjects. For *boat* in sentence 57, there are changes on Vision, Large, Motion, Shape, Complexity, Weight, Sound, Manipulation, Path and Scene, and event attribute Away, reflecting a large moving object. Evaluation and Emotion attributes of Benefit, Pleasant, and Happy represent the experiential and personal nature of using a boat. Similarly, *car* in sentence 142 shows analogous activation for the same brain areas. Since both belong to the same semantic category, they share similar context-related attributes. However, the distinctive weighting on these attributes sets them apart. The CEREBRA model is thus able to find the effect of similar context on these two concepts across subjects.

This section briefly described a case study where CEREBRA served as an instrumental tool for interpreting fMRI patterns. From this study, neuroscientist could gain a better understanding on how semantic knowledge is

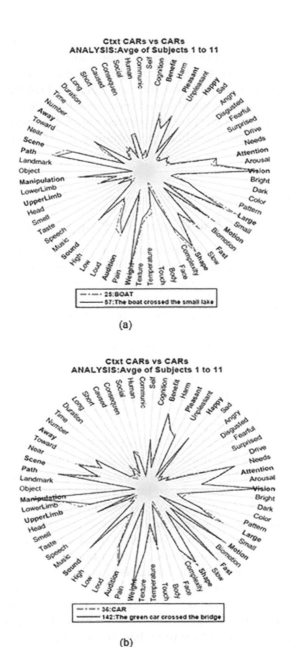

FIGURE 20.3
The effect of similar context for the words (a) *boat* and (b) *car* averaged across subjects. Results are shown for the modified CARs as an average of all subjects. Dotted lines indicate the original CARs, and solid lines specify the context-based representations. Both plots display similar changes, but the different weightings set them apart. One or more concepts share the same context-related attributes, thus forming the vehicle category.

represented, combined, and modified. Overall, the results from this study are expected to advance the development of a unified theory of concepts, of the organization of the semantic space, and of the processes involved in concept representation. In addition, the modified CEREBRA representations might provide enhanced context-based representations for systems such as Siri, Google, and Alexa, to advance grounded natural language understanding systems, supporting embodied applications (e.g., service robots).

20.5 Hybrid Approach to Ground NLP Applications

Language grounding refers to understanding the meaning of words as it applies to the physical world. It assumes that the perceiver is aware of the world, the context, and the communication techniques (e.g., oral, written, visual). However, knowledge acquired experientially is not sufficient to fully account for the brain's semantic representations (Andrews et al., 2009; Anderson et al., 2019). People have knowledge of exotic animals, facts, places, science fiction books, and historic events that they have never experienced. For example, they might know about polar bears, Apollo 11, the Great Wall of China, Frankenstein, and the Jewish Holocaust even though they never encountered them. Yet, distributional approaches are disembodied from the physical world. How, then, can the full semantic space be represented?

To build real-world applications (i.e., NLP systems), it is crucial to use conceptual grounding, and multimodal CEREBRA's grounded representations could be the best option to make such applications more robust. By combining a text-based semantic model and the dynamic representation of experiential-based modified CARs (produced by CEREBRA), current text-based NLP applications (i.e., Siri, Google, and Alexa) could be refined. In addition, such blend might provide some insight on how both types of knowledge interact to produce interesting complex behavior and consequently, benefits future research on brain functions associated with language.

An artificial neural network model that learns to represent context simultaneously from two different sources, such as DSMs representations (linguistic knowledge) and the multimodal CEREBRA representations (experiential knowledge), can be an excellent choice for a natural language understanding system deployed in a service robot (i.e., Agriculture, Medicine, Security). Specifically, it learns from large text corpora of sentence contexts and target words through an artificial neural network architecture. Further, CEREBRA representations are merged with it to provide supplementary knowledge that should produce better performance.

The proposed model consists of two RNNs and a multilayer perceptron (MLP) for learning generic sentence context representations from large text corpora (see Aguirre-Celis 2021 for more details). CEREBRA representations, which provide a different kind of context (i.e., experiential context), can be added to it to provide supplementary knowledge that should improve the natural language understanding process. The dual-knowledge architecture with the CEREBRA extension (red box) is outlined in Figure 20.4.

In this architecture, one RNN reads words from left to right, and another from right to left. The parameters of these two networks are independent,

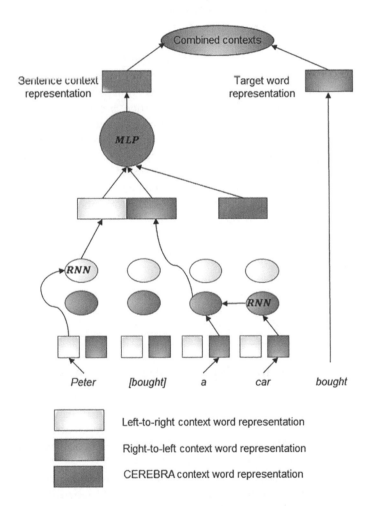

FIGURE 20.4
Dual-knowledge neural network model enhanced with multimodal CEREBRA representations. Context representations for linguistic knowledge are developed using RNNs. One RNN reads words from left to right, and another from right to left. The outputs of the two RNNs are concatenated and fed into an MLP together with the CEREBRA vector. At the same time, the target word is represented with the same dimensionality as that of the sentence contexts. The output of this layer is the representation of the combined sentence context. Thus, the architecture learns generic context representations by integrating linguistic and experiential knowledge for NLP.

including the left-to-right and right-to-left context word representations. For example, to represent the context of a target word in the sentence *Peter [bought] a car*, the architecture concatenates the RNN output vector representing its left-to-right context (*Peter*) with the one representing its right-to-left context (*a car*). This concatenated vector and the CEREBRA vector are fed into an MLP to learn dependencies between the different contexts. At the same time, the target word is represented with the same dimensionality as the sentence contexts. The output of this layer is thus the representation of the combined sentence context.

Basically, the proposed model combines linguistic and experiential information at the contextual level. Thus, the CEREBRA representations would provide the experiential-based data (i.e., concrete words) and the text-based representations would provide the association-based data (i.e., abstract words), leading to a better performance. For instance, agricultural service robots with such representations would have the ability to understand natural language commands (i.e., watering plants), to have encyclopedic knowledge (i.e., to make decisions regarding weed or pest control), to ground language by adapting to the environment (i.e., object recognition and location to plant seeds, prune, or harvest), and by understanding novel concepts (i.e., "rainwater"). For dynamic environments, such robots could accommodate an additional mechanism (e.g., a library of modified CARs) to adapt to new contexts, by using the closest new CAR representation to deliver a possible solution. The next section presents some ideas on how to expand AI algorithms to undertake more brain-like functions (i.e., dynamic learning).

20.6 Challenges for Building More Brain-Like AIs

Today's AI machines, although far from human competence, have shown impressive results across a wide range of tasks in NLP. AI still has fundamental limitations compared to truly intelligent agents that inspired the founders of the field 72 years ago. However, these shortcomings can be seen as challenges to advance the state of the art in AI.

To address these challenges, AI researchers are aiming to recreate cognitive functions such as attention, memory, continual learning, and causal inference. These building blocks are not functioning simultaneously in current AI systems (like they are in humans) but the results are encouraging in many ways, promoting further understanding of how the brain works.

For example, AI researchers discovered that by adding attention mechanisms to AI algorithms has improved their performance. Recent models such as transformer neural networks are designed with a type of attention focusing on specific aspects of the input and ignoring irrelevant information. This approach has improved their ability to process language input considerably.

Another cognitive function that has been a major obstacle for creating brain-like systems with AI is memory (i.e., manipulation of stored concepts; Macpherson et al., 2021). Inspired by biological studies on neural circuits, artificial RNNs have been designed to have memory: they utilize information from prior inputs to influence the values of the current output. RNNs have been successful in many tasks involving sequences of words, such as machine translation, sentiment analysis, image captioning, and summarization. However, even advanced RNNs have limitations while working with long data sequences. By adding an attention mechanism, RNNs can focus on important parts of the input sequence when predicting the output sequence, making such predictions more accurate.

A third research focus is learning. Humans can continually learn, fine-tune, and transfer knowledge throughout their lifetime. This ability is controlled by a set of cognitive mechanisms that contribute to the development and specialization of sensorimotor skills as well as to long-term memory consolidation and retrieval. Continual learning skills are central to AI systems and to autonomous agents interacting in the real world. However, such learning remains a challenge since continual acquisition of information from non-stationary data distributions generally leads to catastrophic forgetting, i.e., loss of earlier information.

Fourth, today's AI systems are data-hungry, training requires labeled data, and the neural networks often learn to associate the labels with minimal or superficial information. For example, a neural network might use the presence of grass to recognize a photo of a cow because cows are typically pictured in fields. As a result, some researchers have been exploring neural networks trained with little or no human-labeled data, suggesting that "The labels come from the data itself." For instance, the network augments images and labels the original and augmented images as a positive pair, different images as negative pair, and pushes the learned features of negative pairs away while moving positive features close (as in learning clusters of similar data points). This process enables the network to learn to group images of similar classes, which further makes tasks like classifications possible, without given any labels.

A fifth research topic is inference. Human cognition has a remarkable ability to learn new concepts from previous knowledge through inductive inference. In contrast, deep learning systems rely on massive amounts of training data to recognize patterns. Recent work in generative adversarial networks (GANs) has started to incorporate brain-inspired inference mechanisms into AI programs. These models can make inferences about a new concept despite little data, and generate new samples from a single example concept. The idea of GAN is to let a generator and a discriminator play an adversarial game with each other. GAN is a generative model that has demonstrated impressive performance. As a result, they can learn to generate distributions of many kinds of data such as images, music, speech, prose, creating novel outputs similar to real-world exemplars (Bishop, 2021).

In sum, the integration of different cognitive functions into AI systems continues to be one of the greatest challenges of the field. Fortunately, the new generation of neural networks is rapidly expanding beyond basic connections between neurons and recreating some of the core building blocks of human intelligence.

20.7 Conclusion

Since the beginning of AI, the brain has been an important inspiration for developing AI mechanisms. While modern neural networks perform at the same level or better than humans in many tasks (i.e., Go, Poker, Gran Tourism, and image classification), it does not mean that they learn and think in the same way. AI mechanisms and behaviors are still remarkably simple compared to those of the human brain.

A promising step was presented in this chapter: the CEREBRA AI model was used to assist neuroscience in understanding how the brain represents and adjust word meanings in the context of a sentence. The approach combines two semantic models: a dynamic representation of modified CARs as a contextual source of the experiential-based knowledge, and a contextual representation of the linguistic knowledge. Such a blend might serve as the next step in understanding how both types of knowledge interact, producing interesting complex behavior and benefiting future research on brain functions associated with language.

Two important differences between AI and brain computations were further identified in this chapter: (i) people learn more from fewer examples and (ii) they acquire richer representations. Humans learn more than what the digital voice assistants collect from the data; people learn a concept in a way that allows it to be applied to many different situations. In addition to recognizing new examples, humans may transform them or even generate new ones. These abilities follow from the acquisition of the underlying concept. Adding cognitive functions, such as Attention, Memory, Life-long learning, Self-Supervised Learning, and Causal Inference to AI algorithms, should help in engineering better solutions to specific tasks.

While the brain has always been an inspiration for AI, it is now also likely for AI scientists to come up with solutions that neuroscientist can test in order to better understand the brain. Thus, the mutually beneficial relationship of neuroscience and AI research is likely to produce most interesting developments of the next decade.

References

Aguirre-Celis, N (a.k.a. Aguirre Sampayo, Nora). From words to sentences and back: Characterizing context-dependent meaning representations in the brain (Doctoral dissertation), Instituto Tecnológico y de Estudios Superiores de Monterrey, Monterrey, Mexico. 2021.

Anderson, A.J., Binder, J.R., Fernandino, L., Humphries, C.J., Conant, L.L., Raizada, R.D., Lin, F., & Lalor, E.C. An integrated neural decoder of linguistic and experiential meaning. *The Journal of Neuroscience: The Official Journal of the Society for Neuroscience*, 39(45): 8969–8987. 2019.

Andrews, M., Vigliocco, G., and Vinson, D. Integrating experiential and distributional data to learn semantic representations. *Psychological Review*, 116(3):463–498. 2009.

Barsalou, L. W. Grounded cognition. *Annual Review of Psychology*, 59:617–645. 2008.

Binder, J. R., Desai, R. H., Graves, W. W., Conant, L. L. Where is the semantic system? A critical review of 120 neuroimaging studies. *Cerebral Cortex*, 19:2767–2769. 2009.

Binder, J. R., Desai, R. H. The neurobiology of semantic memory. *Trends Cognitive Sciences*, 15(11):527–536. 2011.

Binder, J. R. In defense of abstract conceptual representations. *Psychonomic Bulletin & Review*, 23. 2016. doi:10.3758/s13423-015-0909-1.

Binder, J. R., Conant, L. L., Humpries, C. J., Fernandino, L., Simons, S., Aguilar, M., Desai, R. Toward a brain-based Componential Semantic Representation. *Cognitive Neuropsychology*, 33(3–4):130–174. 2016.

Bishop, J. Artificial intelligence is stupid and causal reasoning will not fix it. *Frontiers in Psychology*, 11. 2021. doi:10.3389/fpsyg.2020.513474.

Bruni, E., Tran, N., Baroni, M. Multimodal distributional semantics. *Journal of Artificial Intelligence Research (JAIR)*, 49:1–47. 2014.

Firth, J. R. A synopsis of linguistic theory, 1930–1955. *Studies in Linguistic Analysis*. 1957.

Harris, Z. S. Distributional structure. *Word*, 10:146–162. 1954.

Lake, B. M., Ullman, T. D., Tenenbaum, J. B., Gershman, S. J. Building Machines That Learn and Think Like People. https://arxiv.org/abs/1604.00289 [cs.AI]. 2016.

Macpherson, T., Churchland, A., Sejnowski, T., DiCarlo, J., Kamitani, Y., Takahashi, H., Hikida, T. Natural and Artificial Intelligence: A brief introduction to the interplay between AI and neuroscience research. *Neural Network*. 2021. doi:10.1016/j.neunet.2021.09.018.

Marcus, M., Santorini, B., Marcinkiewicz, M. A. Building a large annotated corpus of English: The Penn Treebank. *Computational Linguistics*, 19(2):313–330. 1993.

Mitchell, T. M., Shinkareva, S. V., Carlson, A., Chang, K.-M., Malave, V. L., Mason, R. A., Just, M. A. Predicting human brain activity associated with the meaning of nouns. *Science*, 320:1191–1195. 2008.

Mitchell, J., Lapata, M. Composition in distributional models of semantics. *Cognitive Science*, 38(8):1388–1439. 2010. doi:10.1111/j.1551-6709.2010.01106.x.

Meteyard, L., Rodriguez Cuadrado, S., Bahrami, B., Vigliocco, G. Coming of age: A review of embodiment and the neuroscience of semantics. *Cortex*, 48:788–804. 2012.

Miller, G. A. Dictionaries in the mind. *Language and Cognitive Processes*, 480(1):171–185. 1986.

Ogden, C. K., Richards, I. *The Meaning of Meaning*. London: Harcourt, Brace and Company, Inc. 33. 1923.

Resnik, P. Using information content to evaluate semantic similarity in a taxonomy. arXiv preprint cmp-lg/9511007. 1995.

Resnik, P. Semantic similarity in a taxonomy: An information-based measure and its application to problems of ambiguity in natural language. *Journal of Artificial Intelligence Research*, 11:95–130. 1999.

Vigliocco, G., Vinson, D. P. Semantic representation. In G. Gaskell (ed.), *Handbook of Psycholinguistics*. Oxford: Oxford University Press. 2007.

Vigliocco, G., Meteyard, L., Andrews, M., & Kousta, S. Toward a theory of semantic representation. *Language and Cognition*, 1(2), 219–247. 2009. doi:10.1515/LANGCOG.2009.011.

Wisniewski, E. J. When concepts combine. *Psychonomic Bulletin & Review*, 4, 167–183. 1997.

Yee, E., Thompson-Schill, S. L. Putting concepts into context. *Psychonomic Bulletin & Review*, 23, 1015–1027. 2016.

21

Adversarial Robustness on Artificial Intelligence

Ivan Reyes-Amezcua
CINVESTAV Unidad Guadalajara

Gilberto Ochoa-Ruiz
Tecnológico de Monterrey

Andres Mendez-Vazquez
CINVESTAV Unidad Guadalajara

CONTENTS

21.1 Recent Advances in Artificial Intelligence

Most recent Artificial Intelligence (AI) advances have revolved around machine learning (ML) methods. Such methods can learn from data and create models that represent a problem (e.g., regression, classification). Traditional ML techniques are useful when working with structured data (i.e., tabular or grid data). During the process of ML, one essential step is to extract features from data, which are then used to learn and improve the model. This is generally done by manually designing the data to look for relationships across independent variables. However, when dealing with unstructured data (e.g., images, text, and sound), ML techniques struggle with extracting effective

DOI: 10.1201/b23345-24

features from the data. This is due to the complexity of the data, which does not permit to extract sufficiently meaningful and powerful representations to feed and train the underlying ML models. Recent strides in AI have happened thanks to a series of factors that have enabled deep learning (DL) models come to the fore of ML research. In particular, DL models have shown an astounding capability for extracting highly efficient features and to learn very complex functions of the features for a great variety of talks, in an end-to-end and automatic manner.

Artificial neural networks are the basis of DL methods. They are basically a neural network, where the fundamental building block are single artificial neurons (perceptron). These neurons take a set of *inputs* to be multiplied by a set of *weights* and get a sum of all of them. The learning and training process of DL models consists in to optimize a *loss function* by adjusting the *weights* of the neural network. The optimization of the *weights* is iteratively modified by using *gradient descent* across several iterations or epochs. The gradient computes how small the adjustment to each parameter θ that will affect the loss function. For deep neural networks, the gradient is computed via backpropagation.

In recent years, both ML and DL applications have been increased in popularity in the research and industrial areas. Examples of successful applications can be found in image and video classification, facial recognition, malware detection, autonomous driving, and natural language processing, among many others. Moreover, constant and rapid data generation has been a key factor in the explosion of ML and DL. It is expected that the rate of data generation will increase in the next years due to the transformation in the industry with smart devices and IoT.

On the other hand, when dealing with model deployment on critical production systems, it should be expected that models do not have vulnerabilities or weaknesses, but in most cases, this is not true. Models can be fooled in several ways, from altering their training data to produce a biased model, to craft adversarial inputs to create a confident incorrect model. Several adversarial attacks can be found in the state-of-the-art, and most of them attempt to use model gradients to maximize a loss function with respect to the input x to create an adversarial example. In contrast, adversarial defenses are focused on preventing some type of attack by inserting adversarial examples into the model training phase. However, there is still a trade-off between model benign performance and adversarial performance. For such reasons, opportunities on adversarial robustness are emerging for both academic and industry purposes to enhance ML models (Figure 21.1).

21.2 Model Vulnerabilities

As the ML model gets increasingly better in terms of performance, the tolerance to possible hacking attacks must be improved in the same manner.

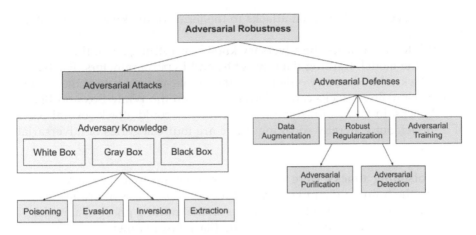

FIGURE 21.1
An overview of the taxonomy of adversarial robustness attacks and defense methods. We broadly categorize attacks into three categories according to the degree of knowledge of an adversary (White, Gray, and Black Box), and by the four types of attacks (poisoning, evasion, inversion, and extraction). In defense methods, we mainly categorize into five types of robust defenses.

Security in critical applications shall be a priority when an AI model is responsible for taking decisions that can affect the safety of people. Furthermore, if an AI model can take decisions without human supervision, they can lead to mistrust if there is uncertainty about their lack of robustness. For instance, autonomous driving is a critical AI application on which a system must control autonomously the status of a vehicle. In such systems, it must be integrated several rigorous tests for model architecture and process of the pipeline to validate robustness to vulnerabilities. Therefore, the need for methods and techniques that guarantee the correct functionality of a model is of utmost importance.

The development and deployment of ML systems in real-world scenarios induce special attention not only to good performance most of the time but also in a robust manner. The term "robustness" has gained common usage in recent times, referring to the ability of classifiers to withstand adversarial attacks or perturbations in input data during testing, thereby preventing the model from being fooled. This concept of robustness is attracting great interest in academia and industry to address the shortcomings of advanced ML systems, in particular DL models.

21.2.1 Adversarial Robustness

When speaking of vulnerabilities of ML/DL models, the adversarial ones are the most important to deal with. Exploitations of such vulnerabilities attempt to find what would be the inputs of a model that change the prediction in a malicious way. We refer to these inputs as adversarial attacks, and there are three types of adversarial attacks in the terms of the knowledge that an adversary knows.

In White-Box Attacks, the adversary knows everything about the architecture of the model, training data, weights, and hyperparameters. For Black-Box Attacks, the adversary only knows the inputs and outputs of the model. Finally, Gray-Box Attacks can be located between the White-Box and Black-Box Attacks. In terms of taxonomy, an adversary could attack several parts of a model, but generally speaking, there are four categories of adversarial attacks.

1. **Poisoning attacks**: Here, the adversary attempts to make alterations on the training data to create a model that produces a wrong classification. This kind of attack works on the training phase of a model and can be done in a White-Box and Black-Box manner.

2. **Evasion attacks**: One of the most common attacks, for this, an adversary tries to exploit the model by using several attacks to find out where there is a misclassification in the prediction. This is done by injecting or adding a small amount of crafted noise to the inputs. The challenge herein is to generate just a small amount of noise that can lead to a big error in the prediction. To create an adversarial example, the adversary strategy should modify small parts of the inputs to fool a model. The most common data types used for these attacks are images. For such reason, the main security risk is that the adversarial images (malicious ones) are almost imperceptible to the human eye. Furthermore, the adversary could be more precise at selecting the output of the prediction since the attack strategy can be directed or undirected.

3. **Inversion attacks**: The goal of a model inversion attack is to violate user privacy by using model predictions to determine whether or not a certain piece of data was included in the training set. It can be used in both Black-Box and White-Box scenarios. Models that have been trained using sensitive data, such as those that might be included in the training set, are particularly susceptible to this kind of attack. The robustness to this type of perturbations is especially pertinent to models that are applied to sensitive data, such as clinical data, which might require extra security.

4. **Extraction attacks**: An adversary tries to steal the settings of a ML model in a model extraction assault. This kind of attack makes it possible to carry out evasion and/or inversion attacks as well as damage a model's intellectual property and confidentiality. Both a White-Box version and a Black-Box version of this attack are possible. This approach has been used to avoid paying for using Machine Learning as a Service (MLaaS) by stealing the intellectual property of a model. To create a model that is placed in a system

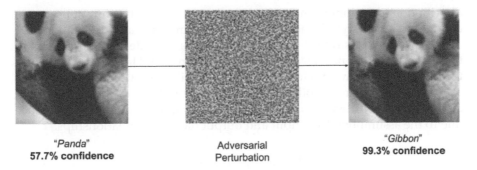

"*Panda*"
57.7% confidence

Adversarial
Perturbation

"*Gibbon*"
99.3% confidence

FIGURE 21.2
A case of adversarial attack [1]. With 57.7% certainty, the original entry identifies the image as a "panda." With a given level of noise added to the image, it receives a 99.3% "gibbon" classification.

that is accessible to the attacker through an API, a legitimate user either trains his model from data or outsources the training to MLaaS providers. Then, a legitimate user trains his model from data or delegates the training to MLaaS services to obtain a model that is deployed in some system, accessible to the attacker through an API. Finally, by making requests to the API, the attacker is able to deduce (to a greater or lesser extent) a model equivalent to the legitimate model (Figure 21.2).

21.3 Limitations of Current Models

It is well known that deep neural networks are susceptible to adversarial and natural image alterations. In particular, when deploying DL systems on critical and high-safety tasks, it is essential to develop models that are robust enough to mitigate vulnerabilities to natural image corruptions. Robustness can have different definitions according to the author and the context of the model task. Most of ML methods and techniques assume that training and testing sets are identical on their distributions [2].

Generally, non-adversarial robustness refers to preserving the model performance under natural manipulations on images [3]. In other words, it is the preservation of the model performance under naturally altered samples generated from the original data distribution, without an explicit adversarial search for these samples. Here, the robustness objective is to deal with transformed samples of the original unaltered image domain distribution. Hence, altered images can be identified as samples of datasets shifts of the original data domain. Therefore, a dataset shift can be thought as a change in the joint distribution of the inputs and outputs of the training and testing subsets, i.e., changes in image style, location, occlusion, zoom, scale, and orientation.

We say a model is robust when the performance under certain metrics in the testing set does not deviate from the training set. However, testing and training data can have distinct probability distributions (Eq. 21.1), producing a shift between them. In other words, the distributions of the training and testing data are different, resulting in a degradation of model performance due to the assumption that input and output have static relationships.

$$P_{\text{train}}(X,Y) \neq P_{\text{test}}(X,Y) \tag{21.1}$$

The shifts that a single distribution could have can differ in terms of the temporal or spatial evolution of data, the data collection/sampling process, or be even unknown due to the underlying behavior of the data. According to Ref. [4], the out-of-distribution (OOD) generalization problem is insolvable unless one makes assumptions about test distribution shifts. For instance, the *covariate shift* is the assumption that marginal distribution of X shifts from training to testing stages, but label generation holds unmodified:

$$P_{\text{train}}(Y \mid X) = P_{\text{test}}(Y \mid X) \rightarrow P_{\text{train}}(X) \neq P_{\text{test}}(X) \tag{21.2}$$

Modeling this process of image generation for understanding dataset shifts is not a trivial task. In fact, it is a problem in real-world domains in which the output may depend on some hidden variables that are not present as predictive features.

On the other hand, adversarial samples can be seen as the worst-case scenario for distributional shifts. This is due to the objective of adversarial attacks, which consists of generating inputs visually indistinguishable from benign ones.

21.4 Designing Robust Models for Evasion Attacks

Let a model or hypothesis function be defined as:

$$h_\theta : X \rightarrow \mathbb{R}^k \tag{21.3}$$

which is a mapping function from our inputs (images) to a k-dimensional vector, where k is the number of classes being predicted. Note the output corresponds to a logit space, so these real values can be positive or negative.

Also, θ stands for all the parameters learned by the model, which are what typically optimized during the training of a neural network. Hence, h_θ is our model object.

Then, we define our loss function $\ell : \mathbb{R}^k \times \mathbb{Z}_+ \to \mathbb{R}_+$ as a mapping from the model predictions (logits) $\mathbb{R}^k$ and true labels (index of true class) $\mathbb{Z}_+$ to a non-negative number $\mathbb{R}_+$ +. Thus, $\ell(h_\theta(x), y)$ is the loss for $x \in X$ the input and $y \in \mathbb{Z}$ the true class that a classifier achieves when prediction x assuming the true class is y. The most common loss function for DL-based classifications is the cross-entropy loss.

$$\ell\big(h_\theta(x), y\big) = -\sum_{j=1}^{k} y \log\big(p\big(h_\theta(x)_j\big)\big) \tag{21.4}$$

where $h_\theta(x)_j$ denotes the jth elements of the vector $h_\theta(x)$, and p is the Softmax probability for the jth class.

21.4.1 Adversarial Examples

Adversarial examples are specifically designed to trick the classifier into misclassifying an input as something else. A common approach to train a model is to optimize parameters θ in order to minimize the average loss over the training set $x_i \in X$, $y_i \in \mathbb{Z}$, $i = 1, \ldots m$.

Thus, we can write this as an optimization problem:

$$\underset{\theta}{\text{minimize}} \frac{1}{m} \sum_{i=1}^{m} \ell\big(h_\theta(x_i), y_i\big) \tag{21.5}$$

Next, we solve it by stochastic gradient descent (SGD), for some minibatch $\mathcal{B} \subseteq \{1, \ldots, m\}$ by computing the gradient of the loss with respect to parameters θ and updating θ to its negative direction.

$$\theta := \theta - \frac{\alpha}{|\mathcal{B}|} \sum_{i \in B} \nabla_\theta l\big(h_\theta(x_i), y_i\big) \tag{21.6}$$

where α is the learning rate or step size.

To create an adversarial example is needed to adjust the image x to maximize the loss. The optimization problem is as follows:

$$\underset{\hat{x}}{\text{maximize}}\, \ell\big(h_\theta(\hat{x}), y\big) \tag{21.7}$$

Here, $\hat{x}$ is the adversarial example that is trying to maximize the loss. This is not a trivial task; we cannot change the image completely so that a classifier is being fooled. The adversarial example must ensure that $\hat{x}$ is close to the original input x. This can be done by optimizing over the perturbation to x, which by convention is denoted as δ:

$$\underset{\delta \in \Delta}{\text{maximize}} \, \ell\left(h_\theta(x+\delta), y\right) \tag{21.8}$$

where Δ represents an allowable set of perturbations.

Characterizing the "correct" set of allowable perturbations is quite difficult as matter of fact: in theory, we would like Δ to capture anything that humans visually perceive to be the "same" as the original input x. Δ can include adding slight amounts of noise, rotation, translation, scaling, or transformations on the image locations. It cannot be possible to give mathematically rigorous definitions of all sets of perturbations allowed. We consider a subset of possible space of allowed perturbations, such that the actual semantic of the image content is not changed under this perturbation.

21.4.2 Perturbations Sets

The L_∞ ball is a common perturbation set, defined as:

$$\Delta = \{\delta : \|\delta\|_\infty \leq \epsilon\} \tag{21.9}$$

where the L_∞ norm of a vector z is defined as the maximum norm in z:

$$\|z\|_\infty = \max_i |z_i| \tag{21.10}$$

Hence, the set of perturbations is allowed to have a magnitude between $[-\epsilon, \epsilon]$ in each of its components. Also, we need to ensure that $x+\delta$ is bounded in $[0, 1]$ in order to be considered a valid image. Furthermore, the advantage of L_∞ ball is that for small ϵ, it creates perturbations that add such a small component to each pixel in the image that they are visually indistinguishable from the original image. Therefore, it provides a necessary but definitely not sufficient condition for us to consider a classifier robust to perturbations.

21.4.3 Training Robust Models

After discussing the problem of attacking by evasion DL classifiers in a more formal fashion, let's look at the challenge of training or otherwise modifying existing classifiers to make them more resistant to such attacks.

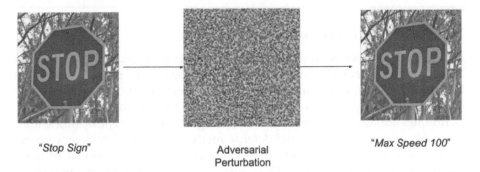

"Stop Sign" Adversarial "Max Speed 100"
Perturbation

FIGURE 21.3
A stop sign with a tiny bit of noise may be misclassified as a "Max Speed 100" by a deep learning model, illustrating the dangers of blindly deploying (even highly accurate) models in critical applications.

The *adversarial risk* of a classifier is the expected loss under the true distribution of data. Rather than experiencing a loss at each sample point, $l(h(x), y)$, we expect the worst-case loss somewhere nearby (Eq. 11).

$$R_{\text{adv}}(h_\theta) = \mathbf{E}_{(x,y)} \sim \mathcal{D}\left[\max_{\delta \in \Delta(x)} \ell(h_\theta(x+\delta)), y\right] \tag{21.11}$$

By using the adversarial risk in an adversarial environment, we can provide an accurate estimate of the expected performance of a classifier. Such environments include spam filters, network intrusion, or traffic signals (Figure 21.3).

A classifier that can resist adversarial attacks (robust) is one that minimizes the empirical adversarial risk. This can be reformulated as an optimization problem. We refer to this step as a min-max robust optimization of adversarial learning.

$$\min_\theta \hat{R}_{\text{adv}}(h_\theta, D_{\text{train}}) = \min_\theta \frac{1}{|D_{\text{train}}|} \sum_{(x,y) \in D_{\text{train}}} \max_{\delta \in \Delta(x)} \ell(h_\theta(x+\delta)), y). \tag{21.12}$$

We solve the above optimization problem by using gradient descent over θ. Furthermore, the gradient of the inner-term function that involves a maximization term is given by the gradient of the function evaluated at the maximum. To put it another way, we iteratively create adversarial examples instead of updating the classifier based on the initial data points. This process, which is now referred to as "adversarial training" in the DL field, is one of the best empirical techniques for developing adversarially resilient models.

21.5 Recent Advances on Robust Models

Although there have been proposals for mitigating adversarial risks for the ML model, there is still a need for enhanced robustness in many settings. Recently, the development and research on adversarial robustness have been focused on creating adversarial attacks to fool a model. In this context, defense techniques are subsequently proposed to mitigate this new type of attack, and so on. Despite many adversarial defense methods proposed in recent years, none of these can guarantee full coverage in terms of security and robustness against all adversarial inputs. Thus, robust adversarial defenses are still a high-demanding and high-priority task for designing trustable ML solutions.

For such reasons, many methods continue to be proposed for creating robust models that are capable of resisting complex adversarial attacks. One of the simplest methods is *Data Augmentation*, including hostile examples in the training set. Compared to the previous approach and its adversarial counterpart, the defender will use these examples to guarantee invariance in the model's output. Also, *Adversarial Logit Pairing* [5] is a technique that ensures that the logits of normal and hostile inputs are invariant. By adding a penalty term to the training objective, this can be accomplished. This approach uses *projected gradient descent* to create a random set of adversarial counterparts x' of a given mini-batch of original samples x.

Furthermore, the adversarial noise from images can be removed by using input transformations. This kind of defense attempts to delete the adversarial perturbations by performing one or more transformations before the input is processed by a model. Examples of such transformations are total variance minimization, image quilting, and JPEG compression. In particular, *PixelDefend* [6] is a defense method that aims to "purify" an image through a generative transformation model. First, a trained generative model (e.g., PixelCNN) on a non-adversarial dataset is used to approximate the distribution of data. Then, an adversarial input is passed through this model to remove the perturbation on the image. These types of defenses are effective against well-known attacks, but they still are vulnerable to recent attacks such as backward pass differentiable approximation (BPDA) [7] and simultaneous perturbation stochastic approximation (SPSA) [8].

On the other hand, regularization methods can be effective for adversarial defense. These methods introduce randomness to the model's gradients during training and inference. This approach is effective due to the additional challenge introduced by randomization, which entails that attackers will need to calculate the exact gradients in order to fabricate an effective attack. Also, another technique to deal with adversaries from the model is to detect and classify clean images from adversarial ones. This is done by training a classifier on adversarial and non-adversarial samples. In particular, Ref. [9] proposes a technique to train an adversarial perturbation detector that takes the output of a classifier at a certain layer as an input. The goal of this method is to differentiate between adversarial and benign inputs. The training is done by including adversarial examples in the detector to take into account future unknown attacks.

Detecting adversarial examples by analyzing the distribution they belong to can be effective to improve model robustness. This category of defenses attempts to estimate the distribution, in order to detect if an input x is OOD. For this, *kernel density estimates* (KDE) [10] proposed to detect if an input is located outside of the class manifold. This can be thought of as an outlier detection approach, where the outliers are adversarial examples. The idea is to determine whether an adversary x' has low $KDE(x')$ or not, which indicates if is detectable. However, this method is vulnerable when x' is close enough to the target manifold.

Trying to defend against unseen attacks is not a trivial task, given that the model does not know what kind of input is expected. For this, *adversarial purification* [11] tries to solve this issue by first adding noise to images through a pipeline (e.g., generative model) and afterward removing this noise by inverting and learning the same pipeline (Figure 21.4). These techniques do not make any assumptions about the type of attack or the classification model; thus, they can protect existing classifiers from dangers that are not yet seen. In other words, a forward diffusion technique is used to initially diffuse an adversarial sample with a little amount of noise, and a reverse generative process is used to recover the clean image.

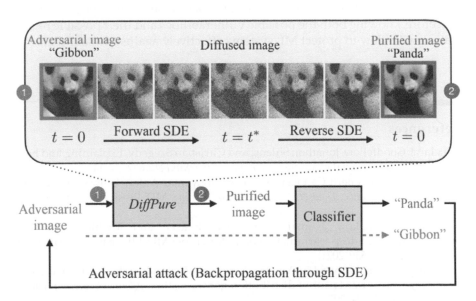

FIGURE 21.4

In order to create diffused images from adversarial photos using a pre-trained diffusion model, DiffPure introduces noise to the images after the forward diffusion process with a short diffusion timestep t.

21.6 Discussion

ML- and DL-based systems currently are useful for a wide domain of applications. However, such systems can be vulnerable to critical scenarios to applications that require high security. Moreover, systems that deal with sensible data (e.g., medical and clinical data) are needed to be resistant to adversarial attacks since the information within such models is highly valuable for adversaries and could be exposed. Having methods to protect against these vulnerabilities is a critical issue in many sectors and applications. There are defensive measures that allow us to protect ourselves against all the attacks described, which are classified into four types (extraction, inversion, poisoning, and evasion).

In Section 21.4, we described how to train robust models to be resistant on evasion attacks. Such attacks allow an adversary to get a model to classify incorrectly. The way to achieve this is to introduce small noise perturbations in the input. These malicious inputs are called adversarial examples, and there are many methods to evade ML models. Most of them have been applied in image classification, although they also exist for video, audio, and text, among others.

Currently, adversarial ML approaches are not sufficient to mitigate all the vulnerabilities of a model. In the near future, adversarial attacks will be adaptive in different scenarios and not only applicable in a certain environment, as there will always be defenses to protect against this specific attack. Therefore, knowing both the possible vulnerabilities and their possible remediation will allow to protect ML systems effectively against these threats.

References

1. Ian J. Goodfellow, Jonathon Shlens, and Christian Szegedy. Explaining and harnessing adversarial examples. *arXiv preprint arXiv:1412.6572*, 2014.
2. Joaquin Quiñõnero-Candela, Masashi Sugiyama, Anton Schwaighofer, and Neil D. Lawrence. *Dataset shift in machine learning*. MIT Press, Cambridge, MA, 2008.
3. Nathan Drenkow, Numair Sani, Ilya Shpitser, and Mathias Unberath. Robustness in deep learning for computer vision: Mind the gap? *arXiv preprint arXiv:2112.00639*, 2021.
4. Zheyan Shen, Jiashuo Liu, Yue He, Xingxuan Zhang, Renzhe Xu, Han Yu, and Peng Cui. Towards out-of-distribution generalization: A survey. *arXiv preprint arXiv:2108.13624*, 2021.
5. Harini Kannan, Alexey Kurakin, and Ian Goodfellow. Adversarial logit pairing. *arXiv preprint arXiv:1803.06373*, 2018.

6. Yang Song, Taesup Kim, Sebastian Nowozin, Stefano Ermon, and Nate Kushman. Pixeldefend: Leveraging generative models to understand and defend against adversarial examples. *arXiv preprint arXiv:1710.10766*, 2017.
7. Anish Athalye, Nicholas Carlini, and David Wagner. Obfuscated gradients give a false sense of security: Circumventing defenses to adversarial examples. In *International Conference on Machine Learning*, pp. 274–283. PMLR, 2018.
8. Jonathan Uesato, Brendan O'donoghue, Pushmeet Kohli, and Aaron Oord. Adversarial risk and the dangers of evaluating against weak attacks. In *International Conference on Machine Learning*, pp. 5025–5034. PMLR, 2018.
9. Jan Hendrik Metzen, Tim Genewein, Volker Fischer, and Bastian Bischoff. On detecting adversarial perturbations. *arXiv preprint arXiv:1702.04267*, 2017.
10. Reuben Feinman, Ryan R. Curtin, Saurabh Shintre, and Andrew B. Gardner. Detecting adversarial samples from artifacts. *arXiv preprint arXiv:1703.00410*, 2017.
11. Weili Nie, Brandon Guo, Yujia Huang, Chaowei Xiao, Arash Vahdat, and Anima Anand-Kumar. Diffusion models for adversarial purification. *arXiv preprint arXiv:2205.07460*, 2022.

Index

Note: **Bold** page numbers refer to tables; *italic* page numbers refer to figures and page numbers followed by "n" denote endnotes.

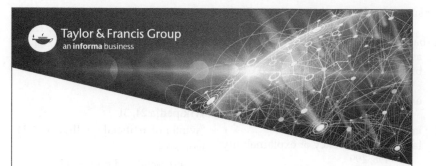